D0217053

Universitext

R. E. Edwards

A Formal Background to Mathematics Ia

Logic, Sets and Numbers

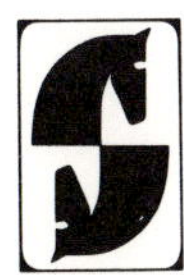

Springer Verlag
New York Heidelberg Berlin

Dr. Robert Edwards
Institute of Advanced Studies
The Australian National University
Canberra, Australia

AMS Subject Classifications: 02–01, 04–01, 06–01, 08–01

Library of Congress Cataloging in Publication Data

Edwards, Robert E
A formal background to mathematics.

(Universitext)
Bibliography: p.
Includes indexes.
Contents: v. 1. Logic, sets and numbers.
1. Mathematics—1961– I. Title.
QA37.2.E38 510 79-15045

Printed in the United States of America.

9 8 7 6 5 4 3 2 1

ISBN 0-387-90431-X Springer-Verlag New York Heidelberg Berlin
ISBN 3-540-90431-X Springer-Verlag Berlin Heidelberg New York

Foreword

§1 Faced by the questions mentioned in the Preface I was prompted to write this book on the assumption that a typical reader will have certain characteristics. He will presumably be familiar with conventional accounts of certain portions of mathematics and with many so-called mathematical statements, some of which (the theorems) he will know (either because he has himself studied and digested a proof or because he accepts the authority of others) to be true, and others of which he will know (by the same token) to be false. He will nevertheless be conscious of and perturbed by a lack of clarity in his own mind concerning the concepts of proof and truth in mathematics, though he will almost certainly feel that in mathematics these concepts have special meanings broadly similar in outward features to, yet different from, those in everyday life; and also that they are based on criteria different from the experimental ones used in science. He will be aware of statements which are as yet not known to be either true or false (unsolved problems). Quite possibly he will be surprised and dismayed by the possibility that there are statements which are "definite" (in the sense of involving no free variables) and which nevertheless can <u>never</u> (strictly on the basis of an agreed collection of axioms and an agreed concept of proof) be either proved or disproved (refuted). In spite of the aforesaid lack of clarity in detail, he will be firmly convinced that mathematics is par excellence a deductive system : one commences from certain statements assumed to be true (axioms) and proceeds to deduce more true statements (theorems) by logical argument. Yet the instances of axiomatics already

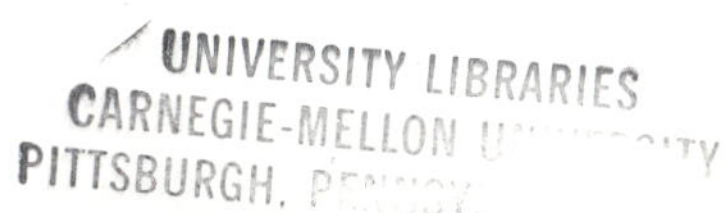

encountered (if any) will clearly fail to get close enough to the bottom of things, principally because they are founded on two things which are left too vague, namely, the concept of set and the (evidently pretty elaborate) logical apparatus for deduction. What he seeks is a clarification of all these matters, embracing in particular more satisfactory answers to the questions listed in the Preface.

Many such questions have "surfaced" in the area of high school mathematics ever since the introduction of the so-called "new mathematics" featuring set language and related topics. On the other hand, the "old mathematics" is also supposedly founded on the same breed of logic and some of the above questions are just as pertinent in connection with it.

These questions, and many others like them, are equally relevant to a great deal of tertiary mathematics. They rarely receive adequate and timely treatment at that level either. Thus teachers and students of tertiary mathematics should also find something of interest to them in this book.

While I willingly concede that no good purpose is served by thrusting questions of this type and attempts to answer them to the forefront at too early a stage or with too much emphasis, I believe that when such questions arise naturally they should be taken seriously. I believe too that everybody intending to be a mathematician, or seeking to understand what mathematics is all about, should at some appropriate stage attend to such questions. Many of the matters discussed in this book may appropriately and with profit be studied by at least some high school teachers, both trainee and in-service, and by the majority of tertiary teachers and students.

It happens (in the Australian Capital Territory, for example) that practising high school teachers are themselves responsible for curriculum design. In view of this alone, some such teachers have themselves stressed the importance of being as well-informed as possible about at least the general aspects of the foundations of their chosen subject.

I am not pressing for any extreme or compulsory measures, merely recommending as highly desirable for an all-round view some steps towards redressing the imbalance represented by an almost total absence of attempts to integrate into the routine treatment of mathematics generally some of the formal aspects of basic material which have developed over the past century, rather than leaving the formal aspect as a specialist study entirely remote from general practice. Such steps should not exclude or be divorced from a complementary study of conventional, informal mathematics. Nor am I suggesting that any more than a minimally adequate degree of formalisation (without some traces of which mathematics has for centuries been almost unthinkable) is a sensible component of a first course on anything for anybody. Almost all beginners at any level (elementary or advanced) and any topic are best served by an informal introduction which is clear and accurate enough for them to learn what it is all about. (It is part of the expositor's job to judge the right amount of formality; either too little or too much results in trouble.) Put another way, it is almost without exception essential that an informal acquaintanceship be developed <u>before</u> a formal approach is attempted, and I am NOT advocating a reversal of this order of events. I suggest merely that, given an informal acquaintanceship, an examination of the formalities in relation to informal practice is desirable.

Most of the best books about formal logic and metamathematics tend to leave aside any detailed consideration of what is done in conventional informal mathematics. On the other hand, most of the best books about informal mathematics tend to take for granted, or offer inadequate and misleading accounts of, the supposed foundations and underlying formalities. This book attempts to fill the gap by providing readers having some familiarity with conventional informal mathematics with a more accurate and detailed description of (one version of) the background formalities, together with commentary which compares the formal and informal aspects of mathematics and draws attention to some of the occasional discrepancies.

Any reader who initially feels overcome by the formalities should perhaps try to give it all the attention he can muster in easy stages and without lingering too long on the first reading of any one topic; see especially the final paragraph in I.0.10. He should at all costs endeavour to keep moving towards Chapters V and VI, where numbers make their entry and form the bridge to more familiar ground. To a large (but not total) extent, numbers are the initial goal of, and provide the earliest motivation for, all the preceding effort. (When a little light, tongue-in-cheek relief seems essential, a glance at Linderholm (1) may be in order!)

The more familiar a reader is with the informal versions of any topic treated, the easier it will be for him to pass to the more formal treatment of that topic. For a reader well-acquainted with the routine treatment, the formal versions may be unnecessarily detailed and finicky. At such points (see, for example, IV.2.2, IV.4.2 and the small print portion of V.5.2) the routinely well-versed reader may indulge in leap-frogging (though always with the slight risk that the formalisation involves details which he may have overlooked).

§2 As has been said, my study groups were initially intended to cover in routinely rigorous fashion topics such as sequences, functions, convergence, calculus, et cetera. Out of this there sprang the logical and foundational questions motivating this book. It soon became clear that no satisfactory answer to the foundational questions raised by the audience in the study group would be possible without a description of formalisation, and I resolved to try that as an experiment.

After casting around for what seemed the most convenient of several formalisations, it seemed to me that the version offered by Bourbaki, modified in a few technical details, was the best for my purposes. As a result, Chapter I of this book is very closely based on Bourbaki's account, though it is (I hope) much better adapted to my special aim and to the needs of my anticipated audience than is the original. (Bourbaki's treatise seems especially valuable to fairly mature mathematicians who already possess an excellent working knowledge of the topics discussed; for most it is too austere to act as an introductory source. For specialists in mathematical logic and/or foundations it has some technical disadvantages; but this is of little significance in the present book.)

Having embarked on a description of a formalisation as a source of answers to the original questions, it was natural to examine in general the link-up between the formal and informal treatments of routine topics. This is not usually done in any detail : if a formal theory is described at all, it is more usual for the description to deal with the formal theory in splendid isolation and to ignore how it relates to informal practice. A closer examination reveals some surprises and troublesome disharmony at various points exemplified by the following:

the use of the terms "not true" and "false" (see I.3.8(ii));

standard remarks about converses (see I.3.8(v));

the coherence and clarity of definitions, and occasional open conflict between the formal and informal styles (see I.3.5, I.4.3 - I.4.5, III.1, III.1.8, IV.1.2, IV.1.4, IV.2.3, IV.10);

the unintentional consequences of some informal definitions (see I.2.9(ix));

the over-hasty and uncritical use of the set builder { : } (see II.3.9);

the use of florid informal language, often quite out of step with the corresponding formalities (see I.3.4(vii));

the use of the phrase "there exists" (see I.3.4(iii), II.4.2(ii) and Remark (ii) following VII.1.3);

careless use of the functional notation (see IV.1.3, IV.1.4, and the end of IV.2.2);

the finer points of proofs by induction (see V.4.1, V.4.3 - V.4.6);

the lack of clarity in the informal statements of results (that is, theorems, theorem schemas and metatheorems) and problems (see, for example, I.4.5(ii) and (iii), V.11.5, VI.10, VII.3.6, IX.2.11).

As has been indicated in the opening paragraph of the Preface, it is unknown whether the chosen formal theory is consistent. (The vast majority of experts believe it to be so, but contradictions may possibly remain hidden.) Should a contradiction be discovered, it is to be expected that attempts would be made to modify the theory in such a way as to block the paths to (render underivable) the contradictions. (Something of this nature has happened before.) However, the broad outlines of the formalistic approach would most probably be retained.

§3 All formal developments have to follow in a strict (though not necessarily unique) logical order. This often results in an appearance which is stark, barren and unfamiliar. In an attempt to counter this, and to stimulate and encourage the reader's patience and interest, informal discussions of certain topics are introduced in advance of their proper logical order. (Notable examples appear in III.2, IV.3, IV.7 and IV.8.) No harm need result from this, provided the derangement is clearly recognised, and provided (above all) that, when the formal developments are resumed, no reliance is placed upon the outcome of such discussions until after these discussions have been formalised.

There is no space in this book for more than a very few such motivating digressions. Many more are desirable and are obtainable by dipping into books such as (for example) Griffiths and Hilton (1), concerning which see Note 1.

The starkness and the strangeness of the formal developments are probably most acutely felt in Chapters I and II, after which familiarity will gradually increase. Even so, such basic mathematical signs as $=$, $\subseteq$, $\cup$, $\cap$ appear only in Chapter II; they, together with a few more set-theoretical signs (such as $\{\ :\ \}$, $P(\ )$, $(\ ,\ )$, pr_1 , pr_2 , $\times$, ...) constitute the sole formal diet throughout Chapters III and IV. Apart from their premature use in Chapters I - IV for illustrative purposes only, the signs 0 , 1 , ..., $+$, which are traditionally among the very first to be encountered in an informal approach to mathematics, do not appear until Chapter V. Integers, rational numbers, and real numbers appear in Chapter VI, and even then in much less than full formal detail. From there on, mathematics in the informal sense moves more rapidly.

It is perhaps safe to assume that many readers with serious intentions will find it difficult to come to terms with the formalities and to resist the feeling that the earlier chapters involve a needless over-elaboration. It is indeed quite probable that some inadvertent elaboration does appear as a result of my shortcomings, but most of it seems unavoidable. There are a number of texts which attempt to present an account which is simpler and more elementary in appearance, but (see I.4.2) most of these fail in their aim and can only serve to mislead. It would, after all, be surprising to find it easy to provide a basis which even appears to be both sound and also adequate to support the vast superstructure represented by a major portion of existing mathematics. The reader should be a little sceptical of apparently easy paths.

§4 For those readers who may wish to refer to Bourbaki, I repeat that Chapter I below follows him closely, the major difference being that I subsequently define equality rather than include it in the primitive alphabet. The correspondence remains fairly close in Chapter II, but from Chapter III onward the differences in detail become quite considerable and I move closer to other texts on formal set theory. I have nowhere followed exactly the terminology of the English translation of Bourbaki. For instance, in place of "assembly" , "relation" , "term" I have used "string" , "sentence" , "set" , the change being made partly because the latter terms are closer to normal English usage, and partly (in the last case) because I emphasise from the outset, more than does Bourbaki, one theory (set theory).

In an attempt to avoid overloading the main text of Volume 1 with details (especially those relating to more purely logical and metamathematical matters), I have included a lengthy Appendix broken into several sections.

§5 Many readers of this book may find it helpful to study, in parallel with or as a preparation for this book, one or more of the less formal accounts of logical and foundational material — for example, Wilder (1), Behnke et al. (1), Gleason (1) (especially Chapters I - 3), Stoll (1) (Stoll (2) is more comprehensive and detailed). Bourbaki (1), pp.296 - 346 provides an illuminating historical survey. A carefully written account of (informal) set theoretical language and methods at high school level appears in Görke (1). Special attention is called to Griffiths and Hilton (1), especially Chapters 1 - 5, 8, 39 thereof; see also Note 1.

§6 Because the point is of paramount importance, I risk being repetitious by re-emphasising that this book is NOT intended as a text for anybody making his first acquaintance with numbers, sequences, functions, convergence, limits, derivatives, integrals and so forth. It is intended for teachers who are now, or who expect to be, holding forth on such topics at high school or early tertiary levels, and for senior tertiary students who seek to learn more about what underlies their earlier routine studies. It is agreed that any more formality than is necessary for a provisional degree of precision and clarity is usually quite out of place in what is to be presented to those meeting a routine topic for the first time. Even so, a higher degree of formalism is a useful or even essential ingredient in the diet of those preparing to present routine versions of those topics or rethinking those topics for their own benefit and satisfaction. For example, the traditional way of presenting arithmetic to young children may be the best way; yet the teacher doing that can and should profit from paying some attention to the formal background. Much understandable disenchantment with the "new mathematics" is the result of poor judgement (often on the part of people with no personal experience of formalism) as to when, for whom, and in what degree a formal approach is to be adopted.

(There is, of course, a reverse side of the coin. The formal theory may in principle develop without resort to any prior knowledge of mathematics. It is accordingly conceivable that a reader who is initially totally ignorant of mathematics could learn mathematics via the formal theory alone. However, such a course is almost unheard-of; and I am certainly not suggesting that this course be tried.)

§7 An almost unique feature of this book is the inclusion of some explicit criticisms of a few books on mathematics. In other disciplines, such criticism is commonplace; in mathematics, it is very rare. While the relative objectivity of mathematics, even in its conventional informal guise, goes part of the way towards explaining the rarity, there remains enough vagueness and subjectivity to leave room for some justifiable criticism. This is even more the case when one has a highly formalised version of mathematics in mind. In the main, the criticism refers to misleading claims and confusion relating to basic general concepts (rather than to localised inaccuracies in detail, though one expects a mathematics book to show some evidence of attention having been paid to localised details as well).

No books known to me are entirely beyond criticism from a formal point of view. (My own earlier books are as objectionable as many others; it is only because they deal with topics remote from those referred to in the present book that they do not receive more criticism. Doubtless, the present book contains several instances of its own peculiar species of blunder; see Note 10, sparing my blushes as you do so.) The criticisms made in the present book refer mainly to fairly widespread practices; particular references are chosen partly for definiteness and partly to make it clear that the practices <u>do</u> occur and thus have the appearance of written authority. In all cases, criticisms are expressed because of the conviction that they are technically called for, and that useful lessons are to be learned from an occasional close analysis of misconceptions and blunders. In some cases, the criticisms are justifiable only from a severely formal standpoint; in other cases, they are justifiable from a conventional and quite informal point of view.

Some of the books cited (for example: Griffiths and Hilton (1), Hewitt and Stromberg (1), Spivak (1), Halmos (1)) have been chosen to typify the conventional informal approach at its best. Others typify a presentation of that same informal approach in much less satisfying fashion.

§8 The majority of the problems are placed near the end of the book, divided according to chapter; but a few appear at appropriate places in the main text. No attempt has been made to present the problems in increasing order of difficulty. Hints are attached to the more difficult problems.

Also collected near the end of the book are a few Notes.

Preface

The aim of this book is to describe one possible formal background to mathematics, and thereafter to discuss its connections with selected but typical topics which form part of informal mathematics up to high school and early tertiary levels. The said formal background is to take the shape of a formal and strictly deductive theory which will, it is hoped, provide a secure foundation for mathematics. For good reasons, I cannot and do not claim that the desired security (which means formal consistency rather than intuitive plausibility or suitability for applications) is in fact achieved; see I.2.7.

The idea for such a book emerged quite unexpectedly from a study group I ran during parts of 1975, 1977 and 1978 for high school teachers. The initial programme for this study group was no more than a routinely rigorous informal treatment of numbers and basic real analysis (sequences, functions, convergence, calculus, and so forth). However, without any prompting from me, the audience raised questions about methods of proof. It thus became clear that at least some teachers of top-level high school courses in mathematics were worried about logical and foundational questions. These teachers were sufficiently concerned to actively seek answers to questions of the following sort:

> What is the nature of proof in (pure) mathematics, as that concept is understood by a competent working mathematician who, although he believes himself to be proceeding logically and rigorously, is not a

specialist in mathematical logic? (What is being sought is the ideal concept, which may and indeed does differ from the conventional working style of proof.)

Why and how do "proofs by contradiction" , "proofs by disjunction of cases" , and "proofs by induction" work?

What is the starting point for a foundational topic such as set theory? By what procedures are other topics founded upon it?

It should be stated without delay and with maximum emphasis that this book is NOT intended as a text for high school students, nor indeed for anybody meeting for the first time any of the routine mathematical concepts (numbers, sequences, functions, limits, convergence, derivatives, integrals, et cetera) which are covered. (More is said on this matter in the Foreword.) Readers of this book should on the contrary have a working acquaintanceship with the topics just mentioned. What is done is to provide a general logical and formal basis, followed by a formal approach to these topics and a comparison of this approach with the conventional informal one.

It is also important to state firmly that this book is not intended as a text on either mathematical logic or metamathematics. Even less is it concerned with the philosophy of mathematics, which is almost totally ignored. (For instance, it seems largely irrelevant to the main purpose of this book to discuss the reduction of mathematics to logic or vice versa.) Formalism is in this book regarded, not as a philosophical doctrine, but merely as a medium allowing (a) a more precise description of (an idealisation of) what most working mathematicians understand by constructions and proofs; and (b) an analysis of and escape from some of the inconsistencies which have emerged in the past. (Others may possibly remain unrecognised.) This is not to claim that these concepts of construction and proof are philosophically satisfactory; interested readers may wish to consult Körner (1) for an introductory survey of philosophical thought on such topics.

Volume 1 contains the more heavily formalistic and foundational material, covering logic, set theory, relations, functions, natural numbers and real numbers.

Volume 2 can, if need demands, be read without detailed knowledge of much of Volume 1 , simply by sticking to the conventional and informal viewpoint and ignoring the links with formalities. In this way Volume 2 may assist high school teachers to present accounts which are at any rate more satisfactory from an informal point of view, and at the same time assist tertiary students in similar fashion. In respect of high school mathematics, it must be reiterated that this book is addressed to teachers; it is entirely up to them to decide to what extent (if any) they wish to modify the content and presentation of their classwork. I hope rather that this book will primarily modify and improve the teacher's general outlook on and understanding of his subject. Such modification in general outlook may aid towards a more confident and effective treatment, although there may be no overt use in class of anything appearing in this book.

The reader may prefer to take up the story at a point other than the beginning I choose. He may, for example, prefer to regard the substance of Chapter I as intuitively-given logical procedure about which nothing needs to be said. I think this is possible without too much loss, though some points of detail will inevitably escape. Similarly, those who wish to handle sets in an informal manner, will be able to move quickly through Chapters II and III, attending merely to the salient informal content of the definitions and basic theorems.

Acknowledgements I am extremely grateful to my two friends and colleagues, Drs John Staples and Jeff Sanders, the former principally for many long and valuable talks, most of them on matters with a logical or metamathematical flavour, and the latter for even more and longer general discussions, plus a great deal of help with the problems. I owe a lot to them both for their generosity. Naturally, I alone am responsible for whatever blunders may appear.

For encouragement and help throughout the whole venture, from beginning to end, I am indebted to my good friend Professor Edwin Hewitt.

I am also grateful to Miss Jane Lake, who read early versions of Chapters

I and II and provided me with her reactions, which led to a number of modifications.

Thanks are due to Mrs Helen Daish for her careful and excellent typing of early drafts of this book, and to Mrs Marion Saville for her similar work on even earlier notes upon which the book has been based. Mrs Lindsay King prepared the final camera-ready copy of the majority of the book (including the whole of Volume 1), an onerous and exacting task discharged in fine style and with enthusiasm. When Mrs King relinquished the work in order to pursue her academic career, my wife took over the completion of the task. My wife also gave me a great deal of assistance with proof-reading and subsequent corrections. To both of them I offer my warmest thanks.

R. E. Edwards,

Canberra, March, 1979

Contents

Appendix

Problems

Notes

Chapter I. Logic and Formal Theories

Most readers of this book will be familiar with the view that sets are the mathematical analogues or models of collections in everyday life, and that the concepts and definitions referring to sets constantly reflect that origin; and perhaps also with the view that set theory should, at the same time, provide a good foundation for almost all the rest of mathematics. It is thus natural to handle sets intuitively as long as this seems both fruitful and legitimate. Limitations to the feasibility of doing this came to be recognised the hard way - by encountering paradoxes (see II.3.9 and II.3.10). Thus forced into attempting some sort of formalisation, one will want sets and everything to do with them to be incorporated into some scheme at least as rigorous and coherent as the portions of mathematics which preceded set theory and which are to be refounded on set theory as a basis. This entails the expectation that sentences will be formed which involve reference to sets; and that one expects to handle and derive (or prove) sentences in much the same way as prevailed in mathematics before the concept of set was imported and gained currency.

At the same time, a practising mathematician sometimes feels the need for something definite and explicit to be said about the type of reasoning habitually used in mathematics (with or without reference to sets) in order to prove theorems. In other words, he seeks clarification on just what logical apparatus lies behind mathematics, traditional or new.

It so happens that most current thought is such as to unite the two quests, in so far as it conjures up a picture of a common logical-mathematical basis for many formal theories; and one instance of such theories, namely set theory, as making an acceptable foundation for a very large portion of everyday mathematics. (On the other hand, I am not intent upon pressing this monolithic picture too far: set theory seems in the main to be currently adequate, but it may be displaced. Even so, the replacement will almost certainly be another formal theory intended to play an exactly similar role.) It must be expected that great effort and care will be needed in founding such a formal theory, which is expected to support so heavy a superstructure. The resort to formalisation is an attempt to make the foundations sound.

It is probably fair to say that set theory, naive or formal in style, is a framework within which almost all current mathematics is expressible; but that such expression (especially that in strictly formal terms) may be judged to be at times neither desirable nor fruitful. See the Preface in Gleason (1), with which the present writer agrees largely but not entirely.

In this chapter, I endeavour to give a description, just adequate for the purposes of this book, of a slightly modified version of the formal system propounded by N.Bourbaki as a common basis for logic and set theory (and thence of almost all current mathematics).

It cannot be over emphasised, however, that this chapter is definitely not intended to be an introduction to current mathematical logic and/or metamathematics per se (which are much wider subjects than this book can indicate). Nor is it likely to meet with the approval of all experts in these areas. I am told that most of mathematical logic these days takes for granted the existence of a going and consistent set theory and is concerned with the interpretation of other theories in set theory (or of one version of set theory in another). My outlook is the restricted one likely to be found in a practising mathematician who is reasonably serious in his concern about the tools he habitually uses and often abuses.

Some readers may find it helpful to read the bird's eye views of axiomatics and formalisation presented in Monro (1) and Halmos (2).

I.0 The Idea of a Formal Language

Introduction Before beginning the detailed description of the formal language and its workings, it may be helpful to the reader to list some features (some of which will be discussed again later at appropriate places) of what one has in view. In this connection, the reader may find it helpful to consult Bourbaki (1), pp.7-13, 296 ff; Gleason (1), pp.7-8; Margaris (1), Chapters 1 and 2; Kleene (1), pp.59 ff; Kleene (2), §§36, 37, 38; Stoll (1), Chapter 3; Stoll (2), Chapter 9; Godement (1), §§0, 1; Hamilton (1).

Gleason's book is especially recommended as a frequent comparative reference, even though its purely technical level is medium to advanced tertiary rather than secondary to early tertiary, because it is an example of a carefully written, conventionally rigorous account of a portion of current mathematics. The degree of formality adopted by Gleason is somewhat higher than is usual in such accounts, yet the author expressly denies (see loc.cit.,p.8, lines 10-15) any attempt at full formalisation. (It is, in other words, very largely concerned with "getting on with the job" of "real" day to day mathematics.) In spite of this disclaimer, the presentation has the virtue of exhibiting to an unusual degree an awareness of and reference to some (unspecified) fully formal scheme; and of discussing from time to time the adopted style (see, for example, loc.cit., pp.22-23). It is the style, rather than the technical content, of Gleason's book which makes it especially appropriate in relation to our programme.

I.0.1 Natural languages It is almost universal to expect (pure) mathematical (see pp.6,7 below) discourse to be more than customarily precise and logical, free from the ambiguities of everyday interchanges and even more precise and logical than technical discussions in, say, chemistry or physics. In spite of this, conventional mathematics is expressed in some natural language (English, in our case) and follows normal usage in many respects. Most of the time, this seems adequate. But occasionally, and most probably during discussions of foundational matters, trouble can arise. This is witnessed by the existence of various paradoxes. The natural language may be compelled to do things which are incompatible with its logical coherence (Tarski); see also I.2.9, II.3.9 and V.4.1. In an attempt to avert such troubles one seeks, at least as an underlying standard by means of which to judge precision, a more specialised formal language with rules of manipulation more closely specified and with a clear statement of the admissible rules of logic.

As to the adequacy of a natural language, doubts may arise the moment one contemplates the abundance of Arts doctoral theses based on the variety of possible interpretations of a few sentences (or conventionally accepted strings of words) in a natural language; see also the interesting discussion in Kleene (2), §27. Few mathematicians desire their key phrases to be open to almost endless debate: their aim is that these key phrases be as unambiguous as can be devised, and then to proceed to what they hope will be irrefutable conclusions. Of course, they do not always succeed; and then debate is unavoidable.

More is sought than merely a mechanism for making statements with maximum clarity. One needs also a notion of truth which, once one has agreed upon precisely stated axioms, will be as objective as can be devised. Given the said axioms, mathematical truth is commonly supposed to be at least as objective as scientific truth. In the case of science, the objectivity is supposedly attained by final appeal to observations of and experiments with portions of the real world. This criterion is not available, or is deemed inappropriate, in the case of mathematics. Conventional informal mathematics is indeed not as objective as it might at first sight seem to be. The pursuit of objectivity in mathematics is certainly aided by a formalisation which will embrace the logical principles which offered the passage from the axioms to the true statements.

This process of formalisation accounts for the mechanical, logical, non-intuitive, objective aspects of mathematics. It leads to fussiness and cumbersomeness. One must never forget the reverse side of the picture - the intuitive, inventive aspect of mathematics which is rarely if ever aided (and is more often handicapped) by too much formalisation, and which is essential to keep the subject alive and developing.

Most natural languages have at least two main functions:

(i) to record and transmit knowledge, facts, information or instructions;

(ii) to record and share emotions, to evaluate and interpret experience, and to entertain.

When the language is being used in role (i), a premium is placed upon precision, lack of ambiguity, objectivity, and adherence to a relatively rigid grammar; words and phrases are expected to retain fixed meanings.

When the language is being used in role (ii), however, a premium is placed upon an ability to transmit the subjectivity and personality associated with the topics being handled, and the grammar is relatively loose and shifting. Many words and phrases are deliberately used metaphorically in an attempt to heighten and dramatise their effect, even though it is recognised that this device may involve a distortion of relevant facts. Words or phrases are often deliberately chosen because they will "sound good" when uttered; they may also be carefully arranged in patterns which are thought to be pleasing or striking when the words are spoken. Many such devices are used to arouse and sustain the reader's interest and to entertain him. He is expected to interpret what is written, and his interpretation is expected to incorporate subjective elements.

A Shavian play is perhaps a good example of the use of a natural language in both roles simultaneously, or at least in rapidly interchangeing roles.

The desirable characteristics of a natural language used in role (i) are those which one expects to be displayed par excellence by a formal language, which should have an exceedingly explicit (and often relatively simple) grammar, to be described as clearly as one can devise in some metalanguage, while the use of the formal language is to be always and absolutely in accord with the grammar. Such a language is expected to be stark and functional; its use strictly according to the rules can soon become a bore, but this is a price that has to be paid for the gains in clarity, precision and (in the case of formal logical mathematics) freedom from fallacious reasoning which its use is intended to secure.

Quite apart from the language used for expression, there is of course a difference of viewpoint between the formal and informal positions. Roughly speaking, intuitive or informal mathematics .... typified by, for example, the development of analysis appearing in Gleason (1) ..... purports to be concerned with meaningful statements about the objects of some (abstract) universe. (Abstract because certainly the physical universe shows no signs of being adequate to accommodate all the mathematical concepts in current use.) These statements, or at least many of them, supposedly correspond to the result of bestowing meanings upon, or interpreting in terms of the said universe, the sentences of the formal language, these formal sentences being syntactically defined as in I.1.6 below and having in themselves no meaning. The truth or falsity of these statements of the informal system is supposedly entirely settled by reference to the adopted meanings of the words and symbols appearing in them, that is, are settled on semantical grounds. The truth or falsity of formal sentences is, on the contrary, determined syntactically in the manner described in I.2.4 below ... which has to be recognised (at least in principle) as a totally different criterion. In the intuitive system, certain meaningful statements, sometimes termed "propositions", are (almost by definition) regarded as being of necessity either true or false; such is not the case with syntactic truth (that is, provability) of formal sentences. (See I.2.9 (ii) and (iv) below.)

I should point out that in this book, "mathematics" means "pure mathematics" unless the contrary is explicitly indicated. Many people (see, for example, Preston (1),pp.87-88 and the Preface in Gleason (1)) suggest that it is often good to seek to obliterate the distinction between pure and applied mathematics. (I take the latter to embrace what is otherwise referred to as mathematical physics, mathematical biology, and so on, in addition to things like "classical" mechanics and hydrodynamics, et cetera.) Doubtless they are right in certain contexts. In this book, however, the distinction is vital because the basic criterion of truth in pure mathematics is quite different from that in applied mathematics : "formally provable" versus "in accord with observation, usually in the real world". The difference is illustrated by Dirac's famous book "The Principles of Quantum Mechanics". The mathematical substratum of this book is formally inconsistent but it presumably leads (or has led) to things of great value in mathematical physics: it is formally "bad" pure mathematics but "good" applied mathematics. Even pure mathematicians have profited from it indirectly, precisely because of the challenges it offered to them.

It may be appropriate to recall here Cantor's famous original definition of "set", which was expressed entirely in the appropriate natural language. Translated into English, the definition runs as follows : A set is a collection into a whole of definite distinct objects of our intuition or of our thought. The objects are called the elements (members) of the set. This definition, or others like it, appears to date from the 1870's and apparently served as a working basis for quite elaborate technical developments during a couple of decades. However, by the late 1890's, Cantor himself recognised its unsatisfactory nature. See Fraenkel et al. (1), Chapter I.

I.0.2 Characteristics of formal languages Leaving aside any attempt to discuss the detailed design of formal languages in general, it is fairly safe to say that the preparatory step would be to analyse carefully the substance and reasoning to be formalised, so as to disentangle as well as maybe a minimal collection of key words and phrases (atomic components of reasoning, so to speak). Each of these would then be assigned a symbol or string of symbols, which symbols would appear among the primitive alphabet and axioms of the final theory being constructed. From then on, all developments in the formal theory have to be legitimised by reference to the axioms, and not by reference to meanings and usage in the everyday language (though it is in practice of great importance to rely on the latter as a source of ideas for conceiving and planning formal developments).

The above very rough outline needs modification in detail. But it suffices to make it easy to appreciate certain vital features. Thus, in everyday language (even when used mathematically) one is confronted with the frequent and usually unregulated addition of new terms and phrases, and manipulations of sentences are allowed on the basis of dictionary meanings (which can be manifestly odd from a mathematical point of view - for example : infinite = boundless = unbounded;

true = correct = proper = appropriate = belonging to).

> In this connection, it is instructive to consider the distorted meaning frequently attached in informal mathematics to the phrase "in general" . Instead of meaning (as it should) "without exception" , it has come to be used in cases where exceptions are very well known to exist. Thus, it is often precisely because they are aware of exceptions to the statement "A line and a circle meet in two points." that many authors feel compelled to write "In general, a line and a circle meet in two points.". I shall endeavour to use the phrase "in general" only with its original meaning.

In a formal language, however, both sentences and objects will on the contrary be formed from a preassigned and invariable list of primitive symbols, the processes of formation being always in strict accordance with explicit rules laid down at the outset and thereafter held fixed. Equally explicit and immutable rules will be given to specify the concepts of proof and theorem; and these rules also will be always strictly observed. The rules of manipulation of sentences, that is, the laws of logic, will be incorporated in the shape of axioms (or axiom schemas), likewise prescribed at the outset and once again always obeyed to the letter. These, taken together with the prescription of proof, fix the form of logical reasoning.

Venturing a little more detail, a proof will be a finite sequence of sentences, the first one or more of which are axioms, and each of the rest being related to its predecessors according to fixed rules. A theorem (or true sentence) will be a sentence which appears in some proof, that is, a provable sentence.

The role of the mathematician falls into two main parts.

(i) To select and formulate interesting and significant concepts and to conjecture interesting and significant theorems relating to them.

(ii) To provide, or describe accurately, proofs of the conjectured theorems.

There are few, if any, general rules which help with (i). The procedure is pretty subjective and each individual relies heavily on intuition and imagination, tempered by hard won experience. Mistakes have to be expected; that is, conjectures quite often turn out to be wrong. Alternatively, one may be unable to decide whether the conjecture is right or wrong ("unsolved problems", maybe even "undecidable problems").

Turning to (ii), one will usually start by struggling to conceive in fairly vague terms a proof; the germ of this may be contained within the original act of conjecture. Once such a vague idea of a proof is obtained (by whatever means), the next task is to describe with all due accuracy a more formal proof: this is where relatively objective standards are provided by the formalism. (For reasons which will emerge, strictly formal proofs cannot in practice be displayed: one has to be content with sufficiently accurate descriptions of them.)

In so far as the formalisation of "good" informal proofs is often (though not always) close to being purely mechanical, it might be asked why one cannot leave it all to a machine. The answer is in part that a purely mechanical generation of formal proofs and theorems, backed up by no intuitively based selection as in (i), would be too inefficient and too long. What is more, a machine might even take too long in working step by step through individual proofs (though this would depend a lot on the particular formalism involved). The future may see the production of machines which will handle stage (i) effectively. But until then, (i) will remain an absolutely vital component in keeping mathematics alive and moving forward, and in teaching people how to make use of existing mathematics.

Formalism becomes essential in relation to stage (ii) by clarifying the idea of proof and so making it easier to decide whether a given alleged proof really is a proof, and also by helping one to avoid logical errors. A considerable degree of formalisation seems to be essential to a better understanding of the deductive phases of mathematics. (Incidentally, the existence of proof-generating machines would seem to presuppose an extreme degree of formalisation.)

Were it part of our aim to discuss formal languages in general and in detail (which it is not), the preceding description would have to be amplified and made explicit in respect of the rules for forming objects (or terms), sentences (or well formed formulae), and formal proofs. However, this book attempts no such general discussion : it is concerned with just one formal language for which the relevant details will be described in I.1 and I.2 below.

I.0.3 <u>Formalisation as an ideal</u> For various reasons, some practical and some matters of principle, formalisation (that is, expression in the formal language) is an ideal which, for the most part, is kept in reserve. (By this it is meant that formalisation shall set the desired standard of precision in matters of substance, but that in practice one will often seek to avoid the fussiness which inevitably accompanies it.) Formalisation is not intended as a routine teaching device, still less a tool of discovery. While meanings are foreign to the formal language, they and descriptive informal language (and even doodle type pictures!) will continue to be used in the initial formulation of concepts, conjectures of theorems, and as an aid in the struggle to construct proofs. On this topic, see Lakatos (1) which includes a stimulating study of the methodology of informal, creative mathematics.

The situation may be compared to having a toolkit with a very fussy handbook of instructions: one pays attention to the latter at first, then gains some confidence in one's common sense judgements. Only when something has gone (or is about to go) wrong, does one resort to the book of rules. According to this analogy, the metamathematician considers the theoretical potentialities and limitations of the toolkit in the hands of a competent worker (mathematician). (The analogy is inadequate in so far as development of the formal theory corresponds in some ways to the manufacture of new tools as one goes along, tools which, although affording great economy, do no more than shorten tasks which could be accomplished using only the original kit.)

To adopt a different metaphor, formality and informality are rather like friction and its absence: too much friction, and a mechanism grinds to a halt; too little friction in appropriate places may result in total loss of control. A balance is needed. Without some degree of formalism, hopeless confusion can and sometimes does ensue; with too much too often, intolerable cumbersomeness results.

From the point of view of a mathematician who is not a specialist in logic or metamathematics, formalisation has two principal aims. The first is to enable him to see with as much clarity as possible the rules underlying his activity and to enable him to confirm from time to time that the rules are not being damagingly bent. The second is to secure adequate precision in communication with his pupils and colleagues. In this second connection, the necessary degree of formalisation will depend on the person with whom he is communicating; presumably only judgement based on experience can guide him here.

I.0.4 Formal mathematical language Formalisation of current mathematics can be attempted in various ways; in this book I describe one possibility to the almost total exclusion of all others. Despite this, and as was suggested in I.0.2, most formal languages intended for mathematical use have a number of common elements. Sentences and objects appear as certain strings of basic signs comprising the primitive alphabet and formed according to strict rules. Unlike natural languages, many formal languages have an endless (countably infinite) alphabet made up of an unlimited list of "logical letters"

$$\underline{x}, \underline{y}, \underline{z}, \ldots, \underline{x}', \underline{y}', \underline{z}', \ldots, \ldots,$$

which are used roughly like what are conventionally (and loosely) spoken of as "variables", plus a short list of other signs.

> One can avoid an infinite list of variables by having just one basic symbol, say $\underline{x}$, and agreeing that $\underline{x}$ is a variable, and that if v is a variable, then so is v' ; thus $\underline{x}$, $\underline{x}'$, $\underline{x}''$, $\underline{x}'''$, ... and these only are variables. This notation becomes clumsy in practice, even at a semiformal level.

In the case of the formal language described in this book, the short list consists of one relational sign $\underline{\in}$ (familiar to everybody who has any acquaintance at all with set theory) and four logical signs $\underline{\vee}$, $\neg$, τ, $\square$. The rules of syntax of the formal language are much simpler than those of natural languages and much more precise.

It should be stressed that in this book almost all the concern is with the syntactic aspects of formal theories and hardly ever with semantics. For a less biased point of view, see Behnke (1), Part A.

The formalisation described in this book is such as to prescribe fully a basis for most of one species of logic and most of the concepts of mathematics.

In this way, it is undoubtedly a step toward securing the extreme logicality and precision that is expected of mathematics. More particularly, it makes possible a critical examination of the standard methods of proof (contradiction, disjunction of cases, induction, and so on).

I.0.5 Formal versus informal style (1) As has been indicated in I.0.3, and as will be stressed from time to time, formalism greatly slows (may even hinder) the creation of new mathematics (that is, the discovery of new theorems). The living quality of the subject relies heavily on a judicious disregard of fussiness at appropriate places. On this issue, see the Preface to Gleason (1) and also the extensive discussion in Lakatos (1), which illustrates in graphic style the role of the informal, heuristic approach, and presents some new views and interpretations of certain famous episodes in the growth of mathematics.

Still on the topic of informality versus formality, the reader may find it both instructive and engrossing to read S. M. Ulam's autobiography (1). This book adds confirmation to the belief that, when a creative mathematician believes himself to be hot on the scent of something new and interesting, formal correctness temporarily fades into insignificance; he will instead make full use of any help to be gained from informality and intuition; and he will sometimes indulge in what he will later freely admit was scarcely more than wishful thinking (which occasionally actually leads him into error). If his research reaches a successful conclusion, he will revert to more formal (though almost never fully formal) reasoning by way of a check. Assuming that no errors come to light, any ensuing published account will carefully disguise any guesswork and illogicality (conscious or not) which played a role in discovering the result and/or its proof.

Incidentally, Ulam is a Polish born and trained highly creative pure mathematician, unusual as such because of his additional success as a theoretical physicist. He makes it clear that formalism has no great appeal to him personally. On the other hand, he was a close collaborator and great admirer of John von Neumann who did (among many other things) contribute heavily to formal logic and set theory. It is presumably just this sort of contrast and intermixing which is most fruitful.

I.0.6 Informal or naive axiomatics Although formalisation involves axioms, it is to be distinguished from informal or naive axiomatics, which has become increasingly popular over the last few decades; see Stoll (2), Chapter 5. In informal axiomatics, the axioms are regarded as meaningful propositions which are taken to be true in some a priori sense of the elements of some given set (or sets). This axiomatic method is familiar in connection with so-called axiomatic theories of groups (the very first steps of which appear in XII.2), or fields, or partially ordered sets, and so on. Simpler but maybe less familiar instances are to be found in the treatment of natural numbers appearing in Chapter A of Thurston (1) and Chapter II of Mozzochi (1). These naive axiomatic theories are, so to speak, usually embedded within informal set theory (though sometimes they are preceded by and embedded within some version of formal set theory).

The distinction is that between formal and existential or material axiomatic theories; cf. Kleene (1), p.28; (2),§36.

I.0.7 Metalanguage and formal language From time to time it will be vitally necessary to maintain the distinction between a metalanguage and a formal language; the former is used to describe and discuss the latter. This is a principle which extends outside of mathematics. Consider, for instance, Epimenides' (or Eubulides') paradox (Kleene (1), p.39; (2), p.188) expressed in the form: "The statement I am now making is false." . The paradoxical nature can be seen to be at least partly the result of mixing a metalanguage and a formal language. The predicate "is false" is presumably to be sensibly applied only to a sentence in the formal language; and "The statement I am now making" ought to be, or at least should clearly and unambiguously name, a sentence of the formal language; and this it does not do.

I.0.8 Formal versus informal style (2) The details of this chapter will be very strange to most readers for whom this book is intended. Yet there is little technical difficulty. The unfamiliarity will slowly diminish chapter by chapter, so that Chapter VII onward will be an almost routinely rigorous treatment of things like convergence, limits, continuity, and so on. Also, as will be further stressed in the sequel, almost all mathematics is initially accomplished at an informal level. (The term "partly formal" is almost always more appropriate than "informal", at least as applied to mathematics at high school and tertiary levels.) At this level, almost every mathematician has a mental picture of a "universe of mathematical objects", which picture he uses to guide and motivate his activities. He comes to believe in a "real", separate existence of these objects and treats accordingly the statements made about them. (There are real and profound problems in this circle of ideas, but this is not a book about the philosophy of mathematics.) When he writes about the existence of an object with certain properties, he may well picture a search conducted in this universe and possibly the successful location of such an object. The formalism, however, offers no support for this sort of picture: the formal definition of $\exists$ given in I.1.7 seems to have little or nothing to do with "real existence". This feature is often an unnecessary blockage. One has to try to accept and live with the fact that the formal $\exists$ is linked with other parts of the formal language in a way which does no more than mimic in some ways the behaviour of statements about "real existence" in relation to other statements about "real" things. One may be reminded of the status of the money and property handled in a game of Monopoly: neither are real, but the rules of the game cause them to behave and to be handled in play in ways similar to real money and real property, and the players are not hindered from playing by the lack of reality.

Formalised mathematics is indeed similar in many ways to a very elaborate and meaningless game; as such, it may inspire interest, indifference or boredom. Unlike most games, however, it admits of being clothed with meaning in a great variety of extraneously significant ways, which is how it comes to be of (relatively) practical use. This is why one is always tempted to contrast the formal point of view with the informal, intuitive (common sense) outlook.

See also Gleason (1), pp.151-154.

I.0.9 <u>Heuristic style</u> There is one important point to be made about the very frequent use of the terms "formal" and "formally". Quite often, mathematicians use these terms as synonyms of "heuristic" and "heuristically", implying a temporary emphasis on the element of discovery or creativity or of reaching tentative conclusions, without too much care for rigour (which is supposedly to follow later). This is almost the <u>exact opposite</u> of the sense in which the words "formal" and "formally" are used in this book, where the formal approach corresponds to the one which pays maximum attention to rigorous argument. (According to the Concise Oxford Dictionary, "formally" in a logical context means: "concerned with the form, not the matter, of reasoning" - concerned, that is to say, with reasoning strictly according to the rules of what is referred to in this book as the formal theory involved.) At other times, conventional mathematical usage of "formally" corresponds closely with ours, often when a definition is described informally and then expressed "formally".

An instance of the conventional use in the first sense is as follows. Suppose one is at the stage of knowing that

$$\ln(1 + x) = \sum_{n=1}^{\infty}(-1)^{n-1}n^{-1}x^n$$

is true for small values of x (this might in fact be a definition). One might then conjecture that

$$\lim_{x\to 0} \ln(1 + x) = 0$$

and write in partial support:

Formally,

$$\lim_{x\to 0} \ln(1 + x) = \sum_{n=1}^{\infty}(-1)^{n-1}n^{-1}\cdot\lim_{x\to 0} x^n$$

$$= \sum_{n=1}^{\infty}(-1)^{n-1}n^{-1}\cdot 0 = 0 \quad .$$

the term "formally" intending to excuse the interchange

$$\lim_{x\to 0} \sum_{n=1}^{\infty}(\ldots.) = \sum_{n=1}^{\infty}\lim_{x\to 0}(\ldots.) \quad ,$$

which is recognised as being in need of rigorous justification. In a similar way one might write:

Formally,

$$\frac{d}{dt} \int_0^1 f(x, t)dx = \int_0^1 \frac{d}{dt} f(x, t)dx$$

as a precursor to examining critically conditions under which the alleged equality is in fact true.

I.0.10 Formal versus informal style (3) Returning to a point discussed in I.0.3, I.0.5 and I.0.8, it needs to be stressed again that progress in creating (or discovering) new things in mathematics seems to demand that an informal approach be adopted in appropriate places. The formal approach was never intended to exclude the other, merely to regulate and compensate for excesses of informality. It is indeed almost inconceivable that mathematics could have originated and got under way as or within the framework of a formal system, or that it could develop at the speed it has done and continues to do, while encompassed too strictly in such a system. (See also the related comments on axioms in II.12.) The formal theory, when it emerges, plays the role of a rein on too liberal use of intuition in the formulation of conjectures, and as the source of as near ultimate justification as seems possible of proposed proofs. Formalities, in this sense, are poor tools of discovery. The hypercritical outlook which has led to formalisation has rarely, until modern times, been a customary or natural outlook on the part of mathematicians. It came to be adopted, rather reluctantly, only when experience showed that one totally ignores it only at the risk of sooner or later reaching flat contradictions (which all mathematicians abhor). Moreover, a close scrutiny of a point in one area may take place, and yet comparable difficulties elsewhere may go unremarked or ignored. What may at first appear to be merely a "local" problem may, on close analysis, turn out to be merely a symptom of a malaise which extends back to the very roots of the subject and which, if curable at all, has to be treated at the roots. Even when one is concerned with matters of pure principle (rather than the practical pursuit of everyday mathematics), formalism has been demonstrated (first by Gödel and then by others) to have surprising inherent limitations, a discovery which marked the end of a dream; see §1 of the Appendix and the remarks in Gleason (1), pp.151-155.

This book owes its existence to my belief that both approaches, the informal and the formal, deserve attention; and to the self evident fact that the vast majority of books about mathematics avoid almost completely any attempt to describe a coherent formal background. (I am referring here to books about mathematics for practising mathematicians, not books about formal logic, most of which ignore the details of relating formal logic to conventional mathematics.) Thus, I am trying to redress a very marked imbalance in favour of informality, without in any way seeking to deny the vital role played by informal procedures.

An interesting discussion of the informal versus formal issue, as it applies in a specific situation, is contained in a letter by H.B.Griffiths (1).

More detailed discussions of the occasional dangers associated with the use of the informal style appear in various places (for example: I.2.9, II.3.9, V.4.1 and VIII.6.9).

In this chapter, apart from attempting to describe the formal language and its workings strictly according to rules, I shall attempt to describe also some of the ways it relates to conventional, informal procedures, in the course of which it is frequently distorted and abused. These points are dealt with notably in I.1.9, I.2.9, I.3.4, I.3.5, I.4.2-I.4.5. Such points are also discussed at appropriate places in later chapters. Some readers may wish to concentrate first on the description of the formal theory in isolation, leaving for later study the often rather "messy", rather vague connections with informal style expositions. They should perhaps first concentrate upon subsections I.1.1-I.1.8, I.1.10, I.2.1-I.2.8, I.3.1-I.3.3, I.3.5(i)-(iii), I.3.6-I.3.9, I.4.1, leaving the rest for a second reading. Material appearing in small print may also be left for a second reading. A glance at the summary in I.5 may be helpful at this stage.

I.0.11 Residual scepticism There will surely be some ultra-critical readers (my good friend, Edwin Hewitt, for one) who will, quite legitimately, raise questions about the possibility of describing with adequate accuracy the formalisation to be attempted. How (they may ask) can one be sure of recognising as "the same" two occurences of a symbol in two different places? How can one legitimately refer to listing or counting as part of the description, when this seems to involve ideas yet to be formalised? Does not this amount to "begging the question"? (Philosophical criticism is also to be expected.)

I do not pretend to have anything like complete answers to such questions (perhaps there are none). The first question seems to refer to human psychology and/or physiology and maybe other things as well. The second question is perhaps less troublesome. In this connection, I join Bourbaki ((1),pp.9-10) in sheltering behind the feeling that the type of counting and finitary induction involved at the lowest levels of metamathematics is different from and more primitive than what is being formalised; see the discussion at the outset of §2 of the Appendix. At the same time, however, I think it has to be admitted that much of the higher strata of metamathematics involve arguments which are "deeper" than this and scarcely less "deep" than what is being formalised, and thus probably less convincing; see the Foreword to the Appendix.

I.1 A formal language, constructions and sets.

I.1.1 A primitive alphabet The starting point is an unlimited supply of logical (or formal) letters, $\underline{x}$, $\underline{y}$, $\underline{z}$, ... , $\underline{x}'$, $\underline{y}'$, $\underline{z}'$, ... , a primitive mathematical sign $\underline{\in}$, together with four primitive logical signs. These last are $\underline{\vee}$ (disjunction), $\neg$ (negation), τ and $\square$. (Some authors write $\sim$ in place of $\neg$. τ , which will be called the selector (see I.1.7), is an alias for Hilbert's " ε-symbol ". $\square$ may be termed the bound variable symbol.) Taken together, these are called primitive signs and are said to form the primitive alphabet. A letter may or may not be a variable, depending on the formal theory being considered; this will be explained in I.2.6. The sign $\underline{\in}$ will be thought of intuitively as that indicating (though in a slightly unconventional way) membership or elementhood; $\underline{\vee}$ and $\neg$ are likewise thought of as counterparts of "or" (in the sense of "either or..., or perhaps both") and "not" respectively, but the correspondence is not perfect; see I.2.9(i), Remark (ii) in I.3.2, and I.3.4 (vii). The last two logical signs defy intuitive meaning at the moment, but they will subsequently be used to define concepts which are closer to intuition and through which they may be endowed with some meaning; see I.1.7 below. (This is rather like a scientist's use of concepts ... such as potentials in mathematical physics ... which have but a tenuous direct link with reality but which seem useful in the study of other concepts with more direct links.) I stress again that the meanings are irrelevant to the internal workings of the formal theory, their sole legitimate use being as essential sources of suggestions and conjectures which have to be checked formally and without reference to the adopted meanings.

Except in the case of $\underline{\in}$ and $\underline{\vee}$, which are not formal letters, a single underline applied to letters of the lower case Roman and Greek alphabets, with or without primes or suffices, will indicate formal letters.

Other logical signs, $\Rightarrow$, $\Leftrightarrow$, $\wedge$, $\exists$, $\forall$, will be introduced as abbreviations a little later.

Returning momentarily to the bound variable symbol $\square$, Gleason ((1), pp.26-28) comments upon its use in this role, though mainly as an ad hoc device for avoiding the blunders which sometimes ensue from the use of letters to denote bound variables. Gleason's account does not present $\square$ as a primitive sign of any formal language, and there is no appearance of a selector τ.

I.1.2 Strings By a string is meant an explicitly written (therefore finite) succession of primitive signs written left to right, wherein certain appearances (perhaps none) of the sign τ are joined by overhead "ties" or "bonds" $\sqcap$ with certain appearances of the sign $\square$. Repetitions of the primitive signs are, of course, allowed ad libitum. For instance,

$$\tau\ \underline{v}\ \neg \in \square\ \underline{y} \in \square\ \underline{z} \qquad (1)$$

is a string of very modest length, in which the one appearance of τ is tied with each of the two appearances of $\square$.

(Were it not for the later technical use of the term "sequence" (see IV.7), one might define a string as a finite sequence of primitive signs together with certain ties; as it is, it seems best to avoid the terminology in this context ... as also again in I.1.5 and I.2.4 in defining constructions and proofs. See the comments at the beginning of §2 of the Appendix.)

If A and B denote strings, AB (or A B) will denote the string obtained by first writing the string denoted by A and then, immediately to the right thereof, the string denoted by B . Similarly, if A , B , C denote strings, ABC will denote the string otherwise denoted by A(BC) (which is identical with the string denoted by (AB)C), et cetera. (Here again one might frame the definitions in terms of finite sequences of primitive signs; see once more the outset of §2 of the Appendix.)

If $\underline{x}$ is a letter, there is a string whose one and only term is $\underline{x}$. Usually, this string will also be denoted by $\underline{x}$. (There are occasions where it is necessary to make a notational distinction between the letter $\underline{x}$ and the string $\underline{x}$; in the notation introduced in §2 of the Appendix, the string $\underline{x}$ would be distinguished by use of the vertical stroke $|$ as a suffix, thus : $\underline{x}_|$.)

I.1.3 <u>The use of τ ; replacements</u> There is no intention or expectation that a general string should, now or at any subsequent time, be endowed with any intuitive meaning. On the contrary, all subsequent interest focuses on two sorts of strings, called sentences and sets respectively, which will be carefully described by rules set out below, and which are the strings which may be helpfully endowed with some intuitive content: intuitively, sets represent (mathematical) objects, and sentences represent meaningful (but not necessarily true) assertions about such objects.

Having said this, I remind the reader yet again that the formal development has to proceed without reference to any adopted intuitive meanings. The role of the latter is to assist one in the planning of formal developments, which then have to be checked for admissibility against the rules. As will become apparent, this role is vital in practice.

Two further pieces of notation are needed to facilitate the definitions and subsequent proceedings.

First, if A denotes any string and $\underline{x}$ any letter, $\tau_{\underline{x}}(A)$ will denote the string obtained by writing out the string denoted by A , placing at its left extremity the sign τ , joining by an overhead tie ⌐¬ each appearance of $\underline{x}$ in A with the new initial sign τ , and then replacing $\underline{x}$ at each of these appearances by the sign $\square$. Thus, if A denotes $\underline{\in}\ \underline{x}\ \underline{y}$ then $\tau_{\underline{x}}(A)$ denotes $\tau\ \underline{\in}\ \square\ \underline{y}$; and $\tau_{\underline{y}}(\tau_{\underline{x}}(A))$ denotes $\tau\ \tau\ \underline{\in}\ \square\ \square$.

In certain strings, one or more ties may cross over certain other ties appearing in that string : such crossings are to be totally ignored. Any tie appearing in any given string has the sole purpose of linking an appearance of τ in that string with certain appearances of $\square$ in that string, its position in relation to all other ties being of no significance whatsoever.

Notice that $\underline{x}$ does not appear in the string denoted by $\tau_{\underline{x}}(A)$, despite the fact that it appears in the name of that string. This (possibly unexpected) feature will recur with other strings and the names adopted for, and/or the symbols used to denote them.

> I might interject here that ties appear in the formal language because of or in deference to the desired intuitive meanings attached to the formalism (about which more is said in I.1.7). Intuitively one does not want
>
> $$\tau_{\underline{x}}(\tau_{\underline{y}}(\underline{v}\ \underline{\in}\ \underline{x}\ \underline{y}\ \underline{\in}\ \underline{x}\ \underline{x}))$$
>
> and
>
> $$\tau_{\underline{x}}(\tau_{\underline{y}}(\underline{v}\ \underline{\in}\ \underline{x}\ \underline{y}\ \underline{\in}\ \underline{x}\ \underline{y}))$$
>
> to denote identical strings. If ties were omitted from the preceding definition of $\tau_{\underline{x}}$, however, they would both denote

one and the same string, namely

$$\tau\ \tau\ \underline{V}\ \underline{\in}\ \square\ \square\ \underline{\in}\ \square\ \square\ .$$

On the other hand, the given definition arranges that the strings denoted are respectively

and

$$\tau\ \tau\ \underline{V}\ \underline{\in}\ \square\ \square\ \underline{\in}\ \square\ \square$$

which are visibly different in respect of their ties.

Next, if A and B denote strings and $\underline{x}$ a letter, $(B|\underline{x})A$ will denote the string obtained by replacing the letter denoted by $\underline{x}$, at each of its appearances in the string denoted by A , by the string denoted by B . If the letter denoted by $\underline{x}$ does not appear in the string denoted by A , the string denoted by $(B|\underline{x})A$ is identical with A .

In the sequel, phrases like " $\underline{x}$ denotes a letter" and "the string denoted by A " sill usually be replaced by the phrases " $\underline{x}$ is a letter" and "the string A " , respectively. This is a blunder at the metamathematical level. It amounts to obliterating the distinction between being and denoting, and between a thing and the name for that thing. (On this point, see Kleene (1), §16 and Kleene (2), §38.) It will usually be clear after a momentary thought what is intended. Care is sometimes needed, however. For example, the result of replacing $\underline{x}$ by $\underline{y}$ in $\tau_{\underline{x}}(A)$ is presumably $\tau_{\underline{y}}(A)$; but the result of replacing $\underline{x}$ by $\underline{y}$ in the string denoted by $\tau_{\underline{x}}(A)$ — which is what is denoted by $(\underline{y}|\underline{x})\tau_{\underline{x}}(A)$ — is $\tau_{\underline{x}}(A)$. See also the remarks in I.1.4.

I.1.4 <u>Replacement rules</u> The following mechanical rules (actually instances of metatheorems; see I.2.9(v)) play a frequent role in subsequent developments. In them, A , B and C denote arbitrary strings, $\underline{x}$ and $\underline{y}$ denote distinct letters, and $\underline{x}$ and $\underline{x}'$ denote not necessarily distinct letters, the letters thus denoted being otherwise arbitrary.

(a) If $\underline{x}'$ does not appear in A , then $(B|\underline{x})A$ is identical with

$(B|\underline{x}')(\underline{x}'|\underline{x})A$.

(b) If $\underline{y}$ does not appear in B , then $(B|\underline{x})(C|\underline{y})A$ is identical with $(C'|\underline{y})(B|\underline{x})A$, where C' denotes $(B|\underline{x})C$. In particular, $(B|\underline{x})(C|\underline{y})A$ is identical with $(C|\underline{y})(B|\underline{x})A$ if $\underline{x}$ does not appear in C and $\underline{y}$ does not appear in B .

(c) If $\underline{x}'$ does not appear in A , then $\tau_{\underline{x}}(A)$ is identical with $\tau_{\underline{x}'}((\underline{x}'|\underline{x})A)$.

(d) If $\underline{x}$ does not appear in B , then $(B|\underline{y})\tau_{\underline{x}}(A)$ is identical with $\tau_{\underline{x}}((B|\underline{y})A)$.

In connection with rule (b), the reader should note that, if $\underline{x}$ appears in the string denoted by C , one must complete the replacement of C for $\underline{y}$ in A <u>before</u> making the replacement of B for $\underline{x}$ in the result. For example, suppose A , B , C denote respectively the strings $\underline{x} \underline{\in} \underline{\vee} \underline{y}$, $\underline{z}$, $\underline{\in} \underline{x} \underline{x} \underline{u}$;

RIGHT		WRONG	
$(C\|\underline{y})A$ denotes	$\underline{x} \underline{\in} \underline{\vee} \underline{\in} \underline{x} \underline{x} \underline{u}$	$(C\|\underline{y})A$ denotes	$\underline{x} \underline{\in} \underline{\vee} \underline{C}$
$(B\|\underline{x})(C\|\underline{y})A$ denotes	$\underline{z} \underline{\in} \underline{\vee} \underline{\in} \underline{z} \underline{z} \underline{u}$	$(B\|\underline{x})(C\|\underline{y})A$ denotes	$\underline{z} \underline{\in} \underline{\vee} \underline{C}$
		which denotes	$\underline{z} \underline{\in} \underline{\vee} \underline{\in} \underline{x} \underline{x} \underline{u}$.

In general, iterated replacements need great care; see the discussion in II.1.3(ii).

Alongside (a) - (d) one should place similar but simpler replacement rules involving the primitive signs $\neg$, $\underline{\vee}$, and $\underline{\in}$. Thus, if A , B and C denote strings and $\underline{x}$ a letter, then $(C|\underline{x})(\neg A)$, $(C|\underline{x})(\underline{\vee}AB)$, $(C|\underline{x})(\underline{\in}AB)$ are identical with $\neg A'$, $\underline{\vee}A'B'$ and $\underline{\in}A'B'$, where A' and B' are $(C|\underline{x})A$ and $(C|\underline{x})B$ respectively. Exactly similar replacement rules apply in the case of the abbreviatory signs $\Rightarrow$, $\Leftrightarrow$, and $\wedge$ to be introduced in I.1.7. The replacement rules referring to the further abbreviatory signs $\exists$ and $\forall$ will be listed explicitly at the appropriate place (see (e) and (f) in I.1.8 below).

Perhaps it should be stressed again (see the final paragraph in I.1.3) that one must sometimes take care to distinguish between a replacement in a string and the same replacement in a name for that string: our concern will always be with the former procedure. Thus, if N is the name of a string, $(A|\underline{x})N$ denotes the result of replacing the letter denoted by $\underline{x}$ by the string denoted by A _in the string denoted by_ N , _not_ in the symbol N itself. In particular, $(A|\underline{x})\tau_{\underline{x}}(B)$ denotes $\tau_{\underline{x}}(B)$ (since $\underline{x}$ does not appear in the string denoted by $\tau_{\underline{x}}(B)$) and not $\tau_A(B)$ (which, if A denotes a string which is not a letter, denotes no string); and $(B|\underline{y})\tau_{\underline{x}}(\in \underline{x}\ \underline{y})$ is not always identical with $\tau_{\underline{x}}(\in \underline{x}\ \underline{B})$ (though it is, according to (d) above, if $\underline{x}$ does not appear in the string denoted by B) . In many cases, investigation proves that the distinction is after all unnecessary. In fact, this is often the point of replacement rules; see, for instance II.1.3(ii), (p) in II.1.3(vi), its analogues at the ends of II.2.1 and II.4.1, and analogous statements following many formal definitions appearing in the sequel. But see also the end of II.3.7 and IV.2.3(iii).

As a simple example of another sometimes useful replacement rule, consider the following variant of (b):

(b') $(B|\underline{x})(C|\underline{x})A$ is identical with $(C'|\underline{x})A$ where C' denotes $(B|\underline{x})C$.

To verify this on the basis of (a) and (b) above, choose a letter $\underline{y}$ different from $\underline{x}$ and not appearing in A , B or C ; then $\underline{y}$ does not appear in C' either. So, by (a)

$$(B|\underline{x})(C|\underline{x})A \text{ is identical with } (B|\underline{x})(C|\underline{y})(\underline{y}|\underline{x})A ,$$

which, by (b), is identical with

$$(C'|\underline{y})(B|\underline{x})(\underline{y}|\underline{x})A ;$$

since $\underline{x}$ does not appear in $(\underline{y}|\underline{x})A$, this last is identical with

$$(C'|\underline{y})(\underline{y}|\underline{x})A \quad ;$$

and, finally, (a) affirms that this is identical with

$$(C'|\underline{x})A \quad ,$$

as alleged in (b').

See also Problem I/6.

I.1.5 Strings of the first and second kinds; constructions and constructs Strings are subdivided into two kinds, a string being said to be of the first kind if it is a letter or if it commences with the sign τ , and to be of the second kind otherwise.

By a construction is meant a (finite) list of strings written in some order (say down the page for definiteness) such that, for every string A of the list, one at least of the following is true:

(i) A is a letter;

(ii) preceding A in the list there is a string B of the second kind, such that A is $\neg B$;

(iii) preceding A in the list there are two strings B and C of the second kind, such that A is $\underline{v}BC$;

(iv) preceding A in the list there is a string B of the second kind and a letter $\underline{x}$, such that A is $\tau_{\underline{x}}(B)$;

(v) preceding A in the list there are two strings B and C of the first kind, such that A is $\underline{\in}BC$.

By a (formal mathematical) construct is meant a string which appears (that is, is listed) in some (that is, at least one) construction.

For example, the following list of strings is a construction:

$$\underline{x}$$

$$\underline{y}$$

$$\underline{z}$$

$$\underline{\in}\ \underline{x}\ \underline{y}$$

$$\underline{\in}\ \underline{x}\ \underline{z}$$

$$\neg\underline{\in}\ \underline{x}\ \underline{y}$$

$$\underline{\vee}\ \neg\underline{\in}\ \underline{x}\ \underline{y}\ \underline{\in}\ \underline{x}\ \underline{z}$$

$$\tau\ \underline{\vee}\ \neg\underline{\in}\ \square\ \underline{y}\ \underline{\in}\ \square\ \underline{z} \qquad (1)$$

In this construction, the first three and the last strings are of the first kind; the fourth, fifth, sixth and seventh are of the second kind.

Three further and more interesting examples of constructions are, for convenience, delayed until near the end of I.1.8.

Observe that no construct commences with the sign $\square$; and that no construct commences with a letter unless it ends there (and is thus the string consisting of that one letter).

I.1.6 Sets and sentences A set is defined to be a string of the first kind which appears in some construction; and a (formal) sentence is defined to be a string of the second kind which appears in some construction.

Otherwise expressed:

> A set is a construct which (as a string) is either a letter or commences with the sign τ . A sentence is a construct which (as a string) commences with one or other of the signs $\underline{\in}$, $\underline{\vee}$, $\neg$.

For example, in the construction appearing in I.1.5 above, the first three and the last strings are sets; the fourth, fifth, sixth and seventh strings are sentences. In particular, I.1.5(1) (identical with I.1.2(1)) is a set.

> It is interesting to compare the above formal definition of "set" with Cantor's definition, cited at the end of I.0.1 above. The formal definition is strikingly more explicit and (in a sense) constructive; sets are to be the end products of carefully regulated constructions. One might presume this definition to be much more restrictive than Cantor's, perhaps so much so as to engender a fear that the resulting concept of set may be too special to be adequate for the purposes of mathematics in general. However, the variety and complexity hidden in this seemingly so restrictive formal definition gradually comes to be recognised as quite astonishing.

A construct is either a set (and then either a letter or a string commencing with τ) or a sentence (in which case it is a string commencing with one or other of the signs $\underline{\in}$, $\underline{\vee}$, $\neg$. Note carefully, however, that not every string commencing with τ is a set; and not every string commencing with one of $\underline{\in}$, $\underline{\vee}$, $\neg$ is a sentence.

> It is possible to analyse and describe in combinatorial terms those strings which are sets and those which are sentences (see Bourbaki (1), pp.50-55). This provides an effective procedure for deciding whether an arbitrarily given string is or is not a set or a sentence; the existence of such a procedure is metamathematically significant but of no practical use in mathematics. (Were there a similar procedure for deciding provability or otherwise of arbitrarily given sentences, <u>that</u> would be of supreme significance. But in fact there is <u>no</u> such procedure. That is to say, not only is no such procedure known at this time, but it can be proved metamathematically that no such procedure exists; see Remark (iv) in I.2.4.)

The term "(formal) sentence" is being used to cover both open and closed sentences (that is, ones with and without letters appearing in them). To use the logician's expression, they are the well-formed formulae of the language. Similarly, I use the word "set" to describe what the logician might refer to as a well-formed term of the language. I use the words "set" and "sentence", since they seem adequate for my purposes and are less forbidding. Some writers use the term "proposition" in place of "sentence, though I shall not so use it; see also the end of I.2.4.

Intuitively, one is to think of sets as being the mathematical objects, and sentences as being meaningful statements about such objects.

The reader may be assailed by the feeling that the word "set" would be better reserved for those sets (such as $\emptyset$, (as defined in Chapter II), N (as defined in Chapter V) and certain realisations of R , Z , Q (as defined in Chapter VI)) which are specific (fixed) in the sense that in them appear no letters, or for those sets which are equal to such a set, the word "term" being applied to sets in which letters may appear. Alternatively, since sentences in which one or more letters appear correspond to what most logicians refer to as sentence forms or propositional forms or propositional functions, one might by analogy refer to sets in which one or more letters appear as set forms. (The term set function is already firmly fixed in mathematics in a quite different sense.)

The position is that in a set S (more properly: in the string denoted by S) there may appear one or more variables (see I.2.6). Informally one may then tend to think of S rather as (or as denoting) a "bundle" (or "family") of sets, one for each selection of "values" of the variables appearing in S . For example, if S denotes $N \cup \underline{x}$, then S denotes a set in which the variable $\underline{x}$ appears; one might informally denote it by $S_{\underline{x}}$ and think of particular sets $S_{\emptyset}$, S_1 , S_{π} , and so on: these are accounted for formally as $(\emptyset|\underline{x})S$, $(1|\underline{x})S$, $(\pi|\underline{x})S$ respectively; and in general S_a is $(a|\underline{x})S$ for an arbitrary string a . See Note 2.

Very frequently, logicians would write $S(\underline{x})$ to indicate that $\underline{x}$ appears in the set (or set form) S and so use S(a) to denote $(a|\underline{x})S$ whenever a denotes a set (or set form). In our formalism at any rate, this is a bad notation for reasons which will emerge in IV.1.4(ii) : confusion arises with the notation for functional values. It is preferable in such situations to write $S[\![a]\!]$, in accord with the notation introduced in II.1.3(iv) below. Similar remarks apply to sentences in which there appear one or more variables.

In any case, one often needs to refer to sets in which letters appear in the course of defining sets in which letters do not appear. For instance, if S is a set (in which letters may or may not appear), one later defines the identity function I_S with domain S to be the set (denoted by)

$$\{(\underline{x}, \underline{x}) : \underline{x} \in S\} ,$$

where $\underline{x}$ is a letter not appearing in S ; and here, although $\underline{x}$ does not appear in the set (denoted by) I_S , reference is made to the set (denoted by) $(\underline{x}, \underline{x})$ in which the letter $\underline{x}$ does appear.

In the sequel, I shall denote sentences by $\underline{\underline{A}}$, $\underline{\underline{B}}$, $\underline{\underline{C}}$, ... and sets by X , X' , Y , Y' , Z , Z' , S , S' , T , T' , ... ; but the reader should bear in mind that every letter is a set. Moreover, lower case letters will also come to be used to denote sets, especially when the sets referred to are informally almost always thought of as things not closely related to sets -

numbers, for example.

Remark It has been stated immediately above that "every letter is a set". More precisely :

If $\underline{x}$ denotes a letter, then the string whose sole term is $\underline{x}$ (see the final paragraph in I.1.2) is a set.

is a true metastatement. On the other hand,

$$\underline{x} \text{ is a set} \tag{1}$$

is plainly not itself a formal sentence; what is perhaps more to the point, nowhere in this book is it adopted as a name for any formal sentence. In this respect it is unlike

$\underline{x}$ is an ordered pair,

$\underline{x}$ is a function ;

these and many other similar bastard expressions are later defined to be and adopted as names for certain formal sentences. There is nothing in principle to prevent the adoption of (1) as a name for some formal sentence; the fact is that there is no formal sentence for which it is a "sensible" and usefully suggestive name. (The situation is otherwise in certain other versions of formal set theory.)

I.1.7 Further notation: the quantifiers At this point I hesitate no longer before modifying the notation in a way that is likely to aid intuitive comprehension, repeating that intuition is a vital aid in framing the rules of a formal system designed with some purpose in mind and in guiding the development of such a system.

Accordingly I make the following modifications:

(i) Introduce a new symbol $\in$ with the convention that, if A and B denote arbitrary strings, one will write more familiarly $A \in B$ in place of (as a name for) the string denoted by $\underline{\in}\, A\, B$; and $A \notin B$ in place of the string denoted by $\neg \underline{\in}\, A\, B$.

Although the distinction between $\in$ and $\underline{\in}$ may look artificial, it has its explanation in seeking to avert possible confusions in connection with the metastatements I.1.8(g), (h), (i) to follow.

The temptation is to ignore the distinction. If one does this, however, the entity $\underline{x} \underline{\in} \underline{y}$ may be viewed both as a string and, at the same time, as denoting a different string (to wit, the string $\underline{\in}\ \underline{x}\ \underline{y}$) . Then I.1.8(g) and (i) might be construed so as to lead to the conclusion that $\underline{x} \underline{\in} \underline{y}$ denotes a sentence, and I.1.8(h) might be construed so as to lead to the conclusion that

$$(\exists \underline{x})(\underline{x} \underline{\in} \underline{y}) \equiv (\tau_{\underline{x}}(\underline{x} \underline{\in} \underline{y})|\underline{x})(\underline{x} \underline{\in} \underline{y}) \qquad (1)$$

denotes a sentence. Yet the right hand side of (1) denotes a string which commences with the sign τ , and so (see I.1.5) denotes a string of the first kind. By I.1.6, therefore, the right hand side of (1) denotes a string which cannot be a sentence. An impasse emerges! It is to avert this type of disaster, while at the same time adopting a conventional notation, that the distinction between $\underline{\in}$ and $\in$ is made. (I am grateful to Mr George Harvey for pointing out this potential contretemps.)

Note that $\in$ is not a sign of the formal language; it is rather a semi-formal sign.

(ii) Similarly, introduce a new sign $\vee$ (not a sign of the formal language) and agree to write more familiarly $A \vee B$ in place of $\underline{\vee}\, A\, B$, A and B here denoting arbitrary strings.

(iii) The remaining standard logical signs are introduced as abbreviations, thus:

$A \Rightarrow B$ will be written in place of $\underline{\vee} \neg A\, B$, that is, in place of $(\neg A) \vee B$;

$A \wedge B$ will be written in place of $\neg((\neg A) \vee (\neg B))$;

$A \Leftrightarrow B$ will be written in place of $(A \Rightarrow B) \wedge (B \Rightarrow A)$;

$(\exists \underline{x})A$ will be written in place of $(\tau_{\underline{x}}(A)|\underline{x})A$;

$(\forall \underline{x})A$ will be written in place of $\neg(\exists \underline{x})(\neg A)$;

here A and B denote arbitrary strings.

The above notations are not universally adopted. For example, $\sim$ is often used in place of our $\neg$; and our $\neg A$ may be replaced (in various books) by any one of $\sim A$, $\bar{A}$, A' . Also, $\supset$ may be used in place of our $\Rightarrow$; we later (see II.2.1) use $\supset$ in a quite different (but commonly adopted) sense.

$\exists$ is referred to as the existential quantifier and $\forall$ as the universal quantifier; see the intuitive interpretations explained below.

It is important to notice that $\underline{x}$ appears in neither of the strings denoted by $(\exists \underline{x})A$ and $(\forall \underline{x})A$.

These modified and abbreviatory notations will certainly facilitate the intuitive interpretation of formal sentences, at least for most readers acquainted with informal mathematics. (There is in reality a price to be paid for using this more familiar notation, inasmuch as it frequently demands the use of parentheses to avoid ambiguities. Thus, it is not clear whether $A \Rightarrow A \vee B$ is intended to denote $A \Rightarrow (A \vee B)$ or $(A \Rightarrow A) \vee B$; these possibilities would be distinguished in the primitive language as $\underline{\vee} \neg A \underline{\vee} A B$ and $\underline{\vee} \underline{\vee} \neg A A B$ respectively. Similarly, $(\exists \underline{x})(A)$ and $(\forall \underline{x})(A)$ are often written in place of $(\exists \underline{x})A$ and $(\forall \underline{x})A$ respectively.)

One ought to carefully codify such use of parentheses, but I omit to do this; cf. Margaris (1), §8. Furthermore, such auxiliary metamathematical use of parentheses has to be distinguished from their mathematical use in the semi-formal language, instanced in the notation (a, b) for ordered pairs and in the functional value notation $f(A)$ introduced in III.1.1 and IV.1.2 below.

In most accounts of logic, one or other of $\exists$ and $\forall$ is treated as an independent sign of the primitive alphabet. In the scheme presented here, $\exists$ is defined in terms of the selector τ and $\forall$ is defined in terms of $\exists$. This leads to a number of important differences in detail, especially in the behaviour of the quantifiers in relation to replacements (often referred to as substitutions). In most accounts, $(\exists \underline{x})A$ and $(\forall \underline{x})A$ are themselves viewed as strings, whereas in our scheme they are properly to be regarded as names for strings. (As has been noted in I.1.3 and I.1.4, it is sometimes vital to recognise this distinction between a name and that of which it is a name.) Most accounts are also somewhat hazy about replacements (or substitutions) by (or of) strings of (or for) variables; in our scheme the description is clearer and is regulated in practice by the replacement rules in I.1.4 and I.1.8. It may be helpful to take a simple illustration suggested by the last two lines on p.81 and the footnote * on page 82 of Gordon and Hindman (1). The usual development, as presented loc. cit., leads one to infer that the result of replacing x by y in $\exists y(x \neq y)$ is $\exists y(y \neq y)$. (Notice the differences between their notation and ours, in spite of which there is a clear correspondence.) On the other hand, if $\underline{x}$ and $\underline{y}$ denote distinct letters, in our scheme it is a consequence of definitions appearing in Chapter II, combined with I.1.8(e) and (f), that $(\underline{y}|\underline{x})(\exists\underline{y})(\underline{x} \neq \underline{y})$ is identical with $(\exists\underline{z})(\underline{y} \neq \underline{z})$, where $\underline{z}$ denotes a letter different from $\underline{x}$ and $\underline{y}$; and, by I.1.8(e) again, this is identical with $(\exists\underline{x})(\underline{y} \neq \underline{x})$. However, $(\exists\underline{y})(\underline{y} \neq \underline{y})$ and $(\exists\underline{x})(\underline{y} \neq \underline{x})$ are quite different; in fact, the theorems of Chapter II show that the first is false and the second true.

Turning to intuitive meanings, $\underline{\underline{A}} \Rightarrow \underline{\underline{B}}$ is the formal counterpart of

$\underline{\underline{A}}$ implies $\underline{\underline{B}}$

or

if $\underline{\underline{A}}$, then $\underline{\underline{B}}$

or

$\underline{\underline{A}}$ is a sufficient condition for $\underline{\underline{B}}$,

while $\underline{\underline{A}} \wedge \underline{\underline{B}}$ is the counterpart of

$\underline{\underline{A}}$ and $\underline{\underline{B}}$,

the conjunction of $\underline{\underline{A}}$ and $\underline{\underline{B}}$; and $\underline{\underline{A}} \Leftrightarrow \underline{\underline{B}}$ is the counterpart of

$\underline{\underline{A}}$ if and only if $\underline{\underline{B}}$

or

$\underline{\underline{A}}$ is necessary and sufficient for $\underline{\underline{B}}$.

The phrases "if and only if" and "necessary and sufficient condition" are often abbreviated to "iff" and "NASC" respectively.

The interpretation of the remaining signs calls for somewhat more imagination. If one thinks of the sentence $\underline{\underline{A}}$ in which the letter $\underline{x}$ appears as expressing a property of $\underline{x}$, $\tau_{\underline{x}}(\underline{\underline{A}})$ is to be thought of as a miraculously pre-selected set, if there is any such set, which has that property; if there is no such set, $\tau_{\underline{x}}(\underline{\underline{A}})$ is to be thought of as a set about which one can a priori say nothing (save perhaps that it does not have the property expressed by $\underline{\underline{A}}$). This is the reason why τ is termed the <u>selector</u> and $\tau_{\underline{x}}$ the <u>x-selector</u>. To affirm $(\exists\underline{x})\underline{\underline{A}}$ is thus to affirm that there is (that is, it is possible to find or choose) at least one set having the property expressed by $\underline{\underline{A}}$; about this set one may a priori know nothing more. Since $\neg$ is interpreted as negation or denial, to affirm $(\forall\underline{x})\underline{\underline{A}}$ is accordingly to affirm that there is no set which fails to have the property expressed by $\underline{\underline{A}}$; that is, that every set possesses the property expressed by $\underline{\underline{A}}$. Thus $(\exists\underline{x})\underline{\underline{A}}$ is usually read as

there exists $\underline{x}$, $\underline{\underline{A}}$

or

there exists $\underline{x}$ such that $\underline{\underline{A}}$;

and $(\forall\underline{x})\underline{\underline{A}}$ is usually read as

for all $\underline{x}$, $\underline{\underline{A}}$

or

$\underline{\underline{A}}$ for all $\underline{x}$,

but see the remarks in I.2.9(vii) and (viii), I.3.4(iii) and II.4.4(ii).

Notice that $\tau_{\underline{x}}(\underline{\underline{A}})$ denotes a set, even though $\underline{\underline{A}}$ may be self contradictory; that is, even though $(\forall\underline{x})\neg\underline{\underline{A}}$ and $\neg(\exists\underline{x})\underline{\underline{A}}$ may both be theorems. It might appear that as a result the formal theory embraces sets which would informally be regarded as absurdities. This is a false alarm, however, because there is no assurance (indeed, no reason to expect) that the set denoted by $\tau_{\underline{x}}(\underline{\underline{A}})$ has the intuitively expected property expressed by $\underline{\underline{A}}$, unless $(\exists\underline{x})\underline{\underline{A}}$ is in fact a theorem. This is why many formal definitions to follow, expressed in terms of the selector, are accompanied (before or after) by a discussion of the appropriate existence theorem; see, for example, the opening remarks in II.3.3, II.4.1, II.6.1 and III.2.2.

When $(\exists\underline{x})\underline{\underline{A}}$ is a theorem, there is a set X such that $(X|\underline{x})\underline{\underline{A}}$ is a theorem; there is, moreover, such a set X , namely $\tau_{\underline{x}}(\underline{\underline{A}})$, in which $\underline{x}$ does not appear and in which the only letters appearing are those (if any) which are different from $\underline{x}$ and which appear in $\underline{\underline{A}}$.

Many enlightening comments on the origin of the τ-symbol (alias Hilbert's ε-symbol), and the ι-symbol which preceded it, are to be found in Kneebone (1), pp.91-107. See also Kleene (2), pp.167-172.

Needless to say, the above intuitive interpretations, and others like them, should never be regarded as providing final justification in formal or semi-formal proofs. Their role -- an essential one, to be sure -- is to act as aids in conjecturing significant theorems to be proved and in constructions of proofs.

thereof. Intuition is used in this latter role almost constantly in everyday mathematics (without it, progress would soon grind to a halt); and it is effective almost always. However, there are situations in which intuition leaves one floundering (see II.4.2(iii) and Remark (ii) in II.10.4) or even leads one astray in everyday mathematics (usually because intuition fails to grasp the hidden content of definitions).

Again, there are a few respects in which intuition and formality fail to harmonise fully right from the start. For example, it will soon emerge that the formal theory is such that a sentence $\underline{\underline{A}} \Leftrightarrow \underline{\underline{B}}$ is true whenever the sentences $\underline{\underline{A}}$ and $\underline{\underline{B}}$ are either both true or both false; in particular, every two theorems are equivalent. Both conclusions are, to say the least, intuitively unexpected. See also I.2.9(i)-(iii), Remark (ii) in I.3.2 and Problem I/34.

I.1.8 More replacement rules; construction rules The replacement rules (a) - (d) in I.1.4 lead directly to the following analogues of (c) and (d) for the quantifiers (A, B, $\underline{x}$, $\underline{x}'$, $\underline{y}$ are as in I.1.4).

(e) If $\underline{x}$ and $\underline{x}'$ denote letters, and if $\underline{x}'$ does not appear in A, then $(\exists\underline{x}')(\underline{x}'|\underline{x})A$ is identical with $(\exists\underline{x})A$; and likewise with $\forall$ in place of $\exists$ throughout.

(f) If $\underline{x}$ and $\underline{y}$ denote distinct letters, and if $\underline{x}$ does not appear in S, then $(S|\underline{y})(\exists\underline{x})A$ is identical with $(\exists\underline{x})(S|\underline{y})A$; and likewise with $\forall$ in place of $\exists$ throughout.

See also Problem I/17(i).

Additionally there are three construction rules (or construction methods), listed as (g), (h) and (i) below. Each of these is a simple metatheorem (that is, a true metamathematical statement; cf. I.2.9(v) below) whose purpose is to render it unnecessary to revert to the defining rules I.1.5(i) - (v) when constructions are required, thus greatly shortening the construction process and its description. (In this respect they play vis à vis constructions a role exactly analogous to that played by the proof methods in I.3.2 and I.3.3 in relation to proofs; see the remarks in I.3.1. The use of these construction rules almost always replaces any overt reference to I.1.5(i) - (v).

As with replacement rules, so with construction rules (and proof methods): they are in principle avoidable, but in practice almost indispensable on the grounds of economy.

As metatheorems, the construction rules demand verification; on this point, see §2 of the Appendix.

(g) Any letter is a set; if $\underline{\underline{A}}$ denotes an arbitrary sentence and $\underline{x}$ an arbitrary letter, then $\tau_{\underline{x}}(\underline{\underline{A}})$ denotes a set; if S and T denote arbitrary sets and $\underline{x}$ an arbitrary letter, then $(S|\underline{x})T$ denotes a set.

(h) If $\underline{\underline{A}}$ and $\underline{\underline{B}}$ denote arbitrary sentences, S an arbitrary set, and $\underline{x}$ an arbitrary letter, then $\neg\underline{\underline{A}}$, $\underline{\underline{A}} \vee \underline{\underline{B}}$, $\underline{\underline{A}} \wedge \underline{\underline{B}}$, $\underline{\underline{A}} \Rightarrow \underline{\underline{B}}$, $\underline{\underline{A}} \Leftrightarrow \underline{\underline{B}}$, $(\exists \underline{x})\underline{\underline{A}}$, $(\forall \underline{x})\underline{\underline{A}}$ and $(S|\underline{x})\underline{\underline{A}}$ all denote sentences.

(i) If S and T denote arbitrary sets, then $S \in T$ denotes a sentence.

At the risk of writing what is totally superfluous, I nevertheless indicate to the reader not to confuse the metastatements "(the letter) $\underline{x}$ appears in the set X " and "the sentence denoted by $\underline{x} \in X$ is a theorem". Both meta-atatements are true in special cases; for example if X is $\{\underline{x}, \underline{y}\}$, in which case $\underline{x}$ and $\underline{y}$ are the letters appearing in X , and the sentences denoted by $\underline{x} \in X$ and $\underline{y} \in Y$ <u>are</u> theorems. But the confusion is usually illegitimate. For example, N (defined as in Chapter V) denotes a set in which no letters appear, and $N \setminus \{\underline{x}\}$ denotes a set in which the letter $\underline{x}$ appears; but (assuming the theory to be consistent), $\underline{x} \in N \setminus \{\underline{x}\}$ is not a theorem of set theory; and in any case $\neg(\underline{x} \in (N \setminus \{\underline{x}\}))$ is a theorem.

This subsection terminates with three examples of constructions. (It is at this early stage difficult to provide illustrations which will be significant from the informal point of view.) The first construction leads to the empty set (characterised by having no elements at all). The second leads to a set whose only element is the empty set, later to be denoted by 1 (the first nonzero natural number). The third leads to the set later (see II.3.2) denoted by $\{\underline{x} : A\}$. In all three constructions, $\underline{x}$ and $\underline{y}$ denote distinct letters. In the first construction it is possible to write out the strings involved, though the second column indicates abbreviated and more familiar names. In the second and third constructions, the strings referred to become too long for explicit inscription to be practicable and it is obligatory to use names.

The first construction is as follows, entries in the third column referring to I.1.5 unless otherwise indicated:

$\underline{x}$		((i))
$\underline{y}$		((i))
$\underline{\in}\ \underline{y}\ \underline{x}$	$\underline{y} \in \underline{x}$	((v))
$\neg\ \underline{\in}\ \underline{y}\ \underline{x}$	$\underline{y} \notin \underline{x}$	((ii))
$\neg\ \neg\ \underline{\in}\ \underline{y}\ \underline{x}$	$\neg(\underline{y} \notin \underline{x})$	((ii))
$\tau\ \neg\ \neg\ \underline{\in}\ \square\ \underline{x}$	$\tau_{\underline{y}}(\neg(\underline{y} \notin \underline{x}))$	((iv))
$\neg\ \neg\ \underline{\in}\ \tau\ \neg\ \neg\ \underline{\in}\ \square\ \underline{x}\ \underline{x}$	$(\exists \underline{y})\neg(\underline{y} \notin \underline{x})$	(I.1.8(g) or ((h))
$\neg\ \neg\ \neg\ \underline{\in}\ \tau\ \neg\ \neg\ \underline{\in}\ \square\ \underline{x}\ \underline{x}$	$\neg(\exists \underline{y})\neg(\underline{y} \notin \underline{x})$ or $(\forall \underline{y})(\underline{y} \notin \underline{x})$	((ii))
$\tau\ \neg\ \neg\ \neg\ \underline{\in}\ \tau\ \neg\ \neg\ \underline{\in}\ \square\ \square\ \square$	$\tau_{\underline{x}}((\forall \underline{y})(\underline{y} \notin \underline{x}))$	((iv))

The final string appearing in this list is later (see II.5) denoted by $\emptyset$; it is a string of length 12.

For the description of the second construction it is necessary to assume the definition of = given in II.1.1 and the elementary metamathematical fact that $X = Y$ denotes a sentence whenever X and Y denote sets. Using abbreviating names where necessary, the construction is indicated thus :

$\underline{x}$		((i))
$\underline{y}$		((i))
$\underline{\in}\ \underline{x}\ \underline{y}$		((v))
$\underline{x} = \emptyset$		(II.1.1)
$(\underline{\in}\ \underline{x}\ \underline{y}) \Leftrightarrow (\underline{x} = \emptyset)$	denote this temporarily by $\underline{\underline{S}}$	(I.1.8(h))
$\neg\underline{\underline{S}}$		((ii))
$\tau_{\underline{x}}(\neg\underline{\underline{S}})$		((iv))
$(\tau_{\underline{x}}(\neg\underline{\underline{S}})\|\underline{x})\neg\underline{\underline{S}}$	$(\exists\underline{x})\neg\underline{\underline{S}}$	(I.1.8(g))
$\neg(\tau_{\underline{x}}(\neg\underline{\underline{S}})\|\underline{x})\neg\underline{\underline{S}}$	$(\forall\underline{x})\underline{\underline{S}}$	((ii))
$\tau_{\underline{y}}(\neg(\tau_{\underline{x}}(\neg\underline{\underline{S}})\|\underline{x})\neg\underline{\underline{S}})$	$\tau_{\underline{y}}((\forall\underline{x})(\underline{x} \in \underline{y} \Leftrightarrow \underline{x} = \emptyset))$	((iv))

The set denoted in the last line is a string of length roughly 10,000. It is equal to (but not identical with) the set later denoted by 1 (which denotes a string of much greater length, of which the construction would be much longer than that given above).

In the third construction, now to be described, $\underline{\underline{A}}$ denotes an arbitrary sentence, $\underline{x}$ an arbitrary letter, and $\underline{y}$ any letter distinct from $\underline{x}$ and not appearing in $\underline{\underline{A}}$. (The appearance of the unspecified sentence $\underline{\underline{A}}$ indicates that one is here concerned with a construction schema (cf. the sentence schemas discussed in I.1.10) rather than a single construction.) The construction has to begin with a construction of $\underline{\underline{A}}$ and may be indicated as follows :

⋮	
$\underline{\underline{A}}$	
$\underline{x}$	((i))
$\underline{y}$	((i))

$\in \underline{x}\ \underline{y}$	$\underline{x} \in \underline{y}$	((v))
$\neg \in \underline{x}\ \underline{y}$	$\underline{x} \notin \underline{y}$	((ii))
$\vee \neg \in \underline{x}\ \underline{y}\ \underline{\underline{A}}$	$(\underline{x} \notin \underline{y}) \vee \underline{\underline{A}}$ or $(\underline{x} \in \underline{y}) \Rightarrow \underline{\underline{A}}$; denote this temporarily by $\underline{\underline{B}}$	((iii))
$\neg\underline{\underline{A}}$		((ii))
$\vee \neg \underline{\underline{A}} \in \underline{x}\ \underline{y}$	$(\neg\underline{\underline{A}}) \vee (\underline{x} \in \underline{y})$ or $\underline{\underline{A}} \Rightarrow (\underline{x} \in \underline{y})$; denote this temporarily by $\underline{\underline{C}}$	((iii))
$\neg\underline{\underline{B}}$		((ii))
$\neg\underline{\underline{C}}$		((ii))
$\vee \neg \underline{\underline{B}} \neg \underline{\underline{C}}$	$(\neg\underline{\underline{B}}) \vee (\neg\underline{\underline{C}})$	((iii))
$\neg \vee \neg \underline{\underline{B}} \neg \underline{\underline{C}}$	$\neg((\neg\underline{\underline{B}}) \vee (\neg\underline{\underline{C}}))$ or $\underline{\underline{B}} \wedge \underline{\underline{C}}$ or $(\underline{x} \in \underline{y}) \Leftrightarrow \underline{\underline{A}}$; denote this temporarily by $\underline{\underline{D}}$	((ii))
$\neg\underline{\underline{D}}$		((ii))
$\tau_{\underline{x}}(\neg\underline{\underline{D}})$		((iv))
$(\tau_{\underline{x}}(\neg\underline{\underline{D}})\,\vert\,\underline{x})\neg\underline{\underline{D}}$	$(\exists\underline{x})\neg\underline{\underline{D}}$	(I.1.8(g) or (h))
$\neg(\tau_{\underline{x}}(\neg\underline{\underline{D}})\,\vert\,\underline{x})\neg\underline{\underline{D}}$	$\neg(\exists\underline{x})\neg\underline{\underline{D}}$ or $(\forall\underline{x})\underline{\underline{D}}$ or $(\forall\underline{x})((\underline{x} \in \underline{y}) \Leftrightarrow \underline{\underline{A}})$; denote this temporarily by $\underline{\underline{E}}$	((ii))
$\tau_{\underline{y}}(\underline{\underline{E}})$		((iv))

The terminating string here, namely, that denoted by

$$\tau_{\underline{y}}((\forall\underline{x})((\underline{x} \in \underline{y}) \Leftrightarrow \underline{\underline{A}})) \quad ,$$

is the set which in II.3 will be adopted as the definition of $\{\underline{x} : \underline{\underline{A}}\}$, which set is termed "the set of (all) $\underline{x}$ such that $\underline{\underline{A}}$ " . As will be seen in due course, (see II.3.3 and II.3.10(iii)), this last name or description is usually intuitively appropriate, but may on occasion be misleading.

I.1.9 Regarding informal terminology If $\underline{\underline{A}}$ denotes a sentence and $\underline{x}$ a letter which appears in $\underline{\underline{A}}$ and in which one is temporarily especially interested, one often informally speaks of $\underline{\underline{A}}$ as being (or, better: as expressing) a condition (or restriction) on (or relative to) $\underline{x}$, or to express a property of $\underline{x}$, the condition or property naturally depending upon $\underline{\underline{A}}$. Then, if S denotes a set, the sentence denoted by $(S|\underline{x})\underline{\underline{A}}$ is regarded as expressing the same condition on S or the same property of S . If also the sentence denoted by $(S|\underline{x})\underline{\underline{A}}$ is true (see I.2.4), one may say that " S satisfies the condition on $\underline{x}$ (expressed by) $\underline{\underline{A}}$ " or that " S possesses the property of $\underline{x}$ expressed by $\underline{\underline{A}}$ " . One may even say (more briefly but ambiguously) that " S satisfies $\underline{\underline{A}}$ " or that " S has property $\underline{\underline{A}}$ " .

Similar remarks apply when there appear in $\underline{\underline{A}}$ two or more letters $\underline{x}, \underline{y}, \ldots,$ $\underline{\underline{A}}$ then being thought of as expressing "a relation between $\underline{x}, \underline{y}, \ldots$".

Assuming momentarily the introduction (see II.1.1) of the sign = , if $\underline{\underline{A}}$ has the form $T = U$, where T and U denote sets, in at least one of which the letter $\underline{x}$ appears and in which letter one is temporarily especially interested, it is usual to refer to $\underline{\underline{A}}$ as "the equation $T = U$ (relative to $\underline{x}$) " . And, if $\underline{\underline{A}}$ has the form $\underline{x} \in X \wedge T = U$, where X also denotes a set (in which usually $\underline{x}$ will not appear), one refers to $\underline{\underline{A}}$ as "the equation $T = U$ in X (relative to $\underline{x}$) " . The term "equation" is still applied to a sentence $T = U$, even if no letters appear in either T or U . See also I.3.4(viii) below.

From a formal point of view, the terminology just described is superfluous and may even appear to encourage the type of vagueness which is to be avoided. On the other hand, from a more practical point of view, it sometimes offers sorely needed assistance in keeping in mind what is going on and in suggesting courses of action. The vital thing is that, when such informal language is used, a check should be made in an attempt to ensure that its use does not lead to serious confusion or to types of reasoning which cannot be justified at the formal level.

In this book, I shall initially be very parsimonious in the use of this sort of terminology, but will gradually become more liberal in step with the reader's improved grasp of what lies behind the informality.

I.1.10 Sentence schemas In general terms, a sentence schema is a metamathematical expression or statement in which there appear one or more symbols, each either a letter symbol, set symbol or sentence symbol, with the property that a sentence results whenever one substitutes for each letter symbol (resp. set symbol, resp. sentence symbol) appearing in the schema a letter (resp. set, resp. sentence). The sentence schema may or may not incorporate one or more metamathematical hypotheses restricting the allowed substitutes. Each sentence resulting from an allowed substitution in a sentence schema is said to result from application, or to be an instance, of the sentence schema. (See also I.3.9(ii) below.)

It would be more proper to introduce "metamathematical variables" to denote letter symbols, et cetera. I shall not do this - using $\underline{x}$, $\underline{y}$, $\underline{z}$, ... in the role of letter symbols, X, Y, Z, S, T, ... in the role of set symbols, and $\underline{\underline{A}}$, $\underline{\underline{B}}$, $\underline{\underline{C}}$, ... in the role of sentence symbols. Cf. the final paragraph on p.28; and see Kleene (1), p.81; Kleene (2), p.33.

If no metamathematical hypotheses are attached to a sentence schema, the schema will be adequately described by displaying the appropriate expression in which the letter symbols, et cetera appear. Thus, in I.2.2 below, reference will be made to the sentence schema

$$(\underline{\underline{A}} \Rightarrow \underline{\underline{B}}) \Rightarrow ((\underline{\underline{C}} \vee \underline{\underline{A}}) \Rightarrow (\underline{\underline{C}} \vee \underline{\underline{B}})) \quad . \tag{1}$$

in which only sentence letters appear and in which no mathematical hypotheses are imposed. It is accordingly understood that an instance of (1) is generated by arbitrary choices of sentences substituted for $\underline{\underline{A}}$, $\underline{\underline{B}}$ and $\underline{\underline{C}}$. On the other hand, (II) in I.3.2 refers implicitly to a sentence schema

$$(S|\underline{x})\underline{\underline{A}} \tag{2}$$

restricted by the metamathematical hypothesis that $\underline{x}$ shall be a variable relative to a certain theory (rather than merely a letter; see I.2.6). Again, Problem I/2 refers implicitly to a sentence schema

$$(\forall \underline{x})(\forall \underline{y})(\underline{\underline{A}} \wedge \underline{\underline{B}}) \Leftrightarrow (\forall \underline{x})\underline{\underline{A}} \wedge (\forall \underline{y})\underline{\underline{B}} \quad , \tag{3}$$

restricted by the metamathematical conditions that $\underline{x}$ does not appear in $\underline{\underline{B}}$ and $\underline{y}$ does not appear in $\underline{\underline{A}}$.

In practice, perhaps the most common way of generating sentence schemas is by replacements in a fixed sentence. Thus, if $\underline{\underline{A}}$ denotes a chosen sentence and $\underline{x}$ a chosen letter appearing in $\underline{\underline{A}}$, one may generate the sentence schema with expression

$$(X|\underline{x})\underline{\underline{A}}$$

having one set symbol denoted by X , upon which metamathematical conditions may or may not be imposed. Similarly, if $\underline{\underline{A}}$ denotes a chosen sentence and $\underline{x}$, $\underline{y}$ distinct chosen letters appearing in $\underline{\underline{A}}$, one may generate sentence schemas with expressions

$$(X|\underline{x})(Y|\underline{y})\underline{\underline{A}} \quad ,$$

$$(Y|\underline{y})(X|\underline{x})\underline{\underline{A}} \quad ,$$

and

$$(X, Y \mid \underline{x}, \underline{y})\underline{\underline{A}} \quad ,$$

this last being the result of simultaneous replacement of $\underline{x}$ by X and $\underline{y}$ by Y (see II.1.3(ii)); in each of these sentence schemas there appear two set symbols denoted by X and Y , upon which metamathematical restrictions may or may not be imposed. A fairly complex natural example of this procedure results in the implicit function sentence schema in IV.4.2.

A little care is needed in connection with the generation of sentence schemas in the above fashion. For instance, if $\underline{\underline{B}}$ denotes a sentence, $\underline{x}$ a letter, and $\underline{\underline{B}}[\![X]\!]$ denotes $(X|\underline{x})\underline{\underline{B}}$, the expression

$$(X \text{ is a set}) \Rightarrow \underline{\underline{B}}[\![X]\!]$$

does not generate a sentence schema, the reason being that nowhere is " $\underline{x}$ is a set" or " X is a set" provided with a formal meaning (within the theory to be discussed in this book); that is, " X is a set" will never be the name of a formal sentence. On the other hand, the expression

$$(X \text{ is an ordered pair}) \Rightarrow \underline{\underline{B}}[\![X]\!]$$

generates a sentence schema, because (see the end of III.1.2) " X is an ordered pair" is the agreed name for the formal sentence denoted by

$$(\exists \underline{x})(\exists \underline{y})(X = (\underline{x}, \underline{y})) \quad ,$$

where $\underline{x}$, $\underline{y}$ denote any two distinct letters not appearing in X . (The appropriate replacement rules ensure that the choice of $\underline{x}$ and $\underline{y}$ is otherwise immaterial.)

<u>Remarks</u> The above description of sentence schemas, like the description of replacement rules, is not in principle essential to the prosecution of the formal theory; nor does it have any bearing on the internal workings of the formal theory beyond the point of making quite clear what are the instances of these sentence schemas chosen to be axiom schemas (see I.2.1) of the formal theory. (This is a delicate point; cf. Remark (iv) in I.2.4.) However, it and other metamathematical discussions may appear rather vague by comparison with the formal theory. This is almost inevitable. The only care likely to be regarded as finally satisfactory lies in formalising the metalanguage, which would then become a second formal language whose workings have to be specified in a meta-metalanguage. But where would one call a halt? Short of commencing this process, one can at this stage only do one's best to secure precision and clarity within the confines of the original metalanguage. (There are obvious dangers in trying to sharpen the exposition by introducing mathematical ideas, whose credentials are themselves being scrutinised.)

I.2 Axioms, theories, proofs and theorems

I.2.1 *Axioms and schemas* As will be explained at greater length in I.2.3 below, the specification of a theory amounts to :

(i) the specification of one or more (finitely many) sentence schemas to be referred to as the *axiom schemas*, or briefly *schemas*, of that theory;

and

(ii) the specification of finitely many (maybe none) sentences to be referred to as the *explicit axioms*, or briefly *axioms*, of that theory.

Every instance of every schema of a theory is termed an *implicit axiom* of that theory.

The role of the axioms, explicit or implicit, of a theory is to act as starting points for proofs of (or in) that theory: this will be explained at length in I.2.4. In other words, they act, in principle anyway, as starting points for all logical reasoning in that theory. (In practice, logical reasoning usually starts from theorems which have already been proved; but in principle one could replace every such theorem, step by step, by a proof which begins with an axiom.)

I.2.2 *The logical schemas* The first five schemas shared by all the theories considered in detail in this book are entirely logical in flavour and may be briefly displayed as follows:

S1 $(\underline{\underline{A}} \vee \underline{\underline{A}}) \Rightarrow \underline{\underline{A}}$;

S2 $\underline{\underline{A}} \Rightarrow (\underline{\underline{A}} \vee \underline{\underline{B}})$;

S3 $(\underline{\underline{A}} \vee \underline{\underline{B}}) \Rightarrow (\underline{\underline{B}} \vee \underline{\underline{A}})$;

S4 $(\underline{\underline{A}} \Rightarrow \underline{\underline{B}}) \Rightarrow ((\underline{\underline{C}} \vee \underline{\underline{A}}) \Rightarrow (\underline{\underline{C}} \vee \underline{\underline{B}}))$;

S5 $(S|\underline{x})\underline{\underline{A}} \Rightarrow (\exists \underline{x})\underline{\underline{A}}$

Each of these is a sentence schema free from metamathematical restrictions on the letter symbols, set symbols, and sentence symbols appearing in it. So, for example, the content of S4 is (as has been explained in I.1.10) that the sentence

$$(\underline{\underline{A}} \Rightarrow \underline{\underline{B}}) \Rightarrow ((\underline{\underline{C}} \vee \underline{\underline{A}}) \Rightarrow (\underline{\underline{C}} \vee \underline{\underline{B}}))$$

is to be accepted as an implicit axiom, whatever the sentences substituted for $\underline{\underline{A}}$, $\underline{\underline{B}}$, $\underline{\underline{C}}$.

In terms of other intuitive interpretations mentioned in I.1.7, each of S1 - S5 is either reasonable, or at least involves no conflict with intuition. Thus, S5 is a formal counterpart to the intuitively acceptable conviction that, if one knows of an object S such that $\underline{\underline{A}}$ is satisfied by S (that is, if $(S|\underline{x})\underline{\underline{A}}$ is true), then surely $(\exists\underline{x})\underline{\underline{A}}$ is true (that is, surely there exists at least one object which satisfies $\underline{\underline{A}}$). (In making this comparison, I am overlooking a subtle distinction which has to be made between " $\underline{\underline{A}} \Rightarrow \underline{\underline{B}}$ is true" and "if $\underline{\underline{A}}$ is true, then B is true" ; see I.2.9(i).)

Useful extensions of S5 will appear in Problem I/17(ii) and (iv).

I.2.3 Theories As has been indicated in I.2.1, a theory is a metamathematical entity which is specified by listing its schemas and its explicit axioms. Such a theory is often spoken of as a formal theory or formal system, though other writers use these terms in a less restrictive sense. (See §4 of the Appendix and also Kleene (1), Chapter IV and (2), §37.)

All theories subsequently dealt with in the main text of this book will have S1 - S8 (and they alone) as their schemas, but they will vary in respect of their explicit axioms. (This will soon be made more precise; see I.2.5.)

The logical schemas S1 - S5 have already been listed in I.2.2. Schemas S6 and S7 relate to set theoretic equality and will appear in II.1.2. Schema S8 , which is the "big" set theoretic schema, appears in II.12.1. Each of the schemas S6 - S8 could be listed here, but their expression would be cumbersome and their intuitive content difficult to grasp.

I.2.4 Proofs and theorems Suppose that Θ denotes a theory in the sense described in I.2.3. A proof in (or of) Θ is a finite list of sentences (written in some definite order, say one under the other down a page; see again §2 of the Appendix) such that, for every sentence $\underline{\underline{A}}$ of the list, at least one of the following is the case:

$\underline{\underline{A}}$ is an explicit axiom of Θ ;

or $\underline{\underline{A}}$ results from the application of an axiom schema of Θ (see I.1.10) ;

or preceding $\underline{\underline{A}}$ in the list there are two sentences $\underline{\underline{B}}$ and $\underline{\underline{C}}$ such that $\underline{\underline{C}}$ is identical with $\underline{\underline{B}} \Rightarrow \underline{\underline{A}}$.

The first two alternatives may together be summarised thus: $\underline{\underline{A}}$ is an axiom (explicit or implicit) of Θ .

A proof in Θ of (a sentence) $\underline{\underline{A}}$ is a proof of Θ in which $\underline{\underline{A}}$ appears (as one entry in the list).

Strictly speaking, a proof should be preceded by a so called auxiliary construction, the purpose of which is to make it clear that every string appearing (that is, listed) in the alleged proof is indeed a sentence, which is effected by checking that every string appearing in the alleged proof also appears in the auxiliary construction. In practice, however, it will be assumed that this verification is effected as part of the line by line generation of the proof.

Every sentence which appears in at least one proof of Θ is called a theorem of (or in) Θ . In other words, a theorem of Θ is a sentence of which there is a proof in Θ . A sentence is said to be true (or provable or derivable) in Θ , if and only if it is a theorem of Θ .

A sentence $\underline{\underline{A}}$ is said to be false in Θ , if and only if the negation $\neg\underline{\underline{A}}$ is true in Θ .

Regarding these uses of the terms "true" and "false", see the remarks in I.2.9.

A proof in Θ of $\neg\underline{\underline{A}}$ is often termed a disproof or refutation in Θ of $\underline{\underline{A}}$; $\underline{\underline{A}}$ is said to be refutable in Θ if and only if $\neg\underline{\underline{A}}$ is provable in Θ

(that is, if and only if $\underline{\underline{A}}$ is false in Θ).

Remarks (i) In this connection, it should be said here that I intend "is a theorem" to be interpreted in the sense of "is provable" , rather than "is proven" (the latter being a concept which is plainly time dependent). Bourbaki (1), p.25 intends this latter view; and indeed, in common mathematical parlance, "is a theorem" usually means "is proven" (although the proof referred to in this case is almost never a proof in the strict formal sense, but rather at best a semi-formal proof of the type described in I.3.4 and making full use of the proof methods mentioned in I.3.2 and I.3.3).

More often that not, the distinction between "is provable" and "is proven" reflects no more than the difference that, while a semiformal or routine style proof has already been (or could easily be) explicitly written, a fully formal proof could not (for practical reasons of time and space) be exhibited in similar fashion.

Although there is room for detailed metamathematical discussion of the significance and verifiability of both "is provable" and "is proven" (cf. (iv) below and §2 of the Appendix), I think it is beyond reasonable doubt that every axiom, explicit or implicit, is provable; and that the proof methods described in I.3.2 and I.3.3 provide valid ways of generating new provable sentences from given provable sentences. (They also provide ways of starting from sentences which are proven at one instant and deriving sentences which are proven at a later instant.) In each case, of course, one is neglecting the practically insurmountable problem of explicitly writing in the formal language proofs of any but the dullest sentences.

In any case, less confusion might result, if the term "provable" were used in place of "true" , and " $\neg\underline{\underline{A}}$ is provable" or " $\underline{\underline{A}}$ is refutable" used in place of " $\underline{\underline{A}}$ is false" . But habits die hard.

(ii) It is most important to recognise that a proof is an essentially terminating, finitary process. As a result one should always be a mite suspicious of any alleged proof which issues instructions to repeat a procedure infinitely often, or culminates in a phrase like "and so on indefinitely" , or ends in a sequence of dots. Sometimes this is merely an intuitive way of referring to a valid proof method, and sometimes it is not. See especially the discussion in V.4 below.

(iii) It has been written (Bourbaki (1), p.297) that the notion of proof possessed by Euclid, Archimedes and Apollonius "differs in no respects from ours". If "proof" here refers to the ideal of a fully formal proof, this statement may be broadly true. (It is difficult to be certain exactly what the Greek ideal was; but see loc. cit. pp.306-310, 312 and 315.) But if "proof" refers to what actually appears in writing in conventional texts (see I.3.4), quite substantial changes have taken place (and will presumably continue to take place). It seems obvious that what was presented and accepted as a proof 100 or 200 years ago would not in all cases be judged acceptable now. The style of proof fashionable then was in some ways similar to what prevails in many current high school text books; and they would often now be described as lacking in rigour. There seems little doubt that changes in the degree of rigour considered acceptable here have been brought about, at least in part, by the deliberate study of formal systems. They have also been brought about partly by teaching at various levels: conscientious preparation of a teaching text should always involve a critical review of material to be presented. (The word "should" has to be stressed here.)

(iv) The intrusion into (i) above of the term "provable" prompts the following "aside" remarks. The reader should appreciate that our description of formal theories and related concepts is far from being complete: it is in fact close to rock bottom minimal for our limited purposes. For example, most metamathematicians would concern themselves with the question of the existence of an effective, mechanical procedure (cf. V.13) for deciding whether an arbitrarily given list of sentences is or is not a formal proof; and likewise in relation to deciding whether an arbitrarily given string is a formula or a term, and whether an arbitrarily given formula is an axiom; cf. the remarks in I.1.6. The existence of such a procedure is guaranteed by a condition known as "recursive axiomatisability" (see Mendelson (1), p.147; Stoll (2), p.374), and this condition is usually taken as part of the rigorous definition of the general concept of formal theory or system. (Beware: this restriction is no guarantee of the existence of an effective, mechanical "decision procedure" for deciding whether an arbitrarily given sentence is or is not provable; the most important and familiar formal theories, including those considered in this book, admit no such decision procedure; see, for example, Kleene (1), p.439.) Such matters are typical of metamathematics, but they are of no immediate relevance to the main purpose of this book, which is concerned with instances of proofs and theorems in one theory (or in certain theories very simply related to that one).

(v) A minor point of terminology: terms such as "lemma" and "proposition" are often used in a way which, from a formal point of view, makes them synonymous with "theorem" . They are used in conventional mathematics merely to indicate a (possibly subjective) estimate of the relative importance or significance of the theorems in question, both lemmas and propositions being theorems regarded merely as steps on the way to proving a "main" theorem. In this role, the aliases are undoubtedly an aid to grasping the overall picture. However, lemmas and propositions have to be proved in no less demanding a fashion than a theorem.

On the other hand, as was indicated in I.1.6, the term "proposition" is sometimes used as a synonym for the term "sentence" . The two uses must be distinguished, of course.

In somewhat similar vein, it is occasionally necessary to distinguish between a sentence or statement and an assertion to the effect that that sentence or statement is true or false or undecidable. For example, depending on the context,

31 is a prime

may be intended merely as a name of (that is, merely as denoting) a certain formal sentence, or as the assertion that the said formal sentence is true (in a specified theory, usually set theory). The former intention may be clarified by the use of quotes :

" 31 is a prime" ;

but this device is rarely used systematically.

I.2.5 <u>Set theory</u> Set theory, which in this book will always be denoted by Θ_0 , results from a certain definite choice of explicit axioms, together with the schemas S1 - S8 ; the schemas S1 - S5 have already appeared and the remaining schemas appear in II.1.2 and II.12.1. The explicit axioms of Θ_0 are not conveniently specified here and now and will be found in II.12.2.

In the sense that I am here using the word "theory" , set theory covers virtually the whole of current mathematics. (The latest discipline, called category theory, shows signs of not being adequately handled within the confines of set theory and may in fact supplant the latter as a foundational system; see Griffiths and Hilton (1), Chapter 38 and pp. 598-599. See also the remarks about Solovay's work in IV.5.3.) Conventional subheadings, like "number theory" , "group theory" , "function theory" , and so on, use the word "theory" in a different and rather ill defined fashion. It is true that there is a so called "first order number theory" of great interest for various reasons to logicians (see V.12 and V.13), but this is not what is commonly understood by "number theory" . Also, although there is a corresponding autonomous "first order group theory" , to which relatively very little attention is paid, what is commonly referred to as "group theory" is, so to speak, embedded within set theory; see XII.2 below.

Although all the ultimate interest here lies with set theory, the vital proof methods discussed in I.3 below involve reference to theories Θ which result from Θ_0 by the adjunction of one or more explicit axioms to those of Θ_0 . These are the _only_ theories with which we shall be "officially" concerned in the sequel. See also Note 3.

I.2.6 _Constants (fixed letters) and variables (free letters)_ Given a theory Θ , those letters (if any) which appear in the explicit axioms of Θ (expressed in the primitive language) are termed the _constants_ or _fixed letters_ of Θ ; all other letters are termed _variables_ or _free letters_ of Θ . A letter is said to be _constant in_, or _fixed in_, Θ if and only if it is a constant or fixed letter of Θ ; otherwise it is said to be _variable in_, or _free in_, Θ .

The reader should note carefully that the above use of the terms "constant" and "variable" bears little relationship to the conventional use of the same terms in informal mathematics; see II.3.9(vi), II.4.4(iii), IV.10 and the Foreword to Volume 2. For one thing, although the said conventional usage is vague in many respects, it would almost certainly involve labelling as constants such sets as those (denoted by)

$$0 , 1 , 2^{\frac{1}{2}} , 1 + 2^{\frac{1}{2}} , e , \pi , (0, 1) , N , N \times N .$$

Yet none of these sets is a letter and so cannot be a constant (fixed letter) of any theory.

Most accounts of mathematical logic (see, for example, Mendelson (1), pp.47,48) make essential use of the idea of free and bound occurrences of a variable in a formula. No use is made of these ideas in this book. Our free and fixed letters are determined in relation to a theory, not in relation to a formula.

Furthermore, it is an important fact that the theory Θ_0 has no constants; all letters are variables of Θ_0 . This feature will become apparent in II.12.2.

I.2.7 <u>Contradictory and consistent theories</u> All the theories to be considered in the sequel share the same sets and the same sentences. But they will differ in their proofs and theorems: what is a proof (or a theorem) in one such theory Θ will not in general be a proof (or a theorem) in Θ_0 .

Furthermore, since such a theory Θ results from adjoining at will any (finite) number of explicit axioms to those of Θ_0 , it should come as no surprise to learn that Θ may be <u>contradictory</u> or <u>inconsistent</u>, that is, that there may be at least one sentence $\underline{\underline{A}}$ such that both $\underline{\underline{A}}$ and $\neg\underline{\underline{A}}$ are theorems of Θ . If this is not the case, Θ is said to be <u>(syntactically) consistent</u>. (Semantic consistency is a different concept; see Behnke (1), p.31.)

If Θ is contradictory, <u>every</u> sentence $\underline{\underline{B}}$ is a theorem of Θ . For suppose $\underline{\underline{B}}$ to be an arbitrary sentence; by S2 in I.1.2 and (I) in I.3.2, $(\neg\underline{\underline{A}}) \vee \underline{\underline{B}}$, that is, $\underline{\underline{A}} \Rightarrow \underline{\underline{B}}$, is a theorem of Θ ; a second appeal to (I) shows that $\underline{\underline{B}}$ is a theorem of Θ . See also Problem I/1(12).

It is fervently hoped (and believed by the vast majority of mathematicians) that set theory is consistent. But no "proof" is known and, if fact, metamathematics (Gödel, 1931; see §1 of the Appendix) provides some grounds for the view that no totally convincing "proof" is possible, even in principle. (Any such "proof" would, of course, be a metaproof; see 1.2.9(v) below.)

There are all sorts of ways of apparently reaching the conclusion that Θ_0 is contradictory. This end can be achieved by deliberately or inadvertently bending the rules (perhaps because the rules sometimes look nonintuitive); by misapplying metatheorems; by abandoning one definition in favour of another (which looks similar but is different) in midstream; by mixing different versions of set theory; and by still other devices. Cf. II.3.9 and see Problem II/11, the end of Remark (ii) appearing in IV.1.3, Problems I/15 and II/34, VIII.4.4(iv) and the Remarks following XII.5.5(1). It is not unknown for interested amateurs to tread such a path and announce loudly that they have established the inconsistency of set theory; see Problem II/34. No such claim has so far been substantiated - though it may yet be otherwise. (Meanwhile, one can only hope that such claims will in future come only after the claimant has taken all reasonable steps to secure a good grasp of what is involved.)

Throughout this book it will be assumed, whenever the issue is or appears to be relevant, and nothing is said to the contrary, that set theory is consistent. Inasmuch as formal mathematics is concerned with theorems, this assumption in a sense does no harm: if set theory is contradictory, every sentence is a theorem. But the assumption is often vital on these rare occasions where metamathematics is involved (as for example, in II.5.8) and thus whenever the alleged non truth of sentences is discussed. And it should not be forgotten that what loosely passes as mathematics is sometimes metamathematics; see I.3.5 and I.3.8.

In particular, every metastatement to the effect that a certain formal sentence is not true (or is not false) is subject to the meta-assumption that Θ_0 is consistent. In spite of this, informal texts often make claims that certain formal sentences are not true, when in fact the most they establish is that the sentence is false. (See the opening paragraph in II.3.10 and also XV.5.4.) On the other hand, it is sometimes claimed that a sentence is false, when all that has been established is that, if the theory is consistent, then the sentence is not true; see I.3.8.

To repeat : if Θ_0 is inconsistent, there is in one sense no loss, since everything that has been proved in Θ_0 (often painfully) is trivially provable. But then so too is every sentence, and there remains no point in further study of Θ_0 itself. The moment the inconsistency of Θ_0 came to be convincingly verified, it is almost certain that mathematicians would immediately seek to weaken the theory in such a way as to render unprovable the known contradiction(s), retaining as much as possible as a basis for a new, corrected mathematics (in which - as they would hope - no contradictions would be derivable).

I.2.8 Comparison of theories If Θ and Θ' denote two theories of the type described above, Θ' will be said to be stronger than Θ and Θ weaker than Θ', if and only if every explicit axiom of Θ is a theorem of Θ'. (Recall that the theories share the same schemas S1 - S8 .) It can be verified (see Problem I/9) that, if Θ' is stronger than Θ, then every theorem of Θ is also a theorem of Θ'. In particular, every theorem of Θ_0 will be a theorem of every theory of the type considered in this book; cf. the final sentence in I.2.5.

I.2.9 <u>Informal and formal languages again</u> (i) Some care is needed when translating between informal language and the formal language, as the following general comments show, and as will later be repeatedly confirmed in particular contexts. (When one considers the application of formal logic to real world situations, the translation is even more ambiguous and troublesome; see, for example, the relevant discussion in Strawson (1), especially pp.35 et seq. and pp.78 et seq. Luckily, these matters are of no direct concern to us, since we are concerned solely with the world of pure mathematics.)

In informal language, there is sometimes some ambiguity and uncertainty over double negatives. On the other hand it will appear (see I.3.3(k)) that $\neg\neg\underline{\underline{A}} \Leftrightarrow \underline{\underline{A}}$ is a theorem, which is decisive.

It is easily seen to follow from S2 in I.2.2 above and (I) in I.3.2 below that, if $\underline{\underline{A}}$ denotes a false sentence and $\underline{\underline{B}}$ any sentence, then $\underline{\underline{A}} \Rightarrow \underline{\underline{B}}$ is true. Some people feel this to be a somewhat paradoxical, rather than intuitive, feature of $\Rightarrow$ regarded as a counterpart of "if ....., then ....." .

Another initial stumbling block is the fact that the metastatements

$$\underline{\underline{A}} \Rightarrow \underline{\underline{B}} \text{ is a theorem} \tag{1}$$

and

If $\underline{\underline{A}}$ is a theorem, then $\underline{\underline{B}}$ is a theorem (that is: Either $\underline{\underline{A}}$ is not a theorem, or $\underline{\underline{B}}$ is a theorem) (2)

are not metamathematically equivalent. Certainly, if (1) holds, then (2) holds. (This ensues from (I) in I.3.2.) But if the theory is consistent, (2) may hold and yet (1) not hold; otherwise expressed, it may (in a consistent theory) happen that neither $\underline{\underline{A}}$ nor $\underline{\underline{A}} \Rightarrow \underline{\underline{B}}$ is a theorem. This matter will be discussed at greater length in §1 of the Appendix; see also Problem I/31.

Here I add merely that it turns out (by using (VI) in I.3.2) that if in (1) and (2) "theorem" is understood to mean "theorem of Θ " , where Θ is some theory, then (1) is metamathematically equivalent to the following metastatement (evidently stronger than (2)) :

If Θ' is any theory stronger than Θ , and if $\underline{\underline{A}}$ is a theorem of Θ' , then $\underline{\underline{B}}$ is a theorem of Θ' . (2')

(ii) In everyday situations one is, or one behaves as if one is, tied to "either true or false and never both" . This is partly due to an interpretation of "false" as "not true" . All the same, a closer examination can cast some doubt on this dictum, sometimes because it may turn out that one or more terms or phrases are not sufficiently precisely defined. (Consider the sentence "All water is wet." ; if one tries to decide whether this is in fact true or false, one cannot avoid pondering the precise meaning of "is wet" - and one may well fail to find a sufficiently precise meaning. Again, even more confusion may result from considering a sentence like "Every sentence is true or false." ; note here the role of English as both object language and metalanguage, a point which recurs below.) A formal language is, it might seem, the most likely instrument for eliminating such ambiguities to the maximum extent we can devise. Certainly, a lot of ambiguity is eliminated. But the concept of formal sentence adopted in I.1.6 is still too broad for "always true or false" to obtain. In fact, one would really only hope for this to apply to closed sentences (that is, sentences in which no letters appear) - and then only in a consistent theory. (It is easy to exhibit open sentences of set theory - assumed to be consistent - which are neither true nor false: the open sentence $\underline{x} = \emptyset$ is an instance; see also I.3.8, Problem 1/5 and §1 of the Appendix.) One may still hope - and with considerable reason - that every closed sentence is either true or false. Again one is disappointed, though it is much more difficult to give counterexamples. In a consistent theory there may well be closed sentences which are undecidable, that is, neither true nor false. See again §1 of the Appendix.

As soon as it is recognised that the decidability of a sentence $\underline{\underline{A}}$ entails that

> If it is impossible to prove $\neg\mathbf{A}$ (that is, if there is no proof of $\neg\mathbf{A}$), then it is possible to prove $\mathbf{A}$ (that is, there is a proof of $\mathbf{A}$)

the decidability of $\mathbf{A}$ may well seem a priori to be more easily questionable.

Again, it has to be recognised that the non-truth of a sentence $\mathbf{A}$ (that is, the circumstance that $\mathbf{A}$ is not true) is no guarantee that $\mathbf{A}$ is false (that is, that $\neg\mathbf{A}$ is true); see Problem I/33.

To this it must be added that, in a contradictory theory, every sentence is both true and false.

It cannot be stressed too often or too strongly that the terms "true" and "false" , as applied to formal sentences, are relative in nature and somewhat in the nature of courtesy titles, chosen on the basis of imperfect analogy with their everyday usage. It is necessary to be prepared at all times to recall that "true" means "provable" ; that "provable" applied to a formal sentence $\mathbf{A}$ signifies the ability in principle to write (or to describe in acceptably precise terms <u>how</u> to write) a finite list of formal sentences which <u>is</u> a proof and in which $\mathbf{A}$ appears; and that " $\mathbf{A}$ is false" signifies " $\neg\mathbf{A}$ is provable" .

It is due to this finitary interpretation of "provable" that the occurrence in alleged proofs of expressions like "and so on (indefinitely)" or "......" ought always to rouse suspicions: this is a point which will be examined closely in Chapter V in connection with proofs by induction. Even if one <u>could</u> "continue indefinitely" , the result could hardly be a proof.

(iii) Another feature which is somewhat strange is that (as was remarked at the end of I.1.7) any two theorems of a given theory Θ are equivalent in Θ . (If $\mathbf{B}$ is true, and if $\mathbf{A}$ is any sentence, then $\mathbf{A} \Rightarrow \mathbf{B}$ is true as a consequence of S2 and S3 in I.2.2 above, and (I) in I.3.2 below.) Of course, one is usually interested in the equivalence of two sentences at a stage where at least one of the sentences is (often both are) not known to be theorems.

In spite of this, mathematicians will often expend not a little effort on providing a special proof of the equivalence of two (or more) especially important known theorems $\underline{\underline{A}}$ and $\underline{\underline{B}}$. (A typical instance is discussed in VII.7.3.) In such a case, the proof provided will not be the formally obvious one (based upon S2 and S3 and (I) in I.3.2 and making use of the known truth of $\underline{\underline{A}}$ and $\underline{\underline{B}}$), which is in no way dependent upon and makes no reference to the internal structure of the sentences $\underline{\underline{A}}$ and $\underline{\underline{B}}$, but rather a proof which ignores the truth of $\underline{\underline{A}}$ and $\underline{\underline{B}}$ and which is thought to be instructive and illuminating precisely because it does feature some of the internal structure of $\underline{\underline{A}}$ and $\underline{\underline{B}}$. (It is this internal structure which, in most cases, is the reason why the equivalence of $\underline{\underline{A}}$ and $\underline{\underline{B}}$ seems intuitively plausible; it is the basis for the intuitive "link", if any, between $\underline{\underline{A}}$ and $\underline{\underline{B}}$, so to speak.)

The aim of such an exercise may be to indicate the possibility of replacing one approach to (that is, method of treatment of) a topic by another.

> (iv) Again, even when formal systems are being considered, logicians sometimes use the words "true" and "false" in a sense discordant with the definition given in I.2.4, which deals with syntactical truth. I refer to the idea of a sentence of a formal theory being true or false in relation to a given model of that theory; cf. Behnke (1), pp.23, 24; Mendelson (1), p.51; Margaris (1), §26; Stoll (2), p.231. Sentences can be true in relation to a chosen model, without being theorems.
>
> On the other hand, for certain formal theories it can be shown metamathematically that "true" in the sense of I.2.4 (that is, "provable") comes to the same thing as "true in <u>every</u> model of that theory" ; see Robinson (1), p.12, Theorem 1.5.3; Mendelson (1), p.68; Bell and Slomson (1). I say nothing about models and the associated latter concept of truth, for the simple reason that an exact description of models presupposes a version of set theory and a notion of truth therein. In other words, as far as I can see, the syntactic approach is the only one offering any hope of providing a satisfactory noncircular foundation.

(v) Another point to be made is that many statements which are in every-day mathematics referred to as theorems are not really theorems in the sense of I.2.4. Rather, they are sentence schemas (see I.1.10) which yield a theorem for every allowed substitution of letters, sets and sentences for the letter symbols, set symbols and sentence symbols appearing in them. Thus they are better described as <u>theorem schemas</u> (cf. the axiom schemas S1 - S5 in I.2.2 above). The proof methods discussed in I.3 contain many instances of theorem schemas; see also the discussion in II.1.4 and II.2.2, and the implicit function theorem schema in

IV.4.2.

At other times, the real status of the alleged theorems is not clear, that is, it is not clear whether the author is really referring to a single theorem or to a theorem schema. In most cases, however, it is possible to convey what is intended in the guise of a genuine theorem; see, for example, II.1.4 and II.2.2. See also I.3.8(vi) and VI.10.

Usually (as in the problems at the end of this and succeeding chapters) one speaks of proving a theorem schema. This is not strictly correct because what is being referred to is not a single proof, but rather a class of proofs sharing a common pattern. It might be better to speak of supplying a species of proof pattern, but the inaccuracy is immaterial for our purposes.

In a similar way, one often encounters metastatements; these are statements about (not in) one or more formal theories, expressed in some metalanguage (usually English or some other natural language, possibly supplemented by a few special symbols, the whole being treated informally; cf. Kleene (2), pp.3, 199, 200, 205-206). Metatheorems (of which replacement rules and theorem schemas are instances) are metastatements which are true. ("True" here is to be taken in an intuitive sense, different from that operating within the formal theory or theories.) For instance, the various replacement and construction rules in I.1.4 and I.1.8, and the proof methods (I) - (XI) in I.3.2, are metatheorems; see also the statements in I.3.5(iii) concerning $\equiv$ and $=$.

To put it more graphically : a metastatement is a statement about a formal theory, expressed in the metalanguage and thus made "outside" the formal theory; and a metatheorem is a metastatement (or meta-assertion) whose truth is, or may be, verified by arguments which are likewise "outside" the formal theory and which usually are based on a relatively simple metalogic. (This metalogic may, of course, itself be questioned and made the object of a separate scrutiny; to do that would involve a metametalanguage, et cetera. Usually, however, the metalogic is taken for granted. See the Introduction to the Appendix.) Such a verification is often termed a metaproof.

schema contains a phrase of the type "if X is a set and $\underline{x}$ a letter" or "if X is any set and $\underline{x}$ any letter" , and if there is no explicit indication to the contrary, the phrase is to be understood in the sense of "if X is (or denotes) an arbitrary set and $\underline{x}$ is (or denotes) an arbitrary letter" .

(vi) The preceding remarks indicate that one has from time to time the problem of distinguishing between at least three languages.

First there is the strictly formal language: this consists of absolutely nothing but strings in the primitive signs $\neg$, $\underline{\vee}$, τ , $\square$, $\underline{\in}$ already mentioned. This is the language of formalised mathematics and is utterly impractical because very soon one becomes involved with strings of stupendous lengths. No abbreviations are permitted and so every such string has to be repeated at length over and over again. Such a text, even when written, would be intolerable to read.

Then, for the purpose of describing the workings of the formalised theory, there is a sort of semiformal language in which there may appear strings properly speaking (but very few), plus a great many abbreviatory signs introduced especially to denote certain strings. Thus, apart from the primitive signs (which are instances of strings), there will in due course appear such symbols as $\tau_{\underline{x}}(\underline{A})$, $(\exists\underline{x})A$, $\{\underline{x} : A\}$, $\{\underline{x}\}$, $\{\underline{x}, \underline{y}\}$, $(\underline{x}, \underline{y})$, $P(\underline{x})$, $\cup$, $\cap$, $\underline{x} \times \underline{y}$, ... Further still, there will be conventional signs for specific sets, like $\emptyset$, 0 , 1 , N , Z , ... , π , + , - , ... , Γ , ζ , ... Each of these last denote strings in the formal language, often of astronomical length; see the penultimate paragraph in V.2.1.

Thirdly, there is the informal language of everyday mathematics; see I.0.5, I.0.8, I.0.10. This is what is used whenever one "does mathematics" in the usual sense. It is in truth not precisely defined and every individual feels free to borrow phrases from his natural language (English, say) whenever he thinks their intended meaning is tolerably clear to his readers. In this scheme, the primitive signs (and even some semiformal signs as well) are replaced by suitable English words and phrases: $\neg$ by "not" , $\vee$ by "or" , $\in$ by "belongs to" (or "is an element of"), and so on. The primitive signs τ and $\square$ rarely if ever appear or receive mention when everyday mathematical language is used: this is because their major role is played in formulating formal definitions of the abbreviatory signs $(\exists\underline{x})\ldots$, $(\forall\underline{x})\ldots$ and $\{\underline{x} : \ldots.\}$ (the last being rendered as "the set of $\underline{x}$ such that ...") and others like them, having done which they disappear from everyday mathematical language. Similarly, informal mathematics usually makes no overt and explicit use of logical (formal) letters $\underline{x}$, $\underline{y}$, $\underline{z}$, ... : one will find simply letters a , b , ... , x , y , z , α , β , ... , υ , Γ , Δ , ... of various alphabets which sometimes denote logical letters and sometime more general sets (in which one or more logical letters may appear). In this book, such licences appear by the time Chapter VI is reached; see especially VI.10, which indicates the sorts of doubts which may be raised by these ambiguities.

This third language is, of course, a bastard in so far as it will be a mixture of the formal language and a metalanguage. Its use is typical of informal or partly formal mathematics and is to a large extent practically unavoidable. Nevertheless, its use is sometimes dangerous and ought always to be viewed with some suspicion, especially if infinite collections or processes are involved. See Problem I/15 and the comments in II.3.9, V.4.1 and VIII.6.9.

(vii) Use of this informal bastard language often leads to replacements typified (but far from exhausted) by the following list (cf. Kleene (2), §27) :-

 or	If $\underline{\underline{A}}$, then $\underline{\underline{B}}$ Suppose (assume) that $\underline{\underline{A}}$. Then $\underline{\underline{B}}$	 or	$\underline{\underline{A}} \Rightarrow \underline{\underline{B}}$ $\underline{\underline{A}} \Rightarrow \underline{\underline{B}}$ is true
	If $\underline{\underline{A}}$, $\underline{\underline{B}}$, ..., $\underline{\underline{D}}$, then $\underline{\underline{P}}$		$(\underline{\underline{A}} \wedge \underline{\underline{B}} \wedge \ldots \wedge \underline{\underline{D}}) \Rightarrow \underline{\underline{P}}$
 or or	$\underline{\underline{A}}$ for all $\underline{x}$ for all $\underline{x}$, $\underline{\underline{A}}$ $\underline{\underline{A}}$, $\forall\underline{x}$		$(\forall\underline{x})\underline{\underline{A}}$

$\underline{\underline{A}}$ for all $\underline{x} \in S$ or for all $\underline{x} \in S$, $\underline{\underline{A}}$	$(\forall \underline{x})((\underline{x} \in S) \Rightarrow \underline{\underline{A}})$
$\underline{\underline{B}}$ whenever (<u>or</u> provided) $\underline{x} \in S$ and $\underline{\underline{A}}$.	$(\forall \underline{x})(((\underline{x} \in S) \wedge \underline{\underline{A}}) \Rightarrow \underline{\underline{B}})$
$\underline{\underline{A}}$ for some $\underline{x} \in S$ or There exists $\underline{x} \in S$ such that $\underline{\underline{A}}$	$(\exists \underline{x})((\underline{x} \in S) \wedge \underline{\underline{A}})$
For all $\underline{x} \in S$, there exists $\underline{y}$ such that $\underline{\underline{A}}$ or To every $\underline{x} \in S$, there corresponds (some, at least one) $\underline{y}$ such that $\underline{\underline{A}}$ or Given $\underline{x} \in S$, there exists $\underline{y}$ such that $\underline{\underline{A}}$	$(\forall \underline{x})((\underline{x} \in S) \Rightarrow (\exists \underline{y})\underline{\underline{A}})$

In the situation last indicated, one will often find written "there exists $\underline{y} = \underline{y}(\underline{x})$ " or "there corresponds $\underline{y} = \underline{y}(\underline{x})$ " in place of "there exists $\underline{y}$ " or "there corresponds $\underline{y}$ " , the intention being to remind or warn the reader that the "choice of $\underline{y}$ may (and usually will) depend upon $\underline{x}$ " : this corresponds to the fact that $\underline{x}$ may (and usually will) appear in $\underline{\underline{A}}$ and so also in $\tau_{\underline{y}}(\underline{\underline{A}})$. For an instance, see VII.1.2(ii). The symbolism " $\underline{y} = \underline{y}(\underline{x})$ " is not to be taken literally as part of the formal language; cf. IV.1.3.

Sometimes one sees written :

"It can be proved that $\underline{\underline{B}}$ is true whenever $\underline{\underline{A}}$."

This would be more coherent (or more "homogeneous") if "whenever $\underline{\underline{A}}$ is true." appeared in place of "whenever $\underline{\underline{A}}$." ; and what would then appear might indeed be what is intended. However, the more usual intention is better expressed by

"It can be proved that $\underline{\underline{A}} \Rightarrow \underline{\underline{B}}$."

or

"It can be shown that $\underline{\underline{A}} \Rightarrow \underline{\underline{B}}$ is true."

In this connection, see again the contrast between (1) and (2) in I.2.9(i).

In addition, one will often find written " $\underline{\underline{A}}$ means $\underline{\underline{B}}$ " in place of " $\underline{\underline{A}} \Leftrightarrow \underline{\underline{B}}$ is a theorem" - or even in place of " $\underline{\underline{A}} \Rightarrow \underline{\underline{B}}$ is a theorem" (this notwithstanding the fact that the formal sentences $\underline{\underline{A}}$ and $\underline{\underline{B}}$ have no meaning).

When a certain theory Θ is in mind (though it will often not be mentioned explicitly), it is customary to write (cf. VII.3.3 and VIII.4.3)

"In order that $\underline{\underline{A}}$ be true, it is necessary (resp. sufficient, resp. necessary and sufficient) that $\underline{\underline{B}}$ "

in place of

" $\underline{\underline{A}} \Rightarrow \underline{\underline{B}}$ (resp. $\underline{\underline{B}} \Rightarrow \underline{\underline{A}}$, resp. $\underline{\underline{A}} \Leftrightarrow \underline{\underline{B}}$) is a theorem of Θ " .

(Alternatively, the displayed bastard sentence may (albeit relatively rarely) be intended as no more than a name for the formal sentences $\underline{\underline{A}} \Rightarrow \underline{\underline{B}}$ (resp. $\underline{\underline{B}} \Rightarrow \underline{\underline{A}}$, resp. $\underline{\underline{A}} \Leftrightarrow \underline{\underline{B}}$) .) See also the discussion in VIII.4.4(i).

(viii) Much use of this informal bastard language (mathematical slang, so to speak) is totally conventional and harmless. Indeed, its use is often helpful since it encourages conjectural thinking based on experience and incorporating intelligent guesswork, almost essential for creative processes. In addition to the instances mentioned in (vii), one finds bastard sentences used as surrogates for formal sentences otherwise denoted by semiformal expressions like those appearing in (vi). Instances are:

" $\underline{z}$ is an ordered pair" to denote $(\exists \underline{x})(\exists \underline{y})(\underline{z} = (\underline{x}, \underline{y}))$,

" $\underline{w}$ is a relation" to denote $(\forall \underline{z})((\underline{z} \in \underline{w}) \Rightarrow (\exists \underline{x})(\exists \underline{y})(\underline{z} = (\underline{x}, \underline{y})))$,

where it is understood that $\underline{x}$, $\underline{y}$, $\underline{z}$, $\underline{w}$ denote distinct letters; and

" f is a function" to denote

$$(\text{ f is a relation}) \wedge (\forall \underline{x})(\forall \underline{y})(\forall \underline{z})(((\underline{x}, \underline{y}) \in f \wedge (\underline{x}, \underline{z}) \in f) \Rightarrow (\underline{y} = \underline{z})) ,$$

where $\underline{x}$, $\underline{y}$, $\underline{z}$ denote distinct letters not appearing in (the string denoted by) f .

In spite of this, the use of the informal language occasionally involves dangers of some consequence, a few instances of which will now be discussed.

To begin with, we have chosen to think of $(\exists \underline{x})\underline{\underline{A}}$ as corresponding in some way to "there exists (that is, it is possible to choose) $\underline{x}$ such that $\underline{\underline{A}}$ " ; see I.1.7. This device is suggestive and useful. However, it involves a risk of confusion simply because the phrase "there exists" means many things, not all of which are compatible with the role of $\exists$ in the formal theory. One illustration appears in II.4.2(iii) and another follows immediately below.

It frequently happens in informal mathematics that possession of a combined existence and uniqueness theorem leads to developments which call for some comment in respect of the use of the phrases "there exists" and "there does not exist" . The said theorem will usually be of the form

If $\underline{\underline{H}}$, then there exists a unique $\underline{x}$ such that $\underline{\underline{A}}$,

wherein $\underline{\underline{H}}$ and $\underline{\underline{A}}$ denote formal sentences (or informal counterparts thereof) and $\underline{x}$ a formal letter (or an informal counterpart thereof, usually spoken of vaguely as a "variable"). A more formal expression of the theorem would be

$$\underline{\underline{H}} \Rightarrow ((\exists \underline{x})\underline{\underline{A}} \wedge (\forall \underline{x})(\forall \underline{x}')((\underline{\underline{A}} \wedge (\underline{x}'|\underline{x})\underline{\underline{A}} \Rightarrow (\underline{x} = \underline{x}'))) \quad ,$$

wherein $\underline{x}'$ denotes a letter different from $\underline{x}$ and not appearing in $\underline{\underline{A}}$.

An example, discussed at more length in V.14 and relating to the very basic study of addition of natural numbers, is the theorem

> If m is a positive natural number, then there exists a unique natural number x such that
>
> $$x + 1 = m \quad .$$

Or again, in connection with the basic study of real numbers, the appropriate theorem might read

> If a is a nonnegative real number, then there exists a unique nonnegative real number x such that
>
> $$x^2 = a \quad .$$

Yet another example, more complicated and relating to the study of limits of real valued sequences, is discussed in VII.1.4(ii) - (iv). See also Problem VI/10.

On the basis of the said existence and uniqueness theorem, it would be informal practice to frame a definition

> If $\underline{\underline{H}}$, define Ω to be the unique object such that
>
> $$(\Omega|\underline{x})\underline{\underline{A}}$$

or

> Define Ω to be the unique object, if it exists, such that $(\Omega|\underline{x})\underline{\underline{A}}$.

Thus, in the first example above, Ω might be written $m - 1$, thereby defining subtraction of 1 as a restricted operation on natural numbers. In the second example, Ω might be written $\sqrt{a}$ or $a^{1/2}$, thus defining the square root operation on nonnegative numbers.

Reverting to the general situation, it has to be recognised that for reasons to be described in I.3.5(iv), neither of the above proposed definitions is formally acceptable. To begin with, neither succeeds in defining Ω as a string in the formal language, as a consequence of which the same lack presumably applies to $(\Omega|\underline{x})\underline{\underline{A}}$. Depending upon the choice of $\underline{\underline{H}}$ and $\underline{\underline{A}}$, Ω will usually be some more or less traditional group of symbols adopted as a convenient (but sometimes misleading) name for whatever it is that is uniquely characterised by the said theorem. No precise formal language will be available. As has been indicated in I.1.7, Ω (when it exists) is equal to $\tau_{\underline{x}}(\underline{\underline{A}})$, but the informal exposition has no way of expressing this since it lacks a selector $\tau_{\underline{x}}$.

In spite of such formalist-inspired objections, the informal text will probably move ahead and incorporate bastard phrases of the type

(a) Ω exists

and/or

(b) Ω does not exist .

In view of the ambiguity about precisely what is denoted or named by Ω, there may be some doubt as to the significance of both (a) and (b). In the simplest cases it seems clear that Ω refers to a definite set or set form, rather than a formal letter or unrestricted variable. But then what is to be made of (a) or (b) ? What of the uneasy feeling that there is something odd about naming an entity and then querying the existence of that entity?

On the one hand, (a) is not to be thought of as affirming the existence of $\tau_{\underline{x}}(\underline{\underline{A}})$, about which there is no room for doubt and which is in any case not part of the informal apparatus. Possibly (a) is to be thought of as referring to a formal sentence which features Ω in some way; perhaps $(\exists\Omega)\underline{\underline{A}}$ or $(\exists\Omega)(\Omega|\underline{x})\underline{\underline{A}}$. But this will not do either : neither of these groups of symbols denotes a string. (Recall that $(\exists\underline{x})\underline{\underline{A}}$ is a recognised name for a string, only when $\underline{x}$ is or denotes a formal letter; and as has been said, there is no expectation that Ω does that.)

The fact is that (a) is after all intended, in this context, to signify either the formal sentence $(\exists\underline{x})\underline{\underline{A}}$ or perhaps the metasentence

$$(\exists\underline{x})\underline{\underline{A}} \text{ is true .}$$

Likewise (b) is intended to signify either the formal sentence $\neg(\exists\underline{x})\underline{\underline{A}}$ or perhaps the metasentence

$$(\exists\underline{x})\underline{\underline{A}} \text{ is false .}$$

Neither involves any real reference to Ω which, in (a) or (b), functions merely as a convenient informal figure of speech. Although the use of (a) and (b) is often thought to typify the precision and technicality expected of mathematics, it is more accurately described as a red herring!

(ix) Again, the omission of the phrase "for all" is quite frequent (although no-one dreams of similar omissions of "there exists"), and this often leads to omissions of the universal quantifier $(\forall \underline{x})$, some of which are illegitimate and lead to entirely unwanted ends. Sometimes, of course, (X) in I.3.2 sanctions such omissions, but it does not sanction all omissions. Part of the trouble is that, in spite of (X) , $\underline{\underline{A}} \Rightarrow (\forall \underline{x})\underline{\underline{A}}$ may not be true (this is discussed at some length in §1 of the Appendix). Another source of the trouble is that (see I.3.3(n)) $(\exists \underline{x})\underline{\underline{A}}$, and even $(\forall \underline{y})(\exists \underline{x})\underline{\underline{A}}$, may be true, and yet $(\exists \underline{x})(\forall \underline{y})\underline{\underline{A}}$ be not true (if Θ_0 is consistent). These features may seem very remote, but omissions of universal quantifiers can lead to trouble even in everyday mathematics. Here is an example.

Suppose informal mathematics has been carried to the stage (Chapter IV below) where " u is a sequence" has been defined in acceptable fashion. One may then (see VI.9.1) want to consider those sequences which are "eventually zero" , that is, which vanish from some point onward (this point varying from one sequence to another). Thus intuitively the constant sequence $\underset{\sim}{0}_N$ is eventually zero, as also is the sequence $\underline{n} \rightsquigarrow \sin(10^{-6}\pi\underline{n}!)$, while the constant sequence $\underset{\sim}{2}_N$ is not. In an attempt to make this new concept more precise, there are at least two plausible choices of a more formal definition, namely:

$$(u \text{ is a sequence}) \wedge (\exists \underline{m})(\underline{m} \in N \wedge ((\underline{n} \in N \wedge \underline{n} > \underline{m}) \Rightarrow (u(\underline{n}) = 0))) \qquad (3)$$

and

$$(u \text{ is a sequence}) \wedge (\exists \underline{m})(\underline{m} \in N \wedge (\forall \underline{n})((\underline{n} \in N \wedge \underline{n} > \underline{m}) \Rightarrow (u(\underline{n}) = 0))); \qquad (4)$$

in each case, $\underline{m}$ and $\underline{n}$ denote distinct letters not appearing in u . The difference between (3) and (4) lies solely in the appearance of $(\forall \underline{n})$ in (4) and its absence from (3) . But this difference is important. If (3) be adopted as the definition, it ensues that every sequence u is eventually zero. (This is almost (but not quite) visible on making the replacement indicated by $(\underline{n}|\underline{m})$; and the correct path to this conclusion is discussed in Problem II/31.) Thus (3) is an "intuitively wrong" definition having unwanted consequences. As it happens, (4) is an "intuitively right" definition - that is, it expresses what is intended and is (or is equivalent to) what has been adopted by the mathematical world at large. ("Right" and "wrong" here are in relation to intended meanings. There is no purely formal objection to (3) per se.). See also IV.12.3.

Harmful omissions of the universal quantifier appear

repeatedly in mathematical literature (almost certainly in my own writings). They occur frequently in the hasty interpretation of bastard sentences of the type " $\underline{\underline{B}}$ whenever $\underline{n} \in S$ and $\underline{\underline{A}}$ " (see (vii) above); cf. Problems VII/10 and X/13. Other instances appear in I.4.5(vii), Problems I/19 and I/31, Remark II.1.3(i), Remark (ii) following II.3.8, IV.1.2(iv), IV.12.3, Problem IV/21, V.2.2, VI.10.1, VII.1.2(ii) and VIII.4.4(i).

It is difficult to frame a simple and satisfactory rule governing the permissible omission of universal quantifiers. Gleason (1), p.18 writes

> As a general rule, any unquantified symbol appearing in a theorem or definition except those such as 1, 2, ..., $\emptyset$, ..., which have a fixed significance, should be understood as a variable and universally quantified.

Incidentally, I think that Gleason's use of "general" here is one which cloaks the very opposite; see the appropriate comment on p.8 above.

In somewhat the same vein, Halmos (private communication) suggests the "tacit convention" expressed thus :

> if a free variable appears in an implication then it is to be bound by a universal quantifier whose scope is just the implication .

It seems to me that this suggestion also fails inasmuch as it involves ambiguities which can result in passage from true sentences to false ones. Consider, for example, the sentence

$$(\underline{x} = \emptyset) \Rightarrow \underline{\underline{B}} \quad ,$$

where $\underline{\underline{B}}$ denotes the sentence

$$(\{\emptyset\} \setminus \underline{x} = \emptyset) \Rightarrow (\underline{x} = \emptyset) \quad .$$

If one works in set theory, the first sentence is true. However, Halmos' proposed rule may be interpreted as leading from it to the sentence

$$(\forall \underline{x})((\underline{x} = \emptyset) \Rightarrow (\forall \underline{x})\underline{\underline{B}}) \quad ,$$

which is false (see Problem II/32).

Quite possibly these rules are "good" in many instances in informal mathematics, but they are clearly not wholly reliable in general.

Incidentally, in Gleason's (and most informal) accounts a clear picture of the rules governing the use of quantifiers is complicated by the underlying rather vague assignment of "domains" to variables and proposition schemas (sentences in which one or more variables appear); see Gleason (1), p.15. These domains are said to be sets, but the concept of set is usually left entirely informal. Thus, in his remarks on Russell's Paradox (loc. cit., p.25), Gleason says that the domain of the variable x there appearing is the "collection of all sets" , the term "collection" being used in preference to "set" , presumably precisely because it is intended to be less formal and to express a subtle (but unclarified) distinction.

The <u>correct</u> use of the quantifiers is (as Godement remarks

(1), p.34, footnote[(*)]) not always as simple as expected.

(x) As has been said in (viii) above, it is often convenient and overall satisfactory to employ bastard sentences to denote formal sentences which are otherwise denoted by semiformal expressions involving letters, logical signs, $\in$, quantifiers, braces, parentheses, and various other symbols which may have been already formally defined. Instances have been given in (viii) above.

Sometimes, however, the policy is less successful: what suggests itself initially as a bastardized name proves, on close examination, to be ambiguous or misleading; and attempts to remedy this may be either abortive or result in bastard sentences almost as cumbersome as the more formal ones they are intended to replace. The reader should not be too discouraged when this happens.

Here is an illustration from current mathematical life, (though one which will not appear again in this book); see, for example, Spivak (1), p.384, Problem *18. When studying real-valued functions of a real variable, one sometimes wishes to define concepts referred to bastard-wise as " f is an R-contraction function" and " f is an R-contractive function" . Now common usage might lead the unwary to expect that these concepts are indistinguishable and that the slight variation in the bastard sentences is of no consequence. But mathematics (even at an informal level) intends there to be a distinction, namely, that which is made clear by writing

(f is an R-contraction function) denotes

$$f \in R^{R} \wedge (\exists \underline{c})(\underline{c} \in]0, 1[\wedge (\forall \underline{x})(\forall \underline{y})((\underline{x} \in R \wedge \underline{y} \in R) \Rightarrow (|f(\underline{x}) - f(\underline{y})| \leq \underline{c}|\underline{x} - \underline{y}|))) \tag{5}$$

and

(f is an R-contractive function) denotes

$$f \in R^{R} \wedge (\forall \underline{x})(\forall \underline{y})((\underline{x} \in R \wedge \underline{y} \in R \wedge \underline{x} \neq \underline{y}) \Rightarrow (|f(\underline{x}) - f(\underline{y})| < |\underline{x} - \underline{y}|)) \quad , \tag{6}$$

where $\underline{c}$, $\underline{x}$, $\underline{y}$ denote distinct letters not appearing in R or f (R denoting the set of all real numbers), - a certain function $R \times R \to R$, and A - B being written in place of -((A, B))) . The right hand sides of (5) and (6) will (in due course) specify quite unambiguously certain formal sentences.

In seeking additional clarity, one might bastardize the right hand side of (5) somewhat as follows:

> f is a real-valued function of a real variable and there exists a real number $\underline{c}$ belonging to $]0, 1[$ such that, for all real numbers $\underline{x}$ and $\underline{y}$, $|f(\underline{x}) - f(\underline{y})| \leq \underline{c}|\underline{x} - \underline{y}|$;

and treat the right hand side of (6) likewise. The result is certainly more familiar in appearance, and it may be regarded as good enough. Even so, one often feels bound to add (supposedly to make doubly sure of avoiding confusion) a phrase such as " $\underline{c}$ is to be independent of $\underline{x}$ and $\underline{y}$, though it may depend upon f " , only to find on closer analysis that the added phrase may generate as much haze as it disperses: indeed, since $\underline{c}$, $\underline{x}$, $\underline{y}$ are distinct letters, in exactly what sense could $\underline{c}$ depend on $\underline{x}$ or $\underline{y}$?

Now such points are automatically cared for in the formalism, thus: the set whose existence is postulated by the appearance of $(\exists \underline{c})$ on the right of (5) is $c \equiv \tau_{\underline{c}}(\underline{\underline{B}})$, wherein $\underline{\underline{B}}$ denotes a sentence in which the only letters appearing are $\underline{c}$ and those letters (if any) which appear in f , R or $\overline{\ }$; neither $\underline{x}$ nor $\underline{y}$ is among these, so that neither they nor $\underline{c}$ appear in $\bar{c}$.

This being so, it is quicker and more effective to formalise, at least to the stage of using (5) .

The reader should look again at (viii) above.

Spivak's definition of "contraction" is slightly ambiguous and might easily be interpreted to correspond to the following (equivalent to (6)):

$$f \in R^R \wedge (\forall \underline{x})(\forall \underline{y})((\underline{x} \in R \wedge \underline{y} \in R) \Rightarrow (\exists \underline{c})(\underline{c} \in]0, 1[\wedge (|f(\underline{x}) - f(\underline{y})| \leq \underline{c}|\underline{x} - \underline{y}|))) .$$

Actually, the context indicates that Spivak intends (5) . It can be shown (see Problem VI/30) that the function

$f : \underline{x} \leadsto |\underline{x}|(1 + |\underline{x}|)^{-1}$ with domain R satisfies (6) but not (5) . (The fact that Spivak speaks about functions $[a, b] \to R$ is here quite immaterial.)

Incidentally, Spivak's Problem *18 asks the impossible. A counterexample is provided by the function

$$f : \underline{x} \leadsto \underline{x} - \underline{x}^2 \text{ with domain } [\tfrac{1}{4}, \tfrac{1}{3}]$$

this has no fixed point, despite the fact that

$$|f(\underline{x}) - f(\underline{y})| \leq \tfrac{1}{2}|\underline{x} - \underline{y}|$$

for all $\underline{x}$, $\underline{y} \in [\frac{1}{4}, \frac{1}{3}]$. Spivak has overlooked the extra hypothesis " Ran $f \subset [a, b]$ " . Furthermore, in Spivak's scenario modified by the inclusion of this extra hypothesis, the problem is more easily solved by applying his Theorems 4 and 5 on pp.101 - 102 (our VIII.2.1) to the function $\underline{x} \leadsto f(\underline{x}) - \underline{x}$. On the other hand, there are excellent reasons why Spivak suggests the method he does: it admits of much wider development.

Other possible confusions arising from the simultaneous use of two or more languages are noted in connection with (IV) and (X) in I.3.2 and receive further attention in §1 of the Appendix. Still more are discussed in II.3.9. See also Kleene (1), Chapter III.

Ambiguities stemming from the use of informal bastard languages are widespread. Usually one can rely on experience and intuition to resolve them, but occasionally one feels the need of a formal background to act as an arbiter; see VI.10.

(xi) A distinction has to be maintained between, say the metamathematical statement

$\underline{x}$ and $\underline{y}$ are (or denote) distinct letters (7)

and the statement usually written

$\underline{x}$ and $\underline{y}$ are distinct (8)

or

$\underline{x}$ and $\underline{y}$ are unequal ; (8')

the last two are names for the formal sentence denoted by

$\underline{x} \neq \underline{y}$

or by

$\neg(\underline{x} = \underline{y})$;

see II.1. To add to the possible confusion, in conventional informal mathematics (8) is usually intended as a replacement for the metamathematical statement

$\underline{x} \neq \underline{y}$ is a theorem, (9)

which is, of course, quite different from the metamathematical statement (7) . It is the meaning (8) or (8') which is appropriate when, for example, the definition of f being a function (see IV.1.1) is rendered informally as:

> f is a relation such that no two distinct ordered pairs belonging to f have equal first coordinates.

Similarly, the bastard sentence

$\underline{x}$ is not equal to $\underline{y}$

may be intended as a name for the formal sentence denoted by

$$\underline{x} \neq \underline{y} ,$$

but in conventional informal mathematics it is almost always intended to signify

$\underline{x} \neq \underline{y}$ is a theorem (or a hypothesis).

Yet another possible interpretation is as

$\underline{x} = \underline{y}$ is not a theorem,

but this is rarely appropriate; cf. the opening paragraph in II.3.10.

Usually, considerations of space prompt the author to say what he judges to be enough and otherwise rely on the context to indicate which meaning is intended. As every teacher knows, such judgements present difficulties.

(xii) I should perhaps add here that there is a numerically relatively small number of mathematicians (including some very eminent representatives) who feel that the logical schemas S1-S5 are themselves unacceptable and who, if they were to seek a formalisation of mathematics at all, would adopt a radically different foundation from that presented here. Among these are the so-called "intuitionists" . (This name does not mean that they exclude formalisation, nor that they are any less rigorous in their development of the appropriate formal theory.) Intuitionist mathematics is in many ways far more rigorous and demanding than the conventional variety. See Kneebone (1), Chapter 9; Behnke (1), p.6; Kleene (1), §13.

I.3 Methods of proof

I.3.1 Need for proof methods In principle one could set about proving theorems in any given theory (set theory, in particular) by referring back to the definition of proof in I.2.4 and writing our completely and explicitly a proof of every theorem, and doing this on every occasion on which reference is made to that theorem. In practice, however, this would be quite impossibly lengthy and tedious a procedure. (For one thing, it involves starting from square one for every proof.) What is more, the limitations of memory and patience throw doubt on whether one could find one's way through the resulting welter of explicitly written strings without working to some preconceived plan based upon intuitive interpretations. Even quite basic theorems would demand almost unbelievably vast books to display their proofs.

The only practical solution seems to be that which is parallel to that adopted in the case of constructions (see I.1.8). In the present case, the aim is to formulate once and for all a relatively small number of basic proof methods which typify and summarise what one would normally refer to as "logical reasoning", which appear to preserve the theoretical possibility of reversion to the primitive language and explicitly written proofs, and which appear to ensue from the schemas and axioms by means of an essentially simpler type of reasoning than that used within the theory itself.

One type of proof method is a true statement (metatheorem) of the sort

If ...... is (or are) a theorem (or theorems)

then ...... is also a theorem ;

see for example (I), (III) and (V) in I.3.2. Other proof methods, such as (IV) in I.3.2 and (j) in I.3.3, are true metastatements of the form

(Whatever the sentence (or sentences)) ...... ,

the sentence ...... is a theorem .

These are instances of sentence schemas and theorem schemas (see I.1.10 and I.2.9(v)).

A proof method may also be described as a metatheorem or theorem schema which is a consequence of the definition of "proof" and "theorem" via simple, finitary metamathematical reasoning of the type set out in §2 of the Appendix. Although this reasoning would be accepted as valid and convincing by almost everybody, they do in the last analysis depend on an underlying informal metalogic.

Alternatively, a proof method might be thought of as a proof schema (compare the notions of sentence schema, axiom schema and theorem schema discussed in I.1.10, I.2.1 and I.2.9(v)). See also Note 4.

These proof methods are intended to function somewhat as preassembled components in semiformal proofs (see I.3.4(i) below), thus reducing an otherwise intolerably long text to more manageable proportions and also making it easier to comprehend a proof as an entirety. (A proof would otherwise be an impossibly long and overwhelming labyrinth of details.) On the other hand, the proof methods stand as signposts indicating the way back to strict formalisation, should the need for this arise. They are thus thought to provide a practically almost essential and theoretically acceptable compromise.

Most of the earlier proof methods will appear very primitive; they are taken so much for granted in mathematical reasoning that conventional books about mathematics rarely mention them at all. The most familiar proof method is perhaps the Principle of Mathematical Induction, which appears as (XVII) in V.3.1. It differs from the earlier ones, which are purely logical in appearance, inasmuch as it makes explicit reference to a specific set, the set N of natural numbers. Much work has to be done before one is ready to formally define N , which is why (XVII) makes its debut so late.

The first eleven basic proof methods are listed in I.3.2. Others will be listed later: (XII) in II.3.7, (XIII) in II.3.11, (XIV) in II.12, (XV) and (XVI) in III.1.2, and (XVII) in V.3.1; see also Problem I/11. Still more might be added, if it seems convenient to do so. (Most books on more advanced set theory would include, for example, a Principle of Transfinite Induction.)

I.3.2 Basic proof methods In this section I shall list eleven of the most basic such proof methods. Some of these will surely be recognisable as definitive statements of "principles of logic" vaguely familiar to anybody who has done any high school mathematics. I will state them in relation to an arbitrary theory Θ (which may be Θ_0 itself); the words "theorem" , "theorem schema" , "constant" (that is, "fixed letter") , "variable" (that is, "free letter") are all to be understood in relation to Θ ; $\underline{\underline{A}}$, $\underline{\underline{B}}$, and $\underline{\underline{C}}$ denote arbitrary sentences.

(I) Rule of Detachment (Modus ponens) If $\underline{\underline{A}}$ and $\underline{\underline{A}} \Rightarrow \underline{\underline{B}}$ are theorems, then $\underline{\underline{B}}$ is a theorem.

(II) Rule of Specialisation (first variant) If $\underline{\underline{A}}$ is a theorem, $\underline{x}$ a variable and S a set, then $(S|\underline{x})\underline{\underline{A}}$ is a theorem.

(III) The Syllogism If $\underline{\underline{A}} \Rightarrow \underline{\underline{B}}$ and $\underline{\underline{B}} \Rightarrow \underline{\underline{C}}$ are theorems, then $\underline{\underline{A}} \Rightarrow \underline{\underline{C}}$ is a theorem.

(IV) Law of the Excluded Middle (Tertium non datur) $\underline{\underline{A}} \vee (\neg \underline{\underline{A}})$ is a theorem schema.

(V) If $\underline{\underline{A}} \Rightarrow \underline{\underline{B}}$ is a theorem, then $(\underline{\underline{B}} \Rightarrow \underline{\underline{C}}) \Rightarrow (\underline{\underline{A}} \Rightarrow \underline{\underline{C}})$ is a theorem.

(VI) Criterion of Deduction Consider the theory Θ' obtained from Θ by adjoining $\underline{\underline{A}}$ as an explicit axiom. If $\underline{\underline{B}}$ is a theorem of Θ', then $\underline{\underline{A}} \Rightarrow \underline{\underline{B}}$ is a theorem (of Θ).

(VII) Proof by contradiction Consider the theory Θ' obtained from Θ by adjoining $\neg \underline{\underline{A}}$ as an explicit axiom. If Θ' is contradictory, then $\underline{\underline{A}}$ is a theorem (of Θ).

(VIII) Proof by disjunction of cases If $\underline{\underline{A}} \vee \underline{\underline{B}}$, $\underline{\underline{A}} \Rightarrow \underline{\underline{C}}$ and $\underline{\underline{B}} \Rightarrow \underline{\underline{C}}$ are theorems, then $\underline{\underline{C}}$ is a theorem. In particular, if $\underline{\underline{A}} \Rightarrow \underline{\underline{C}}$ and $(\neg \underline{\underline{A}}) \Rightarrow \underline{\underline{C}}$ are theorems, then $\underline{\underline{C}}$ is a theorem.

(IX) Rule of Existential Elimination Suppose that $\underline{x}$ is a variable not appearing in $\underline{\underline{B}}$, and that $(\exists \underline{x})\underline{\underline{A}}$ and $\underline{\underline{A}} \Rightarrow \underline{\underline{B}}$ are theorems. Then $\underline{\underline{B}}$ is a theorem.

(X) Rule of Generalisation If $\underline{\underline{A}}$ is a theorem, and if $\underline{x}$ is a variable, then $(\forall \underline{x})\underline{\underline{A}}$ is a theorem.

(XI) Rule of Specialisation (second variant) If $\underline{x}$ is a letter and S a set, then $(\forall \underline{x})\underline{\underline{A}} \Rightarrow (S|\underline{x})\underline{\underline{A}}$ is a theorem schema.

Remarks (i) The proof method (VI) is often called the deduction (meta-)theorem; see, for example, Margaris (1), p.55; Kneebone (1), Chapter 3, §5; Kleene (1), §21.

(ii) In relation to (IV), note that it does not affirm that either $\underline{\underline{A}}$ is a theorem or $\neg\underline{\underline{A}}$ is a theorem. As has been mentioned in I.2.9(ii), if consistency is assumed, it is easy to exhibit open sentences $\underline{\underline{A}}$ which are undecidable (that is, neither $\underline{\underline{A}}$ nor $\neg\underline{\underline{A}}$ is a theorem); see also I.3.4(vii), I.3.8 and §1 of the Appendix. In spite of this dampening note, (VII) provides some redress: the second sentence of (VII) affirms what one would expect on the basis of (I), if it were the case that either $\underline{\underline{A}}$ or $\neg\underline{\underline{A}}$ is a theorem. (In writing this, I am in effect hinting at a fallacious verification of (VII) on the basis of (I); see four paragraphs below.)

The false impression of (IV) (to the effect that in all cases either $\underline{\underline{A}}$ or $\neg\underline{\underline{A}}$ is a theorem) may be gained from some seemingly impeccable sources; an instance is Dieudonné's comparison ((1), p.549, right hand column, first complete paragraph) of intuitionism and formalism. (Jean Dieudonné is an extremely eminent French mathematician, an erstwhile active member of the Bourbaki group, who has had a great deal of influence on the development of modern abstract mathematics.)

(iii) In connection with (VII) it is worth pointing out that, in spite of appearances, a "proof by contradiction" in Θ of $\underline{\underline{A}}$ involves no metamathematical assumption that $\underline{\underline{A}}$ is false in Θ . ($\underline{\underline{A}}$ is false in Θ' , but this is no assumption.)

Sometimes (VII) is expressed in the following form:

(VII') If $\underline{\underline{B}}$ and $\neg\underline{\underline{A}} \Rightarrow \neg\underline{\underline{B}}$ are theorems, then $\underline{\underline{A}}$ is a theorem.

Given (I) and (VII), the verification of (VII') is almost immediate; see also Problem I/12 Also, given S2, (I), (IV), (VI), and (VII'), it is simple to verify (VII); see Problem I/13.

Fallacious or misleading verifications of (VII') are sometimes offered. One runs thus: One may assume Θ to be consistent since otherwise every sentence is true (see I.2.7). Suppose $\underline{\underline{A}}$ to be false in Θ ; then, by (I) and the assumed truth of $\neg\underline{\underline{A}} \Rightarrow \neg\underline{\underline{B}}$, $\neg\underline{\underline{B}}$ is true in Θ . Hence Θ is contradictory, which is not so; hence $\underline{\underline{A}}$ cannot be false in Θ ; hence (here is the fallacy) $\underline{\underline{A}}$ must be true in Θ . This argument is acceptable, only if it is known in advance that $\underline{\underline{A}}$ is decidable in Θ ; see (ii) above.

This type of fallacious verification is surprisingly prevalent: Dieudonné ((1), p.543, left hand column, first paragraph) proffers such an argument by way of verification or explanation of proofs by contradiction (in the version (VII')). See also Griffiths and Hilton (1), p.600.

The same fallacy appears in some alleged verifications of (VIII).

(iv) An appeal to (IX) is in practice often expressed somewhat as follows:

> Choose a so that $(a|\underline{x})\underline{\underline{A}}$ is true (a suitable choice of a being $\tau_{\underline{x}}(\underline{\underline{A}})$) . Then $(a|\underline{x})\underline{\underline{A}} \Rightarrow \underline{\underline{B}}$
>
> is true. (This is because $\underline{\underline{A}} \Rightarrow \underline{\underline{B}}$ is true by hypothesis, and implicit use is being made of (II), the appropriate replacement rule, and the hypothesis that $\underline{x}$ does not appear in $\underline{\underline{B}}$.) Hence (by implicit use of (I)) $\underline{\underline{B}}$ is true.

The parenthesised explanations would usually not appear.

(v) As has been said in I.2.6, all letters are variables of Θ_0 ; so "variable" can be replaced by "letter" in (II), (IX) or (X) when Θ is Θ_0 . Similar remarks apply to I.3.3(1) below.

Remember, however, that in practice logical letters and variables are rarely distinguished, neither by explicit mention nor by some particular notational device (comparable with the single underline used for that purpose in earlier chapters of this book).

(vi) Useful extensions of (XI) appear in Problem I/17(iii) and (v).

(vii) Throughout the rest of this book, the words "theorem" and "true" , without further qualification, signify "theorem of Θ_0 " and "true in Θ_0 " respectively.

(viii) In (II) and (X) it is vital that $\underline{x}$ be a variable; and even then it is not always the case that $\underline{\underline{A}} \Rightarrow (S|\underline{x})\underline{\underline{A}}$ or $\underline{\underline{A}} \Rightarrow (\forall \underline{x})\underline{\underline{A}}$ is a theorem. See Problem I/22 and §1 of the Appendix.

(ix) In connection with (VIII), it turns out that one can (see Problem I/16) prove the theorem schema

$$((\underline{\underline{A}} \Rightarrow \underline{\underline{C}}) \wedge (\underline{\underline{B}} \Rightarrow \underline{\underline{C}})) \Leftrightarrow ((\underline{\underline{A}} \vee \underline{\underline{B}}) \Rightarrow \underline{\underline{C}}) ,$$

of which (VIII) is a simple metamathematical consequence.

In routinely presented proofs (see I.3.4(i) below) hidden appeals to (VIII) are frequently masked in the following way. Suppose the objective to be a proof, in a theory Θ , of a certain formal sentence $\underline{\underline{B}}$. The routine proof may embrace a sentence of the type

It may be assumed without loss of generality that $\underline{\underline{A}}$ (1)

which is then followed by a proof, in the theory Θ' obtained by adjoining to Θ the axiom $\underline{\underline{A}}$, of $\underline{\underline{B}}$. In view of (VI), this amounts to a proof in Θ of

$$\underline{\underline{A}} \Rightarrow \underline{\underline{B}} \quad . \tag{2}$$

The routine proof will terminate at this stage.

In such a situation, (1) is usually intended as an indication that there is a proof in Θ of

$$\neg\underline{\underline{A}} \Rightarrow \underline{\underline{B}} \tag{3}$$

which is either relatively trivial or very similar to that of (2), and which is in either case being omitted; the reader can either accept (3) or provide a proof thereof. Once (2) and (3) are proved, an implicit appeal to (VIII) sanctions the desired conclusion (that $\underline{\underline{B}}$ is a theorem of Θ) .

Examples of this procedure appear in the proofs of (13) in V.7.2, VI.7.14 and VI.7.15 (and probably elsewhere as well).

I.3.3 <u>Further proof methods</u> In addition to the proof methods listed in I.3.2 there are many derivative methods which are in constant use (at least implicitly). These are metatheorems and theorem schemas which are metamathematical consequences of the logical schemas S1 - S5 and (I) - (XI) in I.3.2. I list a few of the most useful ones as (j) - (o) below. As in I.3.2, Θ denotes an arbitrary theory, $\underline{\underline{A}}$, $\underline{\underline{B}}$, $\underline{\underline{C}}$ denote arbitrary sentences and $\underline{x}$ and $\underline{y}$ denote arbitrary letters, subject to no metamathematical restrictions other than those stated in the contexts of (l) and (o) ; and "theorem" means "theorem of Θ " .

(j) $\underline{\underline{A}} \Leftrightarrow \underline{\underline{A}}$, $\underline{\underline{A}} \Rightarrow (\underline{\underline{A}} \vee \underline{\underline{B}})$, $\underline{\underline{B}} \Rightarrow (\underline{\underline{A}} \vee \underline{\underline{B}})$,
$\neg(\underline{\underline{A}} \vee \underline{\underline{B}}) \Leftrightarrow (\neg\underline{\underline{A}}) \wedge (\neg\underline{\underline{B}})$, $(\underline{\underline{A}} \wedge \underline{\underline{B}}) \Rightarrow \underline{\underline{A}}$,
$(\underline{\underline{A}} \wedge \underline{\underline{B}}) \Rightarrow \underline{\underline{B}}$, $\underline{\underline{A}} \vee \underline{\underline{B}} \Leftrightarrow \underline{\underline{B}} \vee \underline{\underline{A}}$, $\underline{\underline{A}} \wedge \underline{\underline{B}} \Leftrightarrow \underline{\underline{B}} \wedge \underline{\underline{A}}$,
$\underline{\underline{A}} \wedge (\underline{\underline{B}} \vee \underline{\underline{C}}) \Leftrightarrow (\underline{\underline{A}} \wedge \underline{\underline{B}}) \vee (\underline{\underline{A}} \wedge \underline{\underline{C}})$,
$\underline{\underline{A}} \vee (\underline{\underline{B}} \wedge \underline{\underline{C}}) \Leftrightarrow (\underline{\underline{A}} \vee \underline{\underline{B}}) \wedge (\underline{\underline{A}} \vee \underline{\underline{C}})$
are theorem schemas; if $\underline{\underline{A}}$ and $\underline{\underline{B}}$ are theorems, then $\underline{\underline{A}} \wedge \underline{\underline{B}}$ is a theorem.

(k) $\underline{\underline{A}} \Leftrightarrow \neg\neg\underline{\underline{A}}$ and $(\underline{\underline{A}} \Rightarrow \underline{\underline{B}}) \Leftrightarrow (\neg\underline{\underline{B}} \Rightarrow \neg\underline{\underline{A}})$ are theorem schemas ("double negation rule" and "contrapositive rule"); $\underline{\underline{A}} \Leftrightarrow \underline{\underline{A}} \vee \underline{\underline{A}}$, $\underline{\underline{A}} \Leftrightarrow \underline{\underline{A}} \wedge \underline{\underline{A}}$ are theorem schemas.

(l) If $\underline{\underline{A}} \Rightarrow \underline{\underline{B}}$ is a theorem, and if $\underline{x}$ is a variable (free letter of Θ), then $(\exists\underline{x})\underline{\underline{A}} \Rightarrow (\exists\underline{x})\underline{\underline{B}}$ and $(\forall\underline{x})\underline{\underline{A}} \Rightarrow (\forall\underline{x})\underline{\underline{B}}$ are theorems.

(m) The following are theorem schemas:

$$\neg(\forall\underline{x})\underline{\underline{A}} \Leftrightarrow (\exists\underline{x})(\neg\underline{\underline{A}}) ,$$
$$\neg(\exists\underline{x})\underline{\underline{A}} \Leftrightarrow (\forall\underline{x})(\neg\underline{\underline{A}}) ,$$
$$(\forall\underline{x})(\underline{\underline{A}} \wedge \underline{\underline{B}}) \Leftrightarrow (\forall\underline{x})\underline{\underline{A}} \wedge (\forall\underline{x})\underline{\underline{B}} ,$$
$$(\exists\underline{x})(\underline{\underline{A}} \vee \underline{\underline{B}}) \Leftrightarrow (\exists\underline{x})\underline{\underline{A}} \vee (\exists\underline{x})\underline{\underline{B}} .$$

On the other hand, it can happen that $(\forall\underline{x})(\underline{\underline{A}} \vee \underline{\underline{B}})$ is true and $(\forall\underline{x})\underline{\underline{A}} \vee (\forall\underline{x})\underline{\underline{B}}$ is false; and that $(\exists\underline{x})\underline{\underline{A}} \wedge (\exists\underline{x})\underline{\underline{B}}$ is true and $(\exists\underline{x})(\underline{\underline{A}} \wedge \underline{\underline{B}})$ is false. (This last is so if, for example, $\underline{\underline{A}}$ is $\underline{x} \in \{\emptyset\}$ and $\underline{\underline{B}}$ is $\neg\underline{\underline{A}}$ (see Chapter II); in fact, one has only to choose for $\underline{\underline{A}}$ a sentence expressing a property possessed by at least one, but not by all, objects.) However, see (o) immediately below and Problem I/32.

(n) The following are theorem schemas:

$$(\exists \underline{x})(\exists \underline{y})\underline{\underline{A}} \Leftrightarrow (\exists \underline{y})(\exists \underline{x})\underline{\underline{A}} \quad ,$$

$$(\forall \underline{x})(\forall \underline{y})\underline{\underline{A}} \Leftrightarrow (\forall \underline{y})(\forall \underline{x})\underline{\underline{A}} \quad ,$$

$$(\exists \underline{x})(\forall \underline{y})\underline{\underline{A}} \Rightarrow (\forall \underline{y})(\exists \underline{x})\underline{\underline{A}} \quad .$$

Yet it can happen that $(\forall \underline{y})(\exists \underline{x})\underline{\underline{A}}$ is true and $(\exists \underline{x})(\forall \underline{y})\underline{\underline{A}}$ is false; see Problem II/12.

(o) If $\underline{x}$ is a letter not appearing in $\underline{\underline{A}}$, then

$$(\exists \underline{x})(\underline{\underline{A}} \wedge \underline{\underline{B}}) \Leftrightarrow \underline{\underline{A}} \wedge (\exists \underline{x})\underline{\underline{B}} \quad ,$$

$$(\forall \underline{x})(\underline{\underline{A}} \vee \underline{\underline{B}}) \Leftrightarrow \underline{\underline{A}} \vee (\forall \underline{x})\underline{\underline{B}}$$

are theorems. (See also Problem I/32.)

The metatheorems (I) - (XI) in I.3.2 and (j) - (o) immediately above are all entirely "reasonable" in appearance, at least in so far as they are translatable into informal language. They would normally be viewed as belonging to "pre-mathematics" , inasmuch as they specify some of the most basic modes of reasoning upon which mathematics is founded. Actually, they belong to that portion of formal logic referred to as (first order) predicate calculus; see Kleene (1), Chapter VII; Kleene (2), Chapter II; Kneebone (1), Chapter 3. Our treatment differs in detail from that of these writers, partly because imbedded in our formalism is the selector τ .

Details of the quite lengthy verifications and proofs of (I) - (XI) and (j) - (o) will be found in §2 of the Appendix.

I.3.4 Description and presentation of proofs

(i) There are three roughly distinguishable ways in which proofs may be presented or described. First are the strictly formal proofs, consisting of explicitly written strings and lists of strings. These are totally impossible in practice and <u>never</u> appear in conventional mathematical texts.

Next come what will be referred to as <u>semiformal proofs</u> which, together with the routine style proofs to be described below, correspond to what Margaris (1), p.13) terms "working proofs" . Semiformal proofs are practically realisable descriptions of formal proofs, their practicability being the result of free use of direct appeals, each occupying only a few lines, to metatheorems (like those in I.3.2 and I.3.3) and to theorems and theorem schemas already proved in like style. (In a formal proof, the only appeals allowed are to axioms, and every proof has to begin at "square one" .) Semiformal proofs often contain in addition various explanations of motivation and strategy : these are, of course, expressed in the metalanguage and have no formal counterparts.

The concept of semiformal proof is somewhat elastic. It undergoes change with the growth of the stock of known theorems. If reasonably detailed and explicit references to proof methods are incorporated, a semiformal proof will itself be very lengthy and tedious. Such versions are rarely found in informal texts, which content themselves with routine style proofs.

Inasmuch as a sufficiently explicit semiformal proof appears to raise no difficulties in principle in the path of complete formalisation (see the remarks in the Foreword to the Appendix on the verification of the proof methods), it will be accepted as a valid replacement for the formal proof which it describes.

The first and simplest examples of semiformal proofs appear in I.3.6, though these are logical rather than conventionally mathematical. Numerous more conventionally mathematical examples appear in natural sequence throughout Chapters II - V , after which they become rarer and give way to routine style proofs.

Finally come <u>routine (style) proofs</u>, which are the descriptions used almost exclusively in conventional texts on mathematics. In these, there is rarely any explicit reference to proof methods. Thus, for example, when an appeal to I.3.2(VII) lurks in the background, the text may say merely,

Assume $\neg \underline{A}$

or

Assume $\underline{\underline{A}}$ is false

and proceed to argue on this basis until a contradiction is reached, at which point the text may simply say

This contradiction completes the proof.

Moreover, there will usually be no special notation for logical letters or variables; and restrictions that certain letters are not to appear in certain sets or sentences will almost invariably be taken for granted and regarded as being indicated by intuition or common sense.

See also (ii) immediately below.

It will at times also be necessary to distinguish between proofs (formal, semiformal and routine) and what will be termed verifications; see the Introduction to and §2 of the Appendix. Verifications are arguments, presented as convincingly as one can devise and based on a simple, informal logic, in support of metastatements about one or more formal theories.

A verification of a metastatement thus lies completely "outside" the formal theory or theories to which the metastatement refers. There is often no expectation of formalising either a metastatement or its verification (if there is one); nor, in principle are there grounds for such an expectation. (In spite of this, one of the great steps in metamathematics was Gödel's ingenious device for formalising certain metastatements about certain formal theories, within those same theories; see the references in §1 of the Appendix.)

An exception to this use of terminology is that we shall usually speak of a proof (rather than a verification) of a theorem schema (cf. I.2.9(v)); numerous instances appear in I.3.3, I.3.6, in various problems, and in the chapters to follow. The reason for making this exception is that the development of the formal theory will make use of individual instances of the theorem schema, and the verification of the theorem schema will (in the cases which directly concern us) exhibit a pattern which indicates clearly how one might (in principle and without further ingenuity) construct a formal proof of that instance.

The few particular verifications essential in this book (namely, those of the replacement rules and the metatheorems in I.3.2 and I.3.3 and a few others like them to appear later) are essentially much less sophisticated than many metamathematical arguments. For reasons which are made clearer in the Foreword to the Appendix, they are widely regarded as being beyond reproach and of such a nature that semiformal proofs are (as has been suggested already) acceptable proxies for their formal counterparts. In fact, semiformal proofs of theorems and theorem schemas are very similar in appearance and status to these particular verifications (but essentially different from many much more sophisticated metamathematical verifications, which frequently invoke a highly developed informal set theory).

In spite of these remarks, a thorough-going formalist might insist that, if proof methods are to be invoked at all, they ought to be applied directly only to individual instances of sentence schemas, so that (for example) an appeal to I.3.2(VI) involves the adjunction of a genuine axiom (and not an axiom schema). The issue might be partially avoided by treating, in place of whatever sentence schema is in question, a suitably chosen generating sentence; see I.1.10 and I.3.8(vi). Such a sentence is usually obtained by replacing the relevant set symbols by distinct letters or variables. One would then prove the corresponding theorem, and then recover the desired theorem schema by formal replacement of the letters or variables by the original set symbols. (This is not a complete answer to the problem inasmuch as sentence symbols may also be involved and cannot be eliminated in a similar way.) For example (see II.2.2(2) below) in place of attempting to deal directly with the theorem schema

$$((X \subseteq Y) \wedge (Y \subseteq Z)) \Rightarrow (X \subseteq Z) \quad ,$$

one might aim at a proof of the sentence

$$((\underline{x} \subseteq \underline{y}) \wedge (\underline{y} \subseteq \underline{z})) \Rightarrow (\underline{x} \subseteq \underline{z}) \quad ,$$

or of the sentence

$$(\forall \underline{x})(\forall \underline{y})(\forall \underline{z})(((\underline{x} \subseteq \underline{y}) \wedge (\underline{y} \subseteq \underline{z})) \Rightarrow (\underline{x} \subseteq \underline{z})) \quad ,$$

wherein $\underline{x}$, $\underline{y}$, $\underline{z}$ denote any chosen distinct letters (or variables). Having achieved that aim, the desired theorem schema could be recovered via triple use of (I) and (II) or (XI) in I.3.2, assuming as one may that $\underline{x}$, $\underline{y}$, $\underline{z}$ are chosen not to appear in X , Y or Z .

For another example, see (4), (6), (4') and (6') in II.4.1.

In practice, however, the reverse procedure (the passage from theorems to theorem schemas) is often adopted (see II.1.4, for example). The reason for this may be the feeling that theorem schemas look less forbiddingly formal than the associated theorems.

(ii) The original aim of a semiformal proof is to make explicit every appeal to the theorems and metatheorems in I.3.2, I.3.3, I.3.6 and I.3.7 and later proof methods, plus any other metatheorems, theorem schemas and theorems used on the way. Fulfillment of even this aim soon leads to forbiddingly long and tedious texts: see, for instance, Remark (i) in II.4.4, Remark (i) in III.2.7, the proofs of IV.4.2, and above all the proof of V.5.2 (which is, after all, nothing more than a very basic theorem relating to natural numbers which subsequently plays an essential role over and over again). In the end, and from the point of view of conventional mathematics, even semiformal proofs tend to appear as ideals. Accordingly, beginning in Chapter II, I shall usually content myself with mentioning only those appeals to the theorems and metatheorems in I.3.2, I.3.3, et cetera, which seem to be the most essential and helpful (and even in this I cannot guarantee 100% accuracy!). It will be up to the reader to check the appeals which remain implicit. (Most frequently these implicit appeals will be to the Rule of Detachment (I) and the Syllogism (III).) This is a tedious task, but it can pay off in terms of familiarity and understanding.

In almost all books which concern themselves at all with formal or semiformal proofs, the tendency is to abandon anything remotely approaching the semiformal style in favour of the routine style long before any substantial theorems are encountered. In this book there is, on the contrary, an attempt to remain more faithful to the semiformal style long enough to convince a sceptical reader that something close to semiformal style is workable in connection with a number of substantial and vitally important theorems used repeatedly throughout mathematics. Readers who are more trusting can ignore the semiformal style as and when they wish, reverting to the routine style found in the overwhelming majority of books.

Concerning a more detailed comparison between semiformal proofs and routine proofs, see the proof of (3) in II.2.2, II.5.5, the proof of IV.4.2, Remark (ii) following IV.5.4, the remarks in V.5.2 following the statement of the Recursion Theorem, and V.5.4; see also Problems II/7 and II/8 and Note 5.

(iii) In writing out semiformal proofs, it will often be helpful to inscribe

$$T\Theta \quad : \qquad \underline{\underline{A}} \qquad (1)$$

to indicate that the sentence $\underline{\underline{A}}$ is, or denotes, a theorem (or sometimes that what occupies the place of $\underline{\underline{A}}$ is more properly termed a theorem schema) of Θ. If Θ is Θ_0, I will write simply T in place of $T\Theta_0$. In the case of theorems or theorem schemas of some lasting interest, I will use the labels $T\Theta!$ or $T!$.

The style $\Theta \vdash \underline{\underline{A}}$ is often used in place of (1). Later on, I will often revert to the customary heading "Theorem".

When one speaks rather loosely of a proof of (1), one may be referring to a formal proof of $\underline{\underline{A}}$ (possible in principle); or one may be referring to a semiformal or routine proof of (1), effected by appeal to whatever available proof methods seem appropriate and to theorems or theorem schemas already proven in the same manner; see (ii) above. In conventional texts, it is usually the second meaning that is intended. Thus, "prove" or "prove that" conventionally signifies "write out a semiformal, or routine, style proof of" .

It would perhaps be more proper to write either

Verify that

$T\Theta$: $\underline{\underline{A}}$

or

Prove $\underline{\underline{A}}$ (in Θ) .

In practice, however, both are intended as injunctions to provide, or describe in acceptable style and detail, a proof in Θ of A .

Recall also the remarks in I.2.9(v) concerning proofs of theorem schemas.

When a proof (in a certain theory Θ) is being conventionally written or described, a sentence such as "Since $\underline{\underline{A}}$ and $\underline{\underline{B}}$, therefore $\underline{\underline{C}}$ " almost always stands for "Since $\underline{\underline{A}}$ and $\underline{\underline{B}}$ are true in Θ , therefore $\underline{\underline{C}}$ is also true in Θ ". The "therefore" will often be justified by reference to one or more metatheorems and/or theorems already verified or proven.

When it is used in routine style proofs, the term "implies" demands care in interpretation. For instance, a theorem might be routinely (informally) recorded as

If $\underline{\underline{A}}$, then $\underline{\underline{B}}$,

a more precise formulation being

$T\Theta$: $\underline{\underline{A}} \Rightarrow \underline{\underline{B}}$,

where Θ is a certain specified theory. A routine style proof of this theorem might include a phrase to the effect that $\underline{\underline{P}}$ implies $\underline{\underline{Q}}$. This might quite reasonably be taken to signify that $\underline{\underline{P}} \Rightarrow \underline{\underline{Q}}$ is a theorem of Θ_0 , or that it is a theorem of Θ . However, it is very frequently intended to signify that $\underline{\underline{P}} \Rightarrow \underline{\underline{Q}}$ is true in the theory Θ^* obtained by adjoining to Θ the axiom $\underline{\underline{A}}$, that is (see (VI) and I.3.7(2)) that $(\underline{\underline{A}} \wedge \underline{\underline{P}}) \Rightarrow \underline{\underline{Q}}$ is true in Θ . This interpretation is especially likely to be the intended one, if the proof is being effected by verifying that $\underline{\underline{B}}$ is true in Θ^* and then making an appeal to (VI) .

Phrases such as "as a corollary" , and "it follows from" , used in routine style proofs, have to be interpreted in a similar way.

Again, when in the course of a routine style proof it is intended to appeal to a known theorem of the form

$$(\exists \underline{x})\underline{\underline{A}} \quad , \tag{2}$$

it is customary to write somewhat as follows:

By (or : in view of) (2) , one may (and will) choose X so that $(X|\underline{x})\underline{\underline{A}}$

or simply

Choose X so that $(X|\underline{x})\underline{\underline{A}}$.

In the latter case, there may or may not appear any justification in the form of an explicit reference to (2) . The semiformal counterpart to this procedure would be to define (often on a merely temporary basis; see I.3.5(viii))

$$X \equiv \tau_{\underline{x}}(\underline{\underline{A}}) \quad ,$$

in terms of which (2) is identical with $(X|\underline{x})A$. (There often is no real choice involved; but in all cases the formalism makes the selection from the outset.)

See also II.4.2(iii) and Remark (ii) in VII.1.4, and Note 6.

(iv) The longer routine and semiformal proofs become, the greater the tendency to incorporate within them explanatory remarks (necessarily informal) about "what is going on" . The outcome of this is that often the order in which various items appear in such proofs will be different (sometimes the very reverse) of that in which they or their counterparts would appear in a formal proof. A better correspondence could be restored, but most mathematicians would prefer intuitive clarity to exact logical order. See the Remarks in II.1.5 and II.4.4 and the proof of (8) in II.5.2.

The said explanatory remarks, or commentary, are of course expressed in the metalanguage. They introduce a subjective element which seems inevitable. They also draw attention to the psychological problem involved in recognising instances or special cases of axiom schemas, theorem schemas and definition schemas. That all "normal" readers will think alike in this respect is an assumption. It is also an assumption that the writer will make "once for all" judgements of this sort - again a psychological issue.

In short, these explanatory remarks are practically essential and yet involve an intrusion of subjectivity, presumably inescapable (unless one seeks to formalise the metalanguage).

One is also led to the conclusion that the formal language provides a high level of precision and objectivity, and yet may paradoxically lack overall clarity, often best restored with the aid of auxiliary explanations expressed in an informal metalanguage. This phenomenon is, I suppose, well known to every teacher: success often comes about through a judicious blend of precision on a small scale and clarity on a larger scale, a blend which has to be chosen in a way depending on the audience. It is probably this feature which gives rise to much justifiable criticism of the so called "new Mathematics" , especially at the primary level.

(v) To reinforce what was said in I.1.7 about the use of intuitive meanings, I should now add two remarks. First, the metatheorems and theorem schemas in I.3.2 and I.3.3 are broadly in harmony with intuition. Second, proofs based upon them are almost always originally conceived in terms of intuitive meanings and it is difficult to see how any progress would otherwise be made. The reader should approach the problems in I.3.7 and elsewhere in this book with these remarks in mind. However, intuitions differ from person to person and so the concept of formal proof is needed in the background. Every proposed step in the proof as initially conceived has to be sanctioned, not merely on intuitive grounds, but also on purely formal grounds. If the latter type of sanction remains in doubt, so does the proof.

See also the Remarks following II.3.8 below.

(vi) From a pedagogical point of view, it is as well to say a little about the interpretation of conventionally stated problems. Perhaps the simplest type is that following the pattern:

> Suppose (or assume) that $\underline{\underline{A}}$. Prove that
> (or: give a proof of) $\underline{\underline{B}}$.

Plainly, there is implicit in this a background theory Θ

(usually Θ_0) . To solve the problem may be interpreted in one of at least two ways:

either provide a proof in Θ of $\underline{\underline{A}} \Rightarrow \underline{\underline{B}}$;

or introduce the theory Θ' obtained by adjoining to Θ the axiom $\underline{\underline{A}}$, and provide a proof in Θ' of $\underline{\underline{B}}$.

The second procedure might be described as a *deduction, in* Θ , *of* $\underline{\underline{B}}$ *from* $\underline{\underline{A}}$; see Problem I/4, where the reader is asked to show that the two procedures are metamathematically equivalent.

For examples, see Problems II/15, V/27 VII.2.6(1) - (3).

Problem VII.2.6(4) is somewhat different and less explicit. It typifies the situation in which one is directed to assume $\underline{\underline{A}}$ (as before) and to decide whether $\underline{\underline{B}}$ is true, not true, or false in Θ' (and, by implication, to provide justification for the decision).

See also (b) and (c) in I.4.5(ii).

When $\underline{\underline{A}} \Rightarrow \underline{\underline{B}}$ is a theorem of Θ , one often says that $\underline{\underline{B}}$ *is a corollary (in* Θ *) of* $\underline{\underline{A}}$ or that $\underline{\underline{B}}$ *is deducible* (or *derivable*) *(in* Θ *) from* $\underline{\underline{A}}$.

The construction and solution of problems presents difficulties for the setter and the solver other than the obvious ones. Further brief comments on this matter will be made from time to time. The remarks in (viii) below are relevant; see also (for example) III.1.8 and IV.1.7. Occasionally remarks will be attached to specific problems appearing in this book; see, for example, Problems IV/10 and IV/14. It will become clear that with many problems, even the intention of the problem poser may be less clear than it might initially appear; and that the solver has to adopt a cooperative attitude, if the aim of the problem poser is not to be defeated.

(vii) Suppose that Θ is a certain theory and that $\underline{\underline{A}} \Rightarrow \underline{\underline{P}}$ and $\neg\underline{\underline{A}} \Rightarrow \underline{\underline{Q}}$ are known to be theorems of Θ . It is then often written:

> One of two things must happen: *either* $\underline{\underline{A}}$ is true, in which case $\underline{\underline{P}}$ is also true; *or* $\neg\underline{\underline{A}}$ is true, in which case $\underline{\underline{Q}}$ is true. ,

"true" standing for "true in Θ " .

From a formal point of view, this is misleading embroidery, simply because it may happen that neither $\underline{\underline{A}}$ nor $\neg\underline{\underline{A}}$ is true (in Θ); see again Remark (ii) in I.3.2.

Similar remarks apply when $\underline{\underline{A}} \vee \underline{\underline{B}}$, $\underline{\underline{A}} \Rightarrow \underline{\underline{P}}$ and $\underline{\underline{B}} \Rightarrow \underline{\underline{Q}}$ are theorems of Θ , and one sees written:

> Either $\underline{\underline{A}}$ is true, in which case $\underline{\underline{P}}$ is true; or $\underline{\underline{B}}$ is true, in which case $\underline{\underline{Q}}$ is true.

There is in fact no assurance that either $\underline{\underline{A}}$ is true or $\underline{\underline{B}}$ is true.

A rather more elaborate (but basically similar) misrepresentation featuring in routine style proofs is illustrated at the end of II.3.10(ii) below.

Again, it is sometimes suggested (see, for example, Gordon and Hindman (1), p.293) that the normal or expected procedure for proving a disjunction $\underline{\underline{P}} \vee \underline{\underline{Q}}$ is to prove at least one of $\underline{\underline{P}}$ or $\underline{\underline{Q}}$. In particular, the suggestion is that, if A , B and C denote sets, one should expect to prove $A = B \cup C$ by proving at least one of $A = B$ or $A = C$. (Gordon and Hindman do, however, state in a footnote that "disjunctions and existential statements are frequently proved by contradiction".) Although the suggested procedure is undeniably effective whenever it is possible, it is by no means obligatory, and in many important cases it is impossible (unless the relevant theory is contradictory). In fact, it is probably fair to state that the suggested procedure is possible only in a negligible proportion of cases, many of those being intuitively quite "dull" .

(viii) Proof by exhibition and existential proofs; solutions and solving As a consequence of S5 in I.2.2 and (I) in I.3.2, one has the proof method

> If $\underline{x}$ is a letter, S a set and $\underline{\underline{A}}$ a sentence, and if $(S|\underline{x})\underline{\underline{A}}$ is a theorem, then $(\exists \underline{x})\underline{\underline{A}}$ is a theorem.

This proof method is in a sense "dual" to (XI) in I.3.2. It is frequently used as a method of proving $(\exists \underline{x})\underline{\underline{A}}$ by "exhibiting a set S such that $(S|\underline{x})\underline{\underline{A}}$ (is true)" . This amounts to (a) writing down or indicating a string S , (b) verifying that S denotes a set, and (c) writing down or indicating a proof of $(S|\underline{x})\underline{\underline{A}}$. Steps (a) and (b) will make use of all those signs of the formal language which seem necessary, together with whatever abbreviating signs which have been defined up to the point in question and which seem relevant (see IV.12 for a list appropriate to one stage of development), and will in the last analysis rely on I.1.5 and the construction rules in I.1.8. Step (c) will make use of the proof methods (I) - (XI) and those of their successors which happen to be available. One then speaks of a constructive proof, or a proof by exhibition, of $(\exists \underline{x})\underline{\underline{A}}$. Examples appear in connection with (35) and (36) in II.3.10 and again in II.3.5, II.5.1, IV.4.2 and V.2.1. See also the discussion in Gleason (1), pp.163, 164.

Inasmuch as " $(\exists \underline{x})\underline{\underline{A}}$ is a theorem" signifies (informally) that there exists a set S such that " $(S|\underline{x})$ is a theorem" , but usually affords no guide as to how to "find" such a set S , it may seem a retrograde step to use proof by exhibition in the above manner. However, luck or sound intuition may make proof by exhibition the quickest way of proving an existential theorem.

Related to the above proof method is the injunction

$$\text{Exhibit (a set) } S \text{ such that } (S|\underline{x})\underline{\underline{A}} \ . \qquad (3)$$

which often appears in the statements of problems (see III.1.8 for the discussion of an example); alternative phrasing might be:

$$\text{Exhibit } \underline{x} \text{ such that } \underline{\underline{A}} \ .$$

or

$$\text{Give an example of } \underline{x} \text{ such that } \underline{\underline{A}} \ .$$

The necessary steps are (a), (b) and (c) above. (Notice that the presence of (c) means that a certain theory is involved, though this theory may not always be made - and very often is not made - explicit by the poser of the problem.)

The instruction (3) may be issued at a stage where $(\exists\underline{x})\underline{\underline{A}}$ is not known to be true, or at a stage where this sentence is known to be a theorem. (In the second case, the idea is that a constructive proof is more demanding than the existential proof and is of some intrinsic interest.) In either case, the intention of the poser of the problem is usually that the selector τ is not to appear explicitly in the solution to the problem. (The most popular versions of naive or formal set theory feature no selector.) For the purpose of this book, the terms and abbreviatory symbols allowed in such exhibitions are those listed in IV.12, together with any others introduced later than that and up to the point at which the problem is posed.

It has to be realised that, when one is working with our formalism, the idea of proof by exhibition makes sense only in a relative fashion and in the limited context of a particular problem. This is because almost all sets are defined, ultimately, by an explicit use of the selector.

In a limited context, one overlooks this and grants the use of a vocabulary (like that listed in IV.12) which embraces a great deal more than the primitive symbols. In certain other versions of formal set theory, a selector is lacking from the outset; and then the idea of proof by exhibition would appear to have a more absolute and more basic significance. (On the other hand, in such a formalism, it may seem questionable whether one is able to satisfactorily and formally define any interesting sets at all!) Basically the problem is then somewhat as follows : whenever $(\exists \underline{x})\underline{\underline{A}}$ is a theorem, it should be possible to exhibit explicitly a set S such that $(S|\underline{x})\underline{\underline{A}}$ is a theorem. In our formalism this is achieved by use of the selector; in the absence of a selector, it is not (to me, anyway) clear how one would proceed. Concerning these matters, see Kneebone (1), pp.91 - 104.

To exhibit a set S such that $(S|\underline{x})\underline{\underline{A}}$ is sometimes also spoken of as solving relative to $\underline{x}$, or finding a solution relative to $\underline{x}$ of , the condition expressed by $\underline{\underline{A}}$ (see I.1.9). If $\underline{\underline{A}}$ has the form $\underline{x} \in X \wedge T = U$, where X , T , U denote sets and $\underline{x}$ appears in at least one of T and U , one speaks in particular of solving relative to $\underline{x}$, in or belonging to X , the equation T = U . See also II.3.2. Such a set S is termed a solution, relative to $\underline{x}$, of the condition (or equation) in question.

As has been hinted above, one often proves $(\exists \underline{x})\underline{\underline{A}}$ without attempting to exhibit an S such that $(S|\underline{x})\underline{\underline{A}}$; sometimes, indeed, there is no known way of exhibiting such an S . One then speaks of a "(purely) existential proof" . (From a philosophical point of view, many people feel much happier with a proof by exhibition than with an existential proof; for at least as long as the consistency of Θ_0 remains in doubt, such feelings may be well founded.) As a simple example, it is an easy deduction from V.9.3(1) that ($\underline{x}$ and $\underline{y}$ denoting distinct letters)

$$(\underline{y} \in N \wedge \underline{y} > 1) \Rightarrow (\exists \underline{x})(\underline{x} \text{ is a prime } \wedge \underline{x} \text{ divides } \underline{y} \text{ in } N) ,$$

or (see I.3.3(m) above)

$$(\exists \underline{x})((\underline{y} \in N \wedge \underline{y} > 1) \Rightarrow (\underline{x} \text{ is a prime } \wedge \underline{x} \text{ divides } \underline{y} \text{ in } N)) , \quad (4)$$

is a theorem. The proof is, relatively speaking, trifling. On the other hand there is (as far as I know) no known way of exhibiting (without explicit use of τ) a set S (in which the letter $\underline{y}$ will assuredly appear) and proving

$$(\underline{y} \in N \wedge \underline{y} > 1) \Rightarrow (S \text{ is a prime } \wedge S \text{ divides } \underline{y}) .$$

In other words, I know of no proof by exhibition of (4) .

The injunction

Exhibit (or give) a counterexample to $(\forall \underline{x})\underline{\underline{A}}$ (or to $\underline{\underline{A}}$) ,

signifies to exhibit an $\underline{x}$ such that $\neg\underline{\underline{A}}$. If this is done, one has a proof by exhibition of $(\exists \underline{x})\neg\underline{\underline{A}}$ and hence a proof of $\neg(\forall \underline{x})\underline{\underline{A}}$. Assuming consistency, this will in turn show that $(\forall \underline{x})\underline{\underline{A}}$ is not a theorem and so that $\underline{\underline{A}}$ is not a theorem. Examples appear in Problems X/1 and X/5.

Referring again to the proof techniques proposed on

p. 292 of Gordon and Hindman (1), the recipe for proving $(\exists\underline{x})\underline{\underline{P}}$ may well mislead; cf. the final paragraph in (vii) immediately above. Translated into our notation, the suggested procedure amounts to this : choose a variable $\underline{y}$ different from $\underline{x}$ and "not quantified in $\underline{\underline{P}}$ " (see loc. cit., pp.53-54); define $\underline{\underline{P}}' \equiv (\underline{y}|\underline{x})\underline{\underline{P}}$; prove $\underline{\underline{P}}'$ and conclude $(\exists\underline{x})\underline{\underline{P}}$. This recipe is effective whenever it is possible, but (as will be seen in a moment) it may be possible only in exceptional cases (unless the relevant theory is contradictory). In fact, one may choose $\underline{y}$ as above and not appearing in $\underline{\underline{P}}$. If $\underline{\underline{P}}'$ is provable, so also is $(\forall\underline{y})\underline{\underline{P}}'$ by I.3.2(X) . On the other hand, according to I.1.8(e), $(\forall\underline{y})\underline{\underline{P}}' \equiv (\forall\underline{x})\underline{\underline{P}}$, and so $(\forall\underline{x})\underline{\underline{P}}$ is provable. Thus the proposed technique can succeed, only if $(\forall\underline{x})\underline{\underline{P}}$ is provable; and in that case the truth of $(\exists\underline{x})\underline{\underline{P}}$ is rarely of much interest. Moreover, it may well happen that $(\exists\underline{x})\underline{\underline{P}}$ is true and $(\forall\underline{x})\underline{\underline{P}}$ is false. (An instance is $\underline{\underline{P}} \equiv \underline{x} \neq \underline{y}$; see Chapter II.)

(ix) As a partial summary: What is conventionally referred to as a theorem is sometimes a theorem, sometimes a theorem schema, and sometimes even a metatheorem. (See II.1.4, IV.2.2, IV.4.2, V.3.1 and V.5.2 for examples.) What is conventionally referred to as a proof is never a formal proof, but rather at most a proof plan based upon appeals to theorem schemas and metatheorems, plus theorems previously proved in a similar fashion. These are what I have referred to as either semiformal proofs or routine proofs. (There is no precise dividing line between the two.) In place of "prove" in this sense, one may say "demonstrate" or "establish" or "show" . The term "verification" is used differently and covers cases of metatheorems in which there is no expectation (even in principle) of a formal proof. (In §2 of the Appendix, which deals with the basis for metatheorems such as (I) - (XI) in I.3.2, I use only the term "verify" for the sake of clarity.)

The concepts of "sentence" , "set" , "proof" , "true" , "false" , "theorem" in relation to a formal theory are all purely syntactic.

What is normally described as mathematics is very often a melange of mathematics and metamathematics - a commentary on formal mathematics, so to speak. Over and over again, the practising mathematician lights on a sentence which seems to him especially significant and poses the metamathematical question: Is this sentence a theorem (of Θ_0)? If he fancies "Yes" as the answer, he struggles to provide or describe a proof (in some, usually routine, style). An example is afforded by the opening paragraph of X.4, the selected sentence being in this case that denoted by

$$(\forall \underline{i})(\forall \underline{g})(((\underline{i} \text{ is an open interval in } R) \wedge (\underline{g} \text{ is a function } \underline{i} \to R)) \Rightarrow (\exists \underline{f})((\underline{f} \text{ is a function } \underline{i} \to R) \wedge (\underline{f}' = \underline{g}))) \quad ,$$

where $\underline{f}$, $\underline{g}$, $\underline{i}$ denote any three distinct letters.

I.3.5 <u>The sign $\equiv$; formal definitions</u> The sign $\equiv$ will be used in either of two ways:

(i) In the form $A \equiv B$ as an abbreviation for the English sentence " A and B denote the same string" , or "the strings denoted by A and B are identical" .

(ii) In the expression of definitions - as, for example, when I write (cf. I.1.7)

$$(\exists \underline{x})A \equiv (\tau_{\underline{x}}(A)|\underline{x})A$$

or

$$\text{there exists } \underline{x} \text{ such that } A \equiv (\tau_{\underline{x}}(A)|\underline{x})A \quad ,$$

meaning that $(\exists \underline{x})A$ or "there exists $\underline{x}$ such that A " is being introduced as a name for the string otherwise and alternatively denoted by $(\tau_{\underline{x}}(A)|\underline{x})A$. The idea here is, of course, simply the making of more convenient names. (I am uncomfortably close to attempting to define "definition" ! There are many thorny problems related to the concept of admissible definitions; see Kleene (1), Chapter III.)

To the extent that definitions do no more than attach names to strings, there is considerable freedom. But the freedom is not total.

On the one hand, there is no harm in attaching more than one name to any one string. This is done repeatedly. Thus " $(\exists \underline{x})A$ " and " $(\tau_{\underline{x}}(A)|\underline{x})A$ " are both adopted as names for one string.

On the other hand, common sense alone leads one to anticipate potential danger in attaching one and the same name to different strings .... something which can inadvertently take place of care is not exercised. See II.1.3(i) and II.4.2(i) for instances and consequences of doing this; see also the final paragraph in VI.6.12 and VIII.4.4(iv).

This warning is to be heeded, even though one soon comes to realise that in the formal theory (though not necessarily in the metamathematics) there is no ultimate harm in attaching the same name to equivalent sentences; nor (by virtue of the schema S6 in II.1.2) in attaching the same name to equal sets.

(iii) In either interpretation, $\equiv$ is not a sign of the formal theory. It must be firmly distinguished from the sign $=$, to be introduced in II.1.1 as a semiformal mathematical sign. If A and B denote strings, $A \equiv B$ denotes a sentence in the metalanguage, whereas $A = B$ will denote a string in the formal language (a formal sentence if A and B denote sets).

Nevertheless, there is a very simple metatheorem which should be noted, namely:

If A and B denote strings, and if $A \equiv B$, then

(a) A denotes a sentence (resp. set) if and only if B denotes a sentence (resp. set);

(b) A denotes a theorem if and only if B denotes a theorem;

(c) if A and B denote sentences, then $A \Leftrightarrow B$ denotes a theorem.

From this it will in due course follow that, if X and Y denote sets, and if $X \equiv Y$, then $X = Y$ denotes a theorem. On the other hand, it can happen that $X = Y$ denotes a theorem and yet X be not identical with Y .

In addition to this, note that a formal definition is not a part of formalised mathematics (in which there appear nothing but explicitly written strings). Rather, they belong to the description of formalised mathematics, and it is in this guise that they appear in routine mathematics which, as is indicated in I.3.4, is usually a commentary on formalised mathematics. They may, nevertheless, be said to be formal inasmuch as their use in the description should not involve difficulties in translating the description into something which can in principle take place in the formal theory. In other words, a definition is formal when and only when it is made in a way which is compatible with the strict development of the formal theory.

(iv) This last proviso leads to an important principle concerning the nature of formally acceptable definitions, which is in conflict with conventional informal practice.

Every set and every (formal) sentence is a string. Accordingly, no acceptable definition of a set or a sentence can legitimately (or even sensibly) be made conditional upon the truth of one or more sentences. In brief, formally acceptable definitions must be <u>unconditional</u>.

In spite of this, informal mathematics abounds in conditional definitions. To illustrate, suppose an informal treatment of natural numbers to be under way. Having proved a theorem informally rendered as

> If n is a natural number and $n \neq 0$,
> there exists a unique natural number m
> such that $m + 1 = n$

the next step might be to frame the definition

> If n is a natural number and $n \neq 0$,
> $n - 1$ is defined to be the unique
> natural number m such that $m + 1 = n$.

This definition is overtly conditional by virtue of its opening phrase. As such, it is formally unacceptable; a formally acceptable definition of $n - 1$ has to be effective for every string n (or at least for every set n). This example will be discussed again in V.14.

Another conditional definition (this time of x/y) might read

$$(y \neq 0) \Rightarrow (z = x/y \Leftrightarrow yz = x) \quad ; \qquad (1)$$

this example is even worse in that it is manifestly less explicit. It might more properly be regarded as a theorem, provided x/y

has already been acceptably defined. Cf. Suppes (1), p.18 and pp. 136-137. See also the Remark at the end of II.3.3 below.

To accept and employ conditional definitions is to imperil the complete formalisability in principle of the allowed procedures. It may no longer be clearly possible in principle to subsequently eliminate all appearances of the signs introduced via such definitions in such a way as to revert to the primitive language. For instance, having accepted (1), how does one eliminate the division sign from the sentence $1/0 \neq 2$?

As has been indicated, conditional definitions are the norm in informal mathematics, where they are almost always regarded as tolerable "working definitions". This is so, partly because to avoid them entirely often involves considerable labour, and partly because in any case they often seem intuitively appropriate and sensible. Moreover, they seem to work well enough! In due course, they will appear in this book. However, I will often pause to discuss unconditional replacements; see, for example, IV.1.4(v), the outset of V.6.2, V.14, VI.8, VII.1.4(ii) - (iv), VII.5.5, the Remark following VIII.1.1, VIII.4.4 and VIII.6.9. The differences between the two styles sometimes have consequences more significant that might be anticipated.

If one starts from a given informal and conditional definition, there will usually be a variety of formal and unconditional replacements for it, and often no evident grounds for choosing between them. In other words, the procedure of formalisation is not unique.

To illustrate, consider the definition of functional values, to be discussed in IV.1.3. The informal and conditional definition might read as follows:

> If f is a function and x is an element of the domain of f, then the functional value $f(x)$ is defined to be the unique object y such that $(x,y) \in f$.

The formal and unconditional replacement actually adopted in IV.1.3 amounts to

$$f(x) \equiv \tau_{\underline{y}}((x,\underline{y}) \in f)\,, \tag{2}$$

$\underline{y}$ denoting any letter not appearing in x or f (which are understood to denote arbitrary strings). An alternative formal and unconditional definition is

$$f(x) \equiv \bigcap\{\underline{y} : (x,\underline{y}) \in f\}\,. \tag{3}$$

Now (2) and (3) are equally acceptable within the intended scope of the conditional definition; this is because (see Problem IV/2(i)), if $\underline{x}$, $\underline{y}$ and $\underline{f}$ denote distinct letters,

$$(\forall\underline{f})(\forall\underline{x})((\underline{f} \text{ is a function} \wedge \underline{x} \in \operatorname{Dom} \underline{f})$$
$$\Rightarrow (\tau_{\underline{y}}((\underline{x},\underline{y}) \in \underline{f}) = \bigcap\{\underline{y} : (\underline{x},\underline{y}) \in \underline{f}\}))$$

is a theorem. On the other hand, (2) and (3) are not unconditionally equivalent; that is to say (see Problem IV/2(ii)),

$$\neg(\forall\underline{f})(\forall\underline{x})(\tau_{\underline{y}}((\underline{x},\underline{y}) \in \underline{f}) = \bigcap\{\underline{y} : (\underline{x},\underline{y}) \in \underline{f}\})$$

is a theorem. Within the scope of the conditional definition, there is nothing to choose between (2) and (3) as formal and unconditional replacements; yet (2) and (3) are, from a formal point of view, quite different definitions. Having selected one of them, one must not in formal developments gaily switch from one to the other when it appears convenient to do so, unless and until the switch can be justified in the context under consideration (which will usually mean proving a theorem for that very purpose).

(v) Definitions of sets are more often than not presented in the style $S = \ldots$ rather than $S \equiv \ldots$. Even though one may have $S = T$ without having $S \equiv T$, the difference is of no ultimate importance in the formal theory: this comes about because of the schema S6 in II.1.2 below, which arranges that equal sets have exactly the same formal (mathematical) properties.

Similarly, many writers (for example, Suppes (1) and Gordon and Hindman (1)) frame what are intended to be formal definitions in the style $\underline{\underline{A}} \Leftrightarrow \underline{\underline{B}}$ (rather than $\underline{\underline{A}} \equiv \underline{\underline{B}}$) , the informal version of which might appear as " $\underline{\underline{A}}$ if and only if $\underline{\underline{B}}$ " (or even merely " $\underline{\underline{A}}$ if $\underline{\underline{B}}$ " or " $\underline{\underline{A}}$ means $\underline{\underline{B}}$ " ; see, for example, the second sentence in XII.2.3). The latter informal version is that which appears almost all the time in conventional texts and sometimes later in this book (see, for example, VII.1.1). From the point of view of our formalism, this procedure is slightly objectionable (though possibly harmless in the end): this is because, if $\underline{\underline{B}}$ is a given sentence, " $\underline{\underline{A}} \Leftrightarrow \underline{\underline{B}}$ " does not determine uniquely a sentence $\underline{\underline{A}}$. It also means that "if and only if" will sometimes denote " $\Leftrightarrow$ " and sometimes " $\equiv$ " .

A typical example of this variation in style is the following proposed definition of $f(x)$, taken from Suppes (1), p.87 (he uses $\leftrightarrow$ in place of our $\Leftrightarrow$):

$$f(x) = y \Leftrightarrow S ,$$

where S is a certain formal sentence (the precise nature of which is here irrelevant). Suppes writes:

> The definition is so framed that the notation 'f(x)' has a definite meaning for any set f and any object x .

This is debatable: the ambiguity involved in using $\Leftrightarrow$ rather than $\equiv$ makes it difficult to see how the proposed definition specifies any definite string to be denoted by $f(x)$. Besides this, it does not make clear what string is denoted by, for example, $y \notin g(y)$, which appears loc. cit. p.97, line 3 up without further explanation. Presumably this last is to denote

$$(\exists \underline{x})(\underline{x} = g(y) \wedge y \notin \underline{x})$$

where $\underline{x}$ is a letter not appearing in y or g. See IV.1.3 and IV.1.4(v).

(vi) In line with what has been written in (v), the typical routine style definition is instanced by

> A perfect square is a number of the form x^2 , where $x \in N$.

or

> y is a perfect square if and only if $y = x^2$, where $x \in N$.

Another instance is (cf. VII.3.1)

> v is a subsequence of u if and only if $v = u \circ s$, where s is a strictly increasing $\dot{N}$ valued sequence.

More generally, one repeatedly encounters intended definitions phrased in terms of a preselected sentence $\underline{\underline{A}}$ and a preselected set T somewhat as follows:

> An oojah is a set (or : number, function, et cetera) of (or : expressible in) the form $T[\![x]\!]$, where $\underline{\underline{A}}[\![x]\!]$ (or : of the form $T[\![x]\!]$ for some x such that $\underline{\underline{A}}[\![x]\!]$) (4)

or, more tersely,

> y is an oojah if and only if $y = T[\![x]\!]$, where $\underline{\underline{A}}[\![x]\!]$ (or : $y = T[\![x]\!]$ for some x such that $\underline{\underline{A}}[\![x]\!]$) (5)

In such circumstances (4) or (5) is usually intended to summarise the following more formal procedure. A letter $\underline{x}$ is preselected, along with the sentence $\underline{\underline{A}}$ and the set T ($\underline{x}$ will usually appear in both $\underline{\underline{A}}$ and T); the definition proceeds by choosing another letter $\underline{y}$, different from $\underline{x}$ and not appearing in $\underline{\underline{A}}$ or T , and making the formal definition

$$y \text{ is an oojah} \equiv (y|\underline{y})(\exists \underline{x})(\underline{\underline{A}} \wedge \underline{y} = T) \quad , \tag{6}$$

applying for arbitrary strings y .

If one chooses a letter $\underline{x}'$, different from $\underline{y}$ and not appearing in y , $\underline{\underline{A}}$ or T , and if $\underline{\underline{A}}[\![S]\!] \equiv (S|\underline{x})\underline{\underline{A}}$ and $T[\![S]\!] \equiv (S|\underline{x})T$ for arbitrary strings S , then I.1.8(e) and (f) show that (6) may be written

$$y \text{ is an oojah} \equiv (\exists \underline{x}')(\underline{\underline{A}}[\![\underline{x}']\!] \wedge y = T[\![\underline{x}']\!]) \quad . \tag{7}$$

Also, if $\underline{x}$ does not appear in y , (6) and I.1.8(f) lead to

$$y \text{ is an oojah} \equiv (\exists \underline{x})(\underline{\underline{A}} \wedge y = T) \quad . \tag{8}$$

Both (7) and (8) are to be compared with (4) and (5).

The originals, (4) and (5) ... especially those versions which terminate with the phrase " ..., where $\underline{\underline{A}}[\![x]\!]$ " .... are objectionable because they do not make plain the role of x and its possible dependence upon y . The formal version terminating in (6) eliminates all such ambiguities. Notice also that the originals lack any distinction between symbols denoting logical letters or variables and symbols denoting more general sets.

The reader should note that there may or may not exist a set of all oojahs. However, in most routinely occurring cases, $\underline{A}$ is of the form $\underline{x} \in X$, where X denotes a set in which the letter $\underline{x}$ does not appear, and then the existence of a set of all oojahs is guaranteed by the axioms; see the discussion in II.12.1.

Similar remarks apply to intended definitions of the type

> z is an oojah if and only if z is of the form $T[\![x,y]\!]$, where $\underline{\underline{A}}[\![x,y]\!]$ (or : $z = T[\![x,y]\!]$ for some x and some y such that $\underline{\underline{A}}[\![x,y]\!]$);

see Problem II/24.

(vii) Even within the sphere of informal mathematics, definitions have received much criticism, especially the so called "impredicative definitions" . When a procedure involves a certain set M and claims or seeks to define an object m in such a way that this alleged definition involves (makes reference to) M and, at the same time, arranges that m is an element of M , the procedure (or the definition) is termed impredicative. Many of the well known paradoxes (see II.3.9, Kleene (1), §11; Kneebone (1) pp.127-128) involve impredicative definitions in a blatant fashion, and seem to collapse if such definitions are banned. It would therefore appear that such a ban should be adopted as an easy way out of the difficulties. Unhappily, however, many such definitions appear in mathematics, where they have been severely criticised, notably by Hermann Weyl. (An instance is the definition of the supremum of a set of real numbers.) The formal theory also involves impredicative definitions. Thus, if M denotes a set such that $M \neq \emptyset$ is a theorem, and if we then define $m \equiv \tau_{\underline{x}}(\underline{x} \in M)$, the definition is impredicative. Yet the formalism bans the known paradoxes in other ways (by its own strictures, so to speak).

(viii) It is part of the customary metamathematical scene that some definitions and names are intended to have, and are in fact accorded, a high degree of permanence; while others (typified by those introduced solely for the purpose of describing the proof of one theorem or theorem schema; see the Remark following the proof of II.3.3(5)) are intended to be transient. This latter category almost always involves confusion : in one proof the symbol " A " will be employed to name one set, and in another proof to name a different set. Potential confusion could, of course, be avoided by adopting at the outset an agreed unending list of name symbols (the usual alphabets are soon exhausted), each to be used to name but one set. However, it is extremely rare to find this device adopted. Instead, one relies on the reader's good sense and vigilance to avoid or to resolve any ensuing confusion. (None of this has any bearing on the formal theory itself, wherein definitions and names do not appear.)

Transient definitions of the sort just described are involved over and over again in informal mathematics, where they are often expressed simply as

$$\text{Let } f = \ldots\ldots\ldots\ ,$$

intended as signifying the adoption of a definition or name

$$f \equiv \ldots\ldots\ldots\ .$$

I shall adopt the convention of writing

$$\equiv_{\text{def}}$$

to indicate a definition or name which is relatively important and which is to be adopted throughout this book. In most cases, such a permanent definition will agree with one of the equivalent definitions widely adopted by other writers. Transient definitions will be indicated by $\equiv$.

(ix) What are often referred to as definitions are very often more properly termed <u>definition schemas</u> (cf. the sentence schemas and theorem schemas discussed in I.1.10 and I.2.9(v); (6) above is an instance). Accordingly, they may well incorporate metamathematical restrictions. An example is the formal definition schema

$$X = Y \equiv_{\text{def}} (\forall \underline{x})((\underline{x} \in X) \Leftrightarrow (\underline{x} \in Y))$$

wherein the letter $\underline{x}$ is subject to the (metamathematical) restriction not to appear in X or Y ; see II.1.1. Such metamathematical hypotheses do not result in conditionality of the type discussed in (iv) above; they are declarative metasentences not intended as names for formal sentences. See also Note 7.

I.3.6 Examples of semiformal proofs There follow four examples illustrating the semiformal style of proof. They are based upon appeals to the axiom schemas S1 - S5 in I.2.2, the metatheorems in I.3.2 and I.3.3(j), together with the first theorem schema listed in I.3.3(k), namely, the double negation rule

$$\underline{\underline{A}} \Leftrightarrow \neg\neg\underline{\underline{A}} \quad ; \tag{1}$$

the other metatheorems in I.3.3(k) are not invoked. Taken together, Examples 1, 2 and 3 provide a semiformal proof of the second theorem schema listed in I.3.3(k), which reappears as (15) below. Notice that, in view of I.3.2(I) and I.3.3(j), (1) implies both the theorem schemas $\underline{\underline{A}} \Rightarrow \neg\neg\underline{\underline{A}}$ and $\neg\neg\underline{\underline{A}} \Rightarrow \underline{\underline{A}}$.

Throughout this subsection, Θ denotes a fixed but arbitrary theory and $\underline{\underline{A}}$, $\underline{\underline{B}}$, $\underline{\underline{C}}$, $\underline{\underline{D}}$ denote arbitrary sentences.

Example 1 I here give a semiformal proof of the theorem schema

$$T\Theta \ : \qquad (\underline{\underline{A}} \Rightarrow \underline{\underline{B}}) \Rightarrow (\neg\underline{\underline{B}} \Rightarrow \neg\underline{\underline{A}}) \tag{2}$$

(Strictly speaking, in this and similar situations below, one is really constructing a semiformal proof of the sentence obtained when definite but arbitrary sentences are substituted for $\underline{\underline{A}}$, $\underline{\underline{B}}$, $\underline{\underline{C}}$ and $\underline{\underline{D}}$; recall the remarks in I.2.9(v).)

Proof Let Θ_1 denote the theory obtained by adjoining to Θ the axiom $\underline{\underline{A}} \Rightarrow \underline{\underline{B}}$, Θ_2 that obtained by adjoining to Θ_1 the axiom $\neg\underline{\underline{B}}$, and Θ_3 that obtained by adjoining to Θ_2 the axiom $\neg\neg\underline{\underline{A}}$. Then

$T\Theta_3$: $\underline{\underline{A}} \Rightarrow \underline{\underline{B}}$, (3)

$T\Theta_3$: $\neg\underline{\underline{B}}$, (4)

$T\Theta_3$: $\neg\neg\underline{\underline{A}}$, (5)

Hence

$T\Theta_3$: $\underline{\underline{A}}$ by (5), (1) and I.3.2(I) , (6)

$T\Theta_3$: $\underline{\underline{B}}$ by (3), (6) and I.3.2(I) ; (7)

note that I.3.2(I) is being applied with Θ_3 in place of Θ .

Comparing (4) and (7) , Θ_3 is seen to be contradictory. Thus

$T\Theta_2$: $\neg\underline{\underline{A}}$ by I.3.2(VII) ,

and hence

$T\Theta_1$: $\neg\underline{\underline{B}} \Rightarrow \neg\underline{\underline{A}}$ by I.3.2(VI) ,

and

$T\Theta$: $(\underline{\underline{A}} \Rightarrow \underline{\underline{B}}) \Rightarrow (\neg\underline{\underline{B}} \Rightarrow \neg\underline{\underline{A}})$ By I.3.2(VI) again. □

Example 2

$T\Theta$: $(\neg\underline{\underline{B}} \Rightarrow \neg\underline{\underline{A}}) \Rightarrow (\underline{\underline{A}} \Rightarrow \underline{\underline{B}})$. (8)

Proof By (2) ,

$T\Theta$: $(\neg\underline{\underline{B}} \Rightarrow \neg\underline{\underline{A}}) \Rightarrow (\neg\neg\underline{\underline{A}} \Rightarrow \neg\neg\underline{\underline{B}})$. (9)

Let Θ_4 be the theory obtained by adjoining to Θ the axiom $\neg\underline{\underline{B}} \Rightarrow \neg\underline{\underline{A}}$, and Θ_5

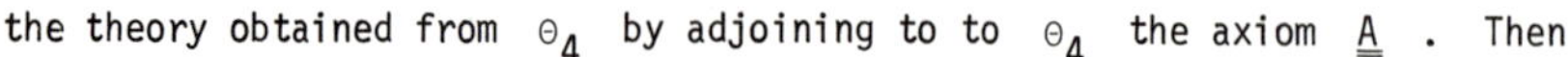

the theory obtained from Θ_4 by adjoining to to Θ_4 the axiom $\underline{\underline{A}}$. Then

$$T\Theta_5 \quad : \qquad \neg\underline{\underline{B}} \Rightarrow \neg\underline{\underline{A}} \quad , \tag{10}$$

$$T\Theta_5 \quad : \qquad \underline{\underline{A}} \quad .$$

This last combines with I.3.2(I) and I.3.3(j) and (1) (with Θ_5 in place of Θ) to give

$$T\Theta_5 \quad : \qquad \neg\neg\underline{\underline{A}} \quad . \tag{11}$$

By (9) , (10) and I.3.2(I) (the last with Θ_5 in place of Θ),

$$T\Theta_5 \quad : \qquad \neg\neg\underline{\underline{A}} \Rightarrow \neg\neg\underline{\underline{B}} \quad . \tag{12}$$

By (11) , (12) and I.3.2(I),

$$T\Theta_5 \quad : \qquad \neg\neg\underline{\underline{B}} \quad .$$

This combines with I.3.2(I) and I.3.3(j) and (1) (with Θ_5 in place of Θ) to give

$$T\Theta_5 \quad : \qquad \underline{\underline{B}} \quad . \tag{13}$$

By (13) and I.3.2(VI) ,

$$T\Theta_4 \quad : \qquad \underline{\underline{A}} \Rightarrow \underline{\underline{B}} \quad . \tag{14}$$

By (14) and I.3.2(VI) ,

$$T\Theta \quad : \qquad (\neg\underline{\underline{B}} \Rightarrow \neg\underline{\underline{A}}) \Rightarrow (\underline{\underline{A}} \Rightarrow \underline{\underline{B}}) \quad ,$$

which is (8) . □

Example 3

$$T\Theta \quad : \qquad (\underline{\underline{A}} \Rightarrow \underline{\underline{B}}) \Leftrightarrow (\neg\underline{\underline{B}} \Rightarrow \neg\underline{\underline{A}}) \, . \tag{15}$$

Proof By (2) , (8) and I.3.3(j) ,

$$T\Theta \quad : \qquad ((\underline{\underline{A}} \Rightarrow \underline{\underline{B}}) \Rightarrow (\neg\underline{\underline{B}} \Rightarrow \neg\underline{\underline{A}})) \wedge ((\neg\underline{\underline{B}} \Rightarrow \neg\underline{\underline{A}}) \Rightarrow (\underline{\underline{A}} \Rightarrow \underline{\underline{B}})) \; ,$$

the sentence denoted here being identical with

$$(\underline{\underline{A}} \Rightarrow \underline{\underline{B}}) \Leftrightarrow (\neg\underline{\underline{B}} \Rightarrow \neg\underline{\underline{A}}) \; .$$

This proves (15) . □

Example 4

$$T\Theta \quad : \qquad (\underline{\underline{A}} \Leftrightarrow \underline{\underline{B}}) \Rightarrow (\neg\underline{\underline{B}} \Leftrightarrow \neg\underline{\underline{A}}) \, . \tag{16}$$

Proof Let Θ_6 denote the theory obtained by adjoining to Θ the axiom $\underline{\underline{A}} \Leftrightarrow \underline{\underline{B}}$. Then, by I.3.3(j) and I.3.2(I),

$$T\Theta_6 \quad : \qquad \underline{\underline{A}} \Rightarrow \underline{\underline{B}} \; .$$

So, by (2) and I.3.2(I) (the last with Θ_6 in place of Θ) ,

$$T\Theta_6 \quad : \qquad \neg\underline{\underline{B}} \Rightarrow \neg\underline{\underline{A}} \; . \tag{17}$$

Similarly, since

$T\Theta_6$: $\underline{\underline{B}} \Rightarrow \underline{\underline{A}}$,

so, by (2) and I.3.2(I) again,

$$T\Theta_6 : \quad \neg\underline{\underline{A}} \Rightarrow \neg\underline{\underline{B}} \ . \tag{18}$$

By (17) , (18) and I.3.3(j) ,

$$T\Theta_6 : \quad (\neg\underline{\underline{B}} \Rightarrow \neg\underline{\underline{A}}) \wedge (\neg\underline{\underline{A}} \Rightarrow \neg\underline{\underline{B}}) \ , \tag{19}$$

which sentence is identical with $\neg\underline{\underline{B}} \Leftrightarrow \neg\underline{\underline{A}}$. By I.3.2(VI), (16) follows from (19) . □

These examples should indicate to the reader how the process of temporary adjunction of axioms functions in practice. Its use in more conventional mathematical situations is exactly the same, as subsequent examples of semiformal proofs will show.

The reader is now urged to attempt at least some of the problems in I.3.7 below; these cover some of the theorem schemas in I.3.3 (see also §2 of the Appendix). Where he finds it helpful, he should consider the intuitive content of the sentences involved as a possible guide to the formulation of semiformal proofs.

As a starting hint, I.3.7(1)(i) is now a relatively simple step from (16) above, combined with the schemas S1 - S5 , the metatheorems in I.3.2 and I.3.3(j), and I.3.6(1).

I.3.7 Problems In these problems one has in mind a fixed but arbitrary theory Θ , while $\underline{\underline{A}}$, $\underline{\underline{B}}$, $\underline{\underline{C}}$ and $\underline{\underline{D}}$ denote arbitrary sentences. Except in Problems 1 and 5, free use may be made of the schemas S1 - S5 and all the metatheorems in I.3.2 and I.3.3.

(1) Using only S1 - S5 in I.2.2, (I) - (XI) in I.3.2, I.3.3(j) and

I.3.6(1), give semiformal proofs of the following theorem schemas:

(i) $(\underline{\underline{A}} \Leftrightarrow \underline{\underline{B}}) \Leftrightarrow (\neg\underline{\underline{A}} \Leftrightarrow \neg\underline{\underline{B}})$;

(ii) $(\underline{\underline{A}} \vee \underline{\underline{B}}) \Leftrightarrow (\underline{\underline{B}} \vee \underline{\underline{A}})$;

(iii) $(\underline{\underline{A}} \wedge \underline{\underline{B}}) \Leftrightarrow (\underline{\underline{B}} \wedge \underline{\underline{A}})$;

(iv) $(\underline{\underline{A}} \Leftrightarrow \underline{\underline{B}}) \Leftrightarrow (\underline{\underline{B}} \Leftrightarrow \underline{\underline{A}})$;

(v) $(\underline{\underline{A}} \Leftrightarrow \underline{\underline{B}}) \Rightarrow ((\underline{\underline{A}} \Leftrightarrow \underline{\underline{C}}) \Leftrightarrow (\underline{\underline{B}} \Leftrightarrow \underline{\underline{C}}))$.

Verify (on the same basis) that if $\underline{\underline{A}} \Leftrightarrow \underline{\underline{B}}$ and $\underline{\underline{B}} \Leftrightarrow \underline{\underline{C}}$ are theorems, then so is $\underline{\underline{A}} \Leftrightarrow \underline{\underline{C}}$; that if $\underline{\underline{A}} \Leftrightarrow \underline{\underline{B}}$ is a theorem, so are $\underline{\underline{A}} \vee \underline{\underline{C}} \Leftrightarrow \underline{\underline{B}} \vee \underline{\underline{C}}$ and $\underline{\underline{A}} \wedge \underline{\underline{C}} \Leftrightarrow \underline{\underline{B}} \wedge \underline{\underline{C}}$; and that if $\underline{\underline{A}}$ is a theorem, then so is $\underline{\underline{B}} \Leftrightarrow \underline{\underline{A}} \wedge \underline{\underline{B}}$.

(2) Give a semiformal proof of the theorem schema

$$(\underline{\underline{A}} \Rightarrow (\underline{\underline{B}} \Rightarrow \underline{\underline{C}})) \Leftrightarrow ((\underline{\underline{A}} \wedge \underline{\underline{B}}) \Rightarrow \underline{\underline{C}}) .$$

(3) Give a semiformal proof of the theorem schema

$$(\underline{\underline{A}} \Rightarrow \underline{\underline{B}}) \Leftrightarrow (\underline{\underline{A}} \Leftrightarrow (\underline{\underline{A}} \wedge \underline{\underline{B}})) .$$

(4) Verify that if $\underline{\underline{A}} \Rightarrow \underline{\underline{C}}$ and $\underline{\underline{B}} \Rightarrow \underline{\underline{C}}$ are theorems, then $(\underline{\underline{A}} \vee \underline{\underline{B}}) \Rightarrow \underline{\underline{C}}$ is a theorem.

(5) Using only S1 - S5 and (I) - (X) in I.3.2, verify that if $\underline{\underline{A}} \Rightarrow \underline{\underline{B}}$ is a theorem, and if $\underline{x}$ is a variable which does not appear in $\underline{\underline{B}}$, then $(\exists \underline{x})\underline{\underline{A}} \Rightarrow \underline{\underline{B}}$ is a theorem. (See also Problem I/31.)

(6) Give semiformal proofs of the theorem schemas

$$(\underline{\underline{A}} \vee \underline{\underline{B}}) \vee \underline{\underline{C}} \Leftrightarrow \underline{\underline{A}} \vee (\underline{\underline{B}} \vee \underline{\underline{C}}) ,$$

$$(\underline{\underline{A}} \wedge \underline{\underline{B}}) \wedge \underline{\underline{C}} \Leftrightarrow \underline{\underline{A}} \wedge (\underline{\underline{B}} \wedge \underline{\underline{C}}) \quad .$$

In view of these theorem schemas, one frequently writes without significant ambiguity

$$\underline{\underline{A}} \vee \underline{\underline{B}} \vee \underline{\underline{C}} \Rightarrow \ldots\ldots\ldots \quad ,$$

$$\ldots\ldots\ldots \Rightarrow \underline{\underline{A}} \vee \underline{\underline{B}} \vee \underline{\underline{C}} \quad ,$$

and so on.

(7) Verify that if $\underline{\underline{A}} \Leftrightarrow \underline{\underline{B}}$ and $\underline{\underline{C}} \Leftrightarrow \underline{\underline{D}}$ are theorems, then so are $(\underline{\underline{A}} \vee \underline{\underline{C}}) \Leftrightarrow (\underline{\underline{B}} \vee \underline{\underline{D}})$ and $(\underline{\underline{A}} \wedge \underline{\underline{C}}) \Leftrightarrow (\underline{\underline{B}} \wedge \underline{\underline{D}})$.

I.3.8 Remarks on theorems and converses The visible products of a mathematician's labour are conventionally displayed (in books and research papers) in the following fashion:

THEOREM ....

PROOF ....

Following the legend "THEOREM" there will almost invariably appear a bastard sentence (see I.2.9(vi)) which is thought to indicate clearly and unambiguously a sentence of the formal language. As will become apparent, however, the indication provided does not always satisfy those demands. For example, it is very often not crystal clear whether the formal sentence referred to is $\underline{\underline{A}}$ or $(\forall \underline{x})\underline{\underline{A}}$ or $(\forall \underline{x})(\forall \underline{y})\underline{\underline{A}}$, et cetera. If the THEOREM in fact asserts no more than the truth of a formal sentence, this type of ambiguity is immaterial (thanks to (X) and (XI) in I.3.2). If, however, the THEOREM incorporates (as it sometimes does) meta-statements asserting the non-truth or falsity of the formal sentence apparently involved or of the converse of that formal sentence, then the presence or otherwise of universal quantifiers may become a vital issue. (See, for example, the discussion of (1) and (2) below.) In such cases, the title "THEOREM" is, strictly speaking, a commonly used misnomer.

Assuming the THEOREM to assert the truth of a certain formal sentence (and to assert that alone), the legend "PROOF" should be and usually is followed by a text of bastard sentences which, taken as a whole, is intended to indicate sufficiently clearly and unambiguously a semiformal or routine style proof (see I.3.4(i)) of the formal sentence indicated in the statement of the THEOREM. (If more is asserted, the PROOF will presumably contain in addition some sort of metamathematical argument as well.) It is obvious that there is room for a great deal of variation here. It is pretty safe to say that what actually appears often has a rather remote chance of satisfying the requirements of all expected readers. This is why anybody reading mathematics (especially proofs) has to be ready to think hard about what does <u>not</u> appear, almost as much as about what does. Consider again what has been said at various places in I.3.4. It is also the reason why writing a mathematics book or paper involves an element of art, as well as a purely technical component.

I turn now to consider some of the comments frequently attached to the statement of theorems, comments which often lead to confusion.

Throughout the rest of this subsection, $\underline{\underline{A}}$, $\underline{\underline{P}}$ and $\underline{\underline{Q}}$ denote formal sentences; letters may or may not appear in $\underline{\underline{A}}$, $\underline{\underline{P}}$ and $\underline{\underline{Q}}$ (not necessarily the same letters appear in any two of these sentences). In addition, $\underline{x}$, $\underline{y}$, $\underline{z}$ denote distinct variables (free letters) of the theory being considered. This theory will be assumed to be consistent.

I proceed to announce and verify a number of very simple metatheorems (based on those in I.3.2 and I.3.3) which sometimes need to be borne in mind.

(i) $\underline{\underline{A}}$ is a theorem, if and only if $(\forall \underline{x})\underline{\underline{A}}$ is a theorem.

This is just a restatement of (II) and (X) in I.3.2. In other words, to announce that $\underline{\underline{A}}$ is true, is metamathematically the same as to announce that $(\forall \underline{x})\underline{\underline{A}}$ is true.

But beware a trap: to say that $\underline{\underline{A}}$ is false, is NOT metamathematically the same as saying that $(\forall \underline{x})\underline{\underline{A}}$ is false. (Explain this.)

(ii) If $\underline{\underline{A}}$ is false, then $\underline{\underline{A}}$ is not true.

But again beware a trap: To say that $\underline{\underline{A}}$ is not true, is NOT the same as saying that $\underline{\underline{A}}$ is false; $\underline{\underline{A}}$ may be neither true nor false (that is, $\underline{\underline{A}}$ may be undecidable; see I.2.9(ii) and Appendix §1).

Now I consider more particularly sentences of the so called conditional form

$$\underline{\underline{P}} \Rightarrow \underline{\underline{Q}} \quad , \tag{1}$$

and compare this with

$$(\forall \underline{x})(\underline{\underline{P}} \Rightarrow \underline{\underline{Q}}) \quad , \tag{2}$$

noting that (by (i)) (1) is true if and only if (2) is true; and (by (i) and (ii)) if (1) is false, then (2) is not true. However, (2) may be not true and (1) be not false. (For example, if $\underline{x}$ appears neither in $\underline{\underline{P}}$ nor $\underline{\underline{Q}}$, then (1) and (2) are identical, and it suffices to choose $\underline{\underline{P}}$ and $\underline{\underline{Q}}$ so that $\underline{\underline{P}} \Rightarrow \underline{\underline{Q}}$ is undecidable; see Appendix, §1.) Also, (2) may be false and (1) be not false. (This will appear from the theorems of Chapter II on taking $\underline{\underline{P}} \equiv \underline{x} \in \{\emptyset, \{\emptyset\}\}$ and $\underline{\underline{Q}} \equiv \underline{x} \in \{\emptyset\}$.)

(iii) If

$$(\exists \underline{x})(\underline{\underline{P}} \wedge \neg\underline{\underline{Q}}) \tag{3}$$

is true, then (2) is false and (1) is not true. However, it can happen that (1) is not true and (3) also is not true.

In fact, if (3) is true, I.3.3(j), (k) and (1) and (I) in I.3.2 show that $(\exists \underline{x})\neg(\underline{\underline{P}} \Rightarrow \underline{\underline{Q}})$ is also true; and then I.3.3(m) and (I) in I.3.2 would show that

$\neg(\forall \underline{x})(\underline{\underline{P}} \Rightarrow \underline{\underline{Q}})$ is true, that is, that (2) is false and therefore (1) is not true. The second statement can be verified on the basis of the theorems in Chapter II, if one takes $\underline{\underline{P}} \equiv \underline{x} = \underline{y}$ and $\underline{\underline{Q}} \equiv \underline{x} = \underline{z}$. (Thus, (1) is not true because, if it were, (I) and (II) in I.3.2 and the theorems of Chapter II would lead to the truth of $\underline{y} = \underline{z}$; and this, by the theorems of Chapter II again, is not true. Further, (3) is not true because, if it were, it could be shown that $\underline{y} \neq \underline{z}$ would be true, which it is not; here again one uses I.3.2, I.3.3 and the theorems of Chapter II.)

The first statement in (iii) is a metatheorem which is used very frequently to show that (1) is not true, an aim which arises in the discussion of converses (see below).

(iv) (1) is false, if and only if $\underline{\underline{P}} \wedge \neg\underline{\underline{Q}}$ is true; and (2) is false, if and only if (3) is true.

These statements follow from I.3.3(j) and (m). However, as has been indicated, there are examples in which (2) is false and (1) is not false; a nonartificial example arises in the opening paragraph of XIII.5 (and there are many similar ones).

The truth of (3) suffices to ensure that (1) is not true and that (2) is false, but not that (1) is false.

(v) I now turn to the consideration of converses. If $\underline{\underline{P}}$ and $\underline{\underline{Q}}$ denote sentences, the <u>converse of</u> (1) is the sentence denoted by $\underline{\underline{Q}} \Rightarrow \underline{\underline{P}}$.

Frequently, having completed a proof of (1), a mathematician will pause to examine the status of the converse, $\underline{\underline{Q}} \Rightarrow \underline{\underline{P}}$. One or more of the following situations may arise:

(a) The converse $\underline{\underline{Q}} \Rightarrow \underline{\underline{P}}$ is also a theorem. In this case it may be affirmed that $\underline{\underline{P}} \Leftrightarrow \underline{\underline{Q}}$ is a theorem.

(b) The converse is not true.

(c) The converse is false.

Now (b) signifies that $\underline{\underline{Q}} \Rightarrow \underline{\underline{P}}$ is not true (that is, that $\neg\underline{\underline{Q}} \vee \underline{\underline{P}}$ is not true), while (c) signifies that $\underline{\underline{Q}} \wedge \neg\underline{\underline{P}}$ is true. These two cases must be carefully distinguished; (c) entails (b) ((ii) above) but (b) does NOT entail (c).

Consider, for example, the case in which $\underline{\underline{P}} \equiv \underline{x} \in \emptyset$ and $\underline{\underline{Q}} \equiv \underline{x} \in \{\emptyset\}$ (again see Chapter II). Then $\underline{\underline{P}} \Rightarrow \underline{\underline{Q}}$ is true, since $\underline{\underline{P}}$ is false: see I.2.9(i) and II.5. The converse, namely,

$$\underline{x} \in \{\emptyset\} \Rightarrow \underline{x} \in \emptyset$$

is not true. (If it were, (I), (II) and the theorems of Chapter II would show that $\emptyset \in \emptyset$ would be true, whereas it will be seen in Chapter II that $\emptyset \in \emptyset$ is false.) The converse is also not false, that is, $\underline{x} \in \{\emptyset\} \wedge \underline{x} \notin \emptyset$ is not true (since $\underline{x} \in \{\emptyset\}$ is not true). The converse is thus undecidable.

Another example arises if $\underline{\underline{P}} \equiv \underline{x} = \underline{y}$ and $\underline{\underline{Q}} \equiv \underline{x} \subseteq \underline{y}$; see II.5.8.

It is indeed rather rare for (c) to be valid, which it is when and only when $\underline{\underline{P}}$ is false and $\underline{\underline{Q}}$ is true.

There is yet another type of confusion to be found in informal accounts. It often happens that a theorem of the form

$$(\underline{\underline{A}} \wedge \underline{\underline{P}}) \Rightarrow \underline{\underline{Q}} \tag{4}$$

is at hand, and there follows a statement of the type

The converse of (4) is not true.

This may indeed be what is intended, but not infrequently the real intention is to state that the converse of

$$(\underline{\underline{A}} \wedge \underline{\underline{P}}) \Rightarrow (\underline{\underline{A}} \wedge \underline{\underline{Q}}) \tag{5}$$

is not true. Notice that (4) and (5) are equivalent sentences (that is, (4) $\Leftrightarrow$ (5) is a theorem schema), but their converses are not necessarily equivalent. (In this connection, see Problems I/23 and IX/19.)

Quite often, the converse of (4) is indeed not true, yet to say so is (in the context in which the discussion takes place) to state what is so obvious as to scarcely merit attention; whereas the non-truth of the converse of (5) is less obvious and thus more noteworthy. Naturally, what is "obvious" , "trivial", "noteworthy" , et cetera is a subjective judgement lying totally outside the formal theory. Such judgements, however, play useful roles in expositions and in teaching.

For illustrations, see Problem V/39 and the discussion in VII.1.13.

One must therefore be on one's guard. Many generally reputable mathematical texts (including my own) blunder by making incorrect assertions of the form

The converse of Theorem .... is false

when only the weaker assertion

The converse of Theorem .... is not true

is justifiable. Furthermore, many assertions of the form

The converse of .... is not true

when made in a conventional framework, are often less clear and informative than they appear to be.

(As an aside, it may be worth calling attention to the metamathematical proofs of (i) - (iv) above. That of (iii) illustrates especially well how these rely on their own style of intuitive logic. Thus, there is one formal logic controlling proofs of sentences in the formal language, and another governing the metaproofs of bastard sentences <u>about</u> proofs of sentences in the formal language. If one doubts these metaproofs, there is need for a meta-metalanguage and a regress has commenced.)

(vi) The preceding remarks arise anew in connection with sentence schemas.

Taking a simple case, suppose that $\underline{\underline{A}}$ is a sentence and $\underline{x}$ a variable (usually appearing in $\underline{\underline{A}}$) . Consider the sentence schema A with the expression $(X|\underline{x})\underline{\underline{A}} \equiv \underline{\underline{A}}[\![X]\!]$, which has just one unrestricted set variable denoted by X ; see I.1.10.

As has been indicated in I.2.9(v), the sentence schema A is said to be true (that is, to be a theorem schema) if and only if the sentence $\underline{\underline{A}}[\![X]\!]$ is true for <u>every</u> choice of the set X , that is (by (X) and (XI) in I.3.2) if and only if the sentence $(\forall\underline{x})\underline{\underline{A}}$ is true.

However, there is a possible ambiguity over the meaning of " A is not true" and " A is false" .

It will be agreed that

> A is said to be not true if and only if the sentence $\neg\underline{\underline{A}}[\![X]\!]$ is true for <u>some</u> (at least one) choice of the set X , that is (using S5 in I.2.2), if and only if the sentence $(\exists\underline{x})\neg\underline{\underline{A}}$ is true.

and that

> A is said to be false if and only if the sentence $\neg\underline{\underline{A}}[\![X]\!]$ is true for <u>every</u> choice of the set X , that is (using (X) and (XI) in I.3.2), if and only if the sentence $(\forall\underline{x})\neg\underline{\underline{A}}$ is true.

It is clear that once again "not true" and "false" must be distinguished; cf. (ii) above.

In addition, if the sentence schema A is not true, then the sentence $\underline{\underline{A}}$ is not true. (For if $\underline{\underline{A}}$ were true and $T \equiv \tau_{\underline{x}}(\neg\underline{\underline{A}})$, then I.3.2(II) would show that $\underline{\underline{A}}[\![T]\!] \equiv (T|\underline{x})\underline{\underline{A}}$ would be true; on the other hand $(\exists\underline{x})\neg\underline{\underline{A}} \equiv (T|\underline{x})\neg\underline{\underline{A}} \equiv \neg(T|\underline{x})\underline{\underline{A}}$ $\equiv \neg\underline{\underline{A}}[\![T]\!]$ is true - since A is not true by hypothesis; but this conflicts with the assumed consistency.) On the other hand, it can happen that the sentence $\underline{\underline{A}}$ is not true and yet the sentence schema A is not not true, that is, both sentences

$\underline{\underline{A}}$ and $(\exists \underline{x})\neg\underline{\underline{A}}$ are not true; see Problem II/28.

Similar remarks apply to sentence schemas having an expression $\underline{\underline{A}}[\![X, Y]\!]$ (the result of simultaneous replacement of $\underline{x}$ by X and $\underline{y}$ by Y in a sentence $\underline{\underline{A}}$, $\underline{x}$ and $\underline{y}$ denoting distinct letters - see Problem I/17 and II.1.3(ii) and (iv)).

What precedes has to be borne in mind whenever one speaks of converses of sentence schemas being not true or false. See also the discussion in III.1.8.

I.3.9 Specialisations and generalisations or extensions As the reader might guess, specialisation (the outcome of which may be described as an "instance" or "special case") refers ultimately to an appeal to (XI), or perhaps (II), in I.3.2. Alongside this one has to consider substitutions in theorem schemas.

In either case, suppose that Θ denotes a theory.

(i) Specialisation on (or of) a variable Specialisation on (or of) (sometimes also spoken of as substitution for) a variable $\underline{x}$ is the term frequently applied to what has hitherto been referred to as replacement for $\underline{x}$ and indicated by $(\ldots|\underline{x})$.

Suppose one has a theorem

$$T\Theta \quad : \qquad \underline{\underline{A}} \ . \tag{1}$$

If $\underline{x}$ is a variable (free letter) of Θ , and if one writes $\underline{\underline{A}}[\![S]\!] \equiv (S|\underline{x})\underline{\underline{A}}$ for arbitrary sets S , then (II) in I.3.2 and (1) yield

$$T\Theta \quad : \qquad \underline{\underline{A}}[\![S]\!] \ . \tag{2}$$

If S is any chosen set, the theorem (2) is said to be obtained from (1) by specialisation on (or of) the variable $\underline{x}$; alternatively, (2) may be said to be a special case of (1) .

Alternatively, if one has the theorem

$$T\Theta \quad : \qquad (\forall \underline{x})\underline{\underline{A}} \quad , \qquad (1')$$

where now the letter $\underline{x}$ need not be a variable of Θ , appeal to (XI) and (I) in I.3.2 again yields (2) , which may therefore also be said to be obtained from (1') by <u>specialisation on</u> $\underline{x}$ or (rather less sensibly) a <u>special case</u> of (1') . (Less sensibly, since $\underline{x}$ does not appear in $(\forall \underline{x})\underline{\underline{A}}$.)

Less directly, if one knows that

$$T\Theta \quad : \qquad \underline{\underline{A}} \Rightarrow \underline{\underline{B}} \quad , \qquad (3)$$

that $\underline{x}$ is a variable of Θ , that S is an arbitrarily chosen set and that

$$T\Theta \quad : \qquad \underline{\underline{A}}[\![S]\!] \quad , \qquad (4)$$

then (II) and (I) in I.3.2 yield

$$T\Theta \quad : \qquad \underline{\underline{B}}[\![S]\!] \quad . \qquad (5)$$

Rather improperly (in view of the essential intervention of (4)), (5) is still sometimes spoken of as a special case of (3) .

For instance, when Θ is Θ_0 , one might have in place of (3) :

$$T \quad : \qquad (\underline{x} \text{ is a real number}) \Rightarrow (\underline{x}^2 \geq 0)$$

and then read off in place of (5) :

$$T \quad : \qquad (\pi - 1)^2 \geq 0$$

and describe this last as a special case. This hides the intervening step (corresponding to (4)) that " $\pi - 1$ is a real number" is true (in Θ).

Similar remarks apply to specialisation of a variable in a theorem schema.

(ii) Substitutions in theorem schemas What are sometimes also termed special cases result from admissible substitutions for some or all of the letter symbols, set symbols and sentence symbols appearing in a theorem schema of Θ . (The simplest cases are instances of theorem schemas, as described in I.1.10 and I.2.9(v).) Such a special case will be a theorem schema or a theorem of Θ .

In the simpler and most frequently occurring cases, a theorem schema will involve one or more unrestricted set symbols X , Y ... , for each of which may be substituted arbitrary sets (see II.1.1(1_1), for example). A so called special case then results from substituting for some or all of X , Y , ... arbitrary chosen sets. (This procedure should not generally be confused with formal replacement of a letter $\underline{x}$ by a set S , as defined in I.1.3 and indicated by $(S|\underline{x})$.)

Similar remarks apply if the theorem schema involves in addition one or more sentence symbols $\underline{\underline{A}}$, $\underline{\underline{B}}$, ... , for each of which may be substituted arbitrary sentences. (An example is II.3.2(2').) Here, of course, one may substitute for X , Y , ... freely chosen sets and for $\underline{\underline{A}}$, $\underline{\underline{B}}$, ... freely chosen sentences. The result will again be a theorem or theorem schema of Θ .

It is, of course, essential to make substitutions only for symbols which denote freely chosen sentences or sets, that is, which are unrestricted sentence symbols and/or set symbols.

For instance in Chapters II and V specific sets denoted by $\emptyset$ and N are introduced. There is then a theorem schema

$$T \ : \quad (X \in N \wedge X \neq \emptyset) \Rightarrow (\emptyset \in X) ,$$

in which X is the only unrestricted set symbol in the sense of I.1.10. While any substitution for X leads to a theorem of Θ_0 , there is no reason to expect that a substitution for $\emptyset$ (or for N) will lead to a theorem schema of Θ_0 . In fact, if one substitutes $\{\emptyset\}$ for $\emptyset$ one obtains the sentence schema

$$(X \in N \wedge X \neq \{\emptyset\}) \Rightarrow (\{\emptyset\} \in X) ,$$

which, if Θ_0 is consistent, is not a theorem schema of Θ_0 . Another instance is discussed in II.5.4.

Usually, specialisations and substitutions are carried out without mention, much less detailed description. It is regarded as enough to indicate at most the theorem or theorem schema in which the specialisations and/or substitutions are to be made.

For example, suppose one were to begin with the theorem schema II.4.1(4), namely:

$$(\forall \underline{z})((\underline{z} \in \{X, Y\}) \Leftrightarrow ((\underline{z} = X) \vee (\underline{z} = Y)))$$

wherein $\underline{z}$ denotes a letter not appearing in X or Y. Then one may first specialise on $\underline{z}$ to obtain the theorem schema

$$a \in \{X, Y\} \Leftrightarrow ((a = X) \vee (a = Y)) \quad ,$$

and then substitute a for X and b for Y to obtain the theorem schema

$$a \in \{a, b\} \Leftrightarrow ((a = a) \vee (a = b)) \quad ;$$

also, specialisation in the theorem II.1.4(2) (which reads $(\forall \underline{x})(\underline{x} = \underline{x})$), yields $a = a$ as a theorem schema; hence (see I.3.3(j) and I.3.2(I) and substitution) $(a = a) \vee (a = b)$ is a theorem schema; hence (via substitutions in (I) and (III) in I.3.2) $a \in \{a, b\}$ is a theorem schema. All of this might routinely be covered by saying simply that $a \in \{a, b\}$ ensues from II.4.1(4)!

Special cases of theorem schemas are more difficult to describe in completely accurate terms than specialisation of a variable in a genuine theorem. One might see in this a case for the banishment of theorem schemas in favour of genuine theorems (cf. the discussion in I.3.4(i)), but axiom schemas are accepted as part of the description of the formal system adopted here, so that theorem schemas seem natural. In addition, as was remarked in I.3.4(i), theorem schemas are often found by most people to have an intuitive appeal lacking in the corresponding theorems which would appear in their place. See also VI.10.2.

(iii) Generalisation or extension A conditional sentence $\underline{\underline{A}}' \Rightarrow \underline{\underline{B}}$ is often said to be a generalisation or extension of a conditional sentence $\underline{\underline{A}} \Rightarrow \underline{\underline{B}}$, if $\underline{\underline{A}} \Rightarrow \underline{\underline{A}}'$ is true. In this case, if $\underline{\underline{A}}' \Rightarrow \underline{\underline{B}}$ is true, then so too is $\underline{\underline{A}} \Rightarrow \underline{\underline{B}}$ (thanks to the syllogism (III) in I.3.2); and then the theorem $\underline{\underline{A}}' \Rightarrow \underline{\underline{B}}$ is termed a generalisation or extension of the theorem $\underline{\underline{A}} \Rightarrow \underline{\underline{B}}$.

In practice, one usually begins with a known theorem $\underline{\underline{A}} \Rightarrow \underline{\underline{B}}$ and seeks interesting or significant generalisations $\underline{\underline{A}}' \Rightarrow \underline{\underline{B}}$. The simplest instances are those in which $\underline{\underline{A}} \equiv \underline{x} \in S$ and $\underline{\underline{A}}' \equiv \underline{x} \in S'$ where S and S' are sets and $S \subset S'$

is true (see II.2). For example,

$$(\underline{x} = 3 \vee \underline{x} = 4) \Rightarrow \underline{x}(\underline{x} + 1) \text{ is even}$$

is a theorem, of which the theorem

$$\underline{x} \in N \Rightarrow \underline{x}(\underline{x} + 1) \text{ is even}$$

is a very significant generalisation.

To be honest, many alleged generalisations are nowhere near as simple as this. In order to squeeze them into the above pattern, all sorts of identifications (see IV.9.4) are involved (but usually left hidden). See also Note 8.

To illustrate this hidden complexity, suppose that R and C denote the sets of real and complex numbers respectively (see Chapters VI and XII). Loosely speaking (in conventional style) the theorem

$$(\underline{x} \in R \wedge \underline{y} \in R \wedge \underline{x}\cdot\underline{y} = 0 \wedge \underline{x} \neq 0) \Rightarrow (\underline{y} = 0) \qquad (6)$$

is said to have as a generalisation the theorem

$$(\underline{x} \in C \wedge \underline{y} \in C \wedge \underline{x}\cdot\underline{y} = 0 \wedge \underline{x} \neq 0) \Rightarrow (\underline{y} = 0) \quad . \qquad (7)$$

In the formalism developed in this book (and indeed in almost any current formalism), this is a distortion of the truth. In the first place (and speaking more strictly) R is not a subset of C ; the 0 in (6) is not the same as the 0 in (7); the $\underline{x}\cdot\underline{y}$ in (6) is not the same as the $\underline{x}\cdot\underline{y}$ in (7) . What is true, is that R may be identified with a subset of C in such a way as to compensate for these differences and so make (7) into a generalisation of (6) . More precisely: there is an injective function f with domain R and range a subset of C , an element 0_C of C , and a function π_C with domain $C \times C$ and range a subset of C such that (in lieu of (7))

$$(\underline{x} \in C \wedge \underline{y} \in C \wedge \pi_C((\underline{x}, \underline{y})) = 0_C \wedge \underline{x} \neq 0_C) \Rightarrow (\underline{y} = 0_C) \qquad (7')$$

and also

$$(\underline{x} \in R \wedge \underline{y} \in R) \Rightarrow (f(\underline{x}\cdot\underline{y}) = \pi_C((f(\underline{x}), f(\underline{y})))) \qquad (8)$$

and

$$f(0) = 0_C \qquad (9)$$

are all true. Then (6) may be deduced from (7') , (8) and (9) (and the stated properties of f and π_C) , so that, strictly speaking, it is (7') (not (7)) which is to be viewed as a generalisation of (6) .

I.4 General comments

I.4.1 Formalism and metamathematics What has been said so far about formal theories is, I think, enough to make it fairly clear that they do attain certain limited objectives; and this will hopefully become clearer by example as the reader progresses. The concept of a formal theory does provide a yardstick by which to judge specific examples of reasoning, that is, it does help to decide whether a specific alleged proof is indeed a proof according to explicitly stated and widely accepted canons. Also, it does help to make clear the basis for standard proof methods used without thought in conventional mathematics. This is certainly an addition to clarity. (However, see the comment at the end of VIII.6.9.)

On the other hand, the concept raises questions of a new order which are no less troublesome and which have come to form part of metamathematics, a subject now comparable in scope and intricacy with mathematics itself. Thus, given a formal theory, it can often happen that one is interested in the provability in that theory of a particular sentence. Supposing that repeated efforts to formulate a proof all fail, and that the same is true of the negation of the said sentence, one casts around for less "bare-handed" techniques for settling the matter. Instances of this sort of question arise whenever one considers the consistency of a formal theory, or the independence of the axioms of a formal theory. Metamathematics is concerned with precisely such questions as these, and with the investigation of methods for settling them (if possible). Fragmentary comments on such matters occur occasionally in this book, as asided or as remarks in the Appendix, but absolutely no attempt has been made to deal systematically with such matters. What is done in this book is to look in some detail at a few of the well established and basic theorems in one formal theory, and to compare their formal and informal versions.

The metamathematical and philosophical aspects of formal theories present some features which are even today very perplexing. Even now the classical portions of metamathematics contain conclusions about the limitations of formal theories which were entirely unexpected. Semantical notions (as opposed to the syntactical ones which are alone featured in the preceding description) are sometimes very puzzling and seem almost always to be made difficult by the hierarchies of metalanguages and concepts of truth. Some little encouragement can be derived from the fact that the considered opinions of even those best qualified to judge, sometimes require modification with the passage of time.

It may seem fair overall to claim that formalism seems sometimes essential in a search for clarity and precision and as an aid in avoiding at least some of the paradoxes; that it has had some significant success in these respects; and that (what was presumably not commonly intended at the outset) it has led to many valuable and exciting advances in our understanding of the powers and limitations of strictly logical reasoning. It is perhaps psychologically interesting that at least some of those who have spent the most time dealing with remote abstractions and formalisation nonetheless cling resolutely to a belief that, behind and independent of all the meaningless formalism, there is a real and objective universe of sets, some aspects only of which can be incorporated into and accounted for in any one formalism.

I.4.2 Other approaches discussed The rest of this section has two principal aims. The first is to explain why the treatment in this book is so abstract compared with alternatives which may look to be, and are sometimes presented as being, both simpler and satisfactory. This will be aided by detailed references to three books, namely, Glicksman and Ruderman (1), Anderson (1), and Dippy and Olyjnik (1), often abbreviated in the rest of this section to (GR), (A) and (DO) respectively. These references and comments combine with the second aim, which is to help the reader to appreciate what is involved by calling attention to some typical specific issues which demand careful handling, some at the basic logical level and some at a later conventionally mathematical level.

Readers who feel replete in these areas can ignore the rest of this section (but see also Note 9).

The first of the books cited above go together, because they both make broadly very similar (and seemingly serious) attempts to deal with the logical substructure of the mathematics to follow. I begin with some general comments on these attempts.

It is not easy to compare the common approach of (GR) and (A) with that I have adopted. Roughly speaking, however, they concern themselves with sentences (nowhere clearly defined) in which there may or may not appear one or more variables (again nowhere clearly defined). Each variable is said to have attached to it a domain, which is a set; cf. II.4.4(iii) below. The term "set" is left unexplained: this in itself is an almost insuperable stumbling block in the path of regarding this approach as a satisfactory foundation for all that is to follow. Sentences in which no variables appear are termed statements; in the contrary case they are termed s-forms (statement forms) by (GR) and conditions by (A). ((A)'s vaguely described combinations of statements are seemingly still statements.) If variables x, y, ... appear in an s-form, substitution for (that is, replacement of) x, y, ... by certain permissible elements of their respective domains is to result in a statement. There are (unexplained) concepts of true and false; and each statement is assumed (questionably) to be either true or false. The logical connectives $\neg$, $\vee$, $\rightarrow$, et cetera are supposed to lead from sentences to sentences; and (explicitly writable, finitary) truth tables prescribe truth values of $\neg s$, $s \vee t$, $s \rightarrow t$ and so on in terms of s and t, at least when s and t denote statements. If, as is sometimes the case, they denote s-forms, the situation is never satisfactorily clarified: as will become clearer in a moment, this lack is a manifestation of a hiatus in their treatment which they neither mark nor bridge and which to all appearances is in a sense unbridgeable (see three paragraphs below).

As long as attention is restricted to statements (as opposed to s-forms), one is in effect concerned with what is usually termed the propositional calculus. Here it is well known that the approach via truth tables is fully satisfactory and equivalent to a syntactic approach of the sort I have adopted. Both (GR) and (A) are mainly adequate up to this point, though even here their accounts need tightening up before they would be really acceptable (compare with Margaris (1), §13 or Kneebone (1), Chapter 2; for a full and careful treatment, see Kleene (1), Chapter VI and Kleene (2), Chapter I) and then they would appear to be in reality no simpler than the syntactic approach. (For one thing, they have need for another symbol $\Rightarrow$, related to but different from their $\to$; it is their $\to$ which corresponds to our $\Rightarrow$, while their $\Rightarrow$ has no counterpart in our scheme.)

So far their treatment is broadly acceptable, and does indeed avoid some of the abstraction of the syntactic approach (though only by sacrificing clarity on a number of vital points).

The real fly in the ointment is that it is generally recognised that the propositional calculus is quite inadequate as a foundation for mathematics; see, for example, Kneebone (1), p. 64. It is necessary to introduce variables (or some counterpart of them), s-forms and quantifiers, characteristic of the so-called predicate calculus; see Margaris (1), Chapter 2; Kneebone (1), Chapter 3; Kleene (1), Chapters VII and XIV; Kleene (2), Chapters II, III and VI. It is also generally recognised that, having proceeded thus, one has to abandon the (finitary) truth table approach in favour of the syntactic one; see Mendelson (1), p.56 and Kleene (2), p.117. (Do not be misled by the informal discussion in Margaris (1), §6.) Both (GR) and (A) gloss over this vital point (cf. I.4.3(ii) and I.4.4(ii) below) or, what is perhaps worse, they try to give the impression that they have surmounted the insurmountable. (An essential part of the difficulty is the difference between the unrestricted quantifiers $(\exists \underline{x})$ and $(\forall \underline{x})$ and the restricted quantifiers $(\exists_X \underline{x})$ and $(\forall_X \underline{x})$ associated with finite sets X; see II.4.4(iii).)

It is important to point out here that the predicate calculus can be "truth tabled", if one allows infinite truth tables; see Kleene (2), especially the comments on p.117. However, while the use of infinite truth tables to verify the validity of any given proof may be of interest to logicians as an application of existing mathematical ideas, it is suspect in any attempt to found mathematics when compared with the syntactic verification described in I.2.4 (which, when applied to any given alleged proof, is a finitary procedure in which the construction itself virtually provides the verification). In addition to this, there is little doubt that almost all working mathematicians (as opposed to logicians) think of proofs being constructed and verified syntactically, not by means of computations with truth tables (finite or infinite). In any case, neither (GR) nor (A) give the slightest hint that infinite truth tables are involved, much less any indication how to handle them.

(Parenthetically it should perhaps be added also that even the predicate calculus is inadequate as a foundation for mathematics; to it has to be added something set theoretical or its equivalent. All this is taken into account in our approach, in which a set theoretical component is built in right from the start.)

In the end, our treatment (which incorporates the predicate calculus from the word "go" and makes no overt reference to the more special propositional calculus) is much more coherent and is, to boot, simpler and smoother when one gets down to details. Moreover, we have no need of an external naive set theory to bolster the discussion of the logical calculus. Leaving aside the use of infinite truth tables, and neglecting the choice of which particular formal system is to be adopted, there appears to be no alternative to the syntactically defined deductive approach (cf. I.2.9(iv)). Its strangeness and abstractness is a price that seemingly has to be paid for something which has any hope of supporting conventional mathematics. Nothing is to be gained by seeking to disguise this state of affairs.

A discussion of a few more specific issues follows.

I.4.3 Discussion of (GR) Beginning with (GR), Chapters 1 - 4, one is encouraged by their stress (pages 1-3) on the distinction between the meta (syntax) language and the formal (object) language, and the mention of various demands to be placed upon definitions. On the other hand, they go only part of the way towards formalisation. This would be acceptable, if it were subsequently shown to function satisfactorily. But, as the following items indicate, shortcomings soon appear.

(i) The sign = is surely intended to be at times part of the formal language, yet it is confusingly used also as part of the metalanguage in the expression of definitions; see (iv) and (vi) below.

(ii) On page 43 one finds $\forall_{x\in A}s(x)$, which is clearly intended to be (or to denote) a formal sentence, defined to be

$s(x)$ is true for all values of x in set A,

which is visibly a sentence in the metalanguage. The proposed definition may be intended to define a truth table for $\forall_{x\in A}s(x)$, but it does not achieve this aim (except perhaps when the set A is finite); even less does it serve to specify a statement in the formal language; cf. Problem I/37. This is one place where the authors seem to encounter the problem of truth tabling the predicate calculus without offering any hint of solution; see the comments in I.4.2 above.

So much for the good intentions expressed on pages 1-3.

(iii) Exercise 35 on page 77 asks "Can there be such a thing as the set of all sets?" (The book contains no answers to exercises, by the way.) Neither at this stage, nor at any other in the book, is the reader really in a position to answer this question. The set theory expounded is naive. Presumably the reader is expected to argue informally towards a version of one of the standard paradoxes and so conclude that the answer is "No" (see II.3.10 below). But see (v) below, which indicates that later developments in the book will suggest to the reader that the answer should (or may) be "Yes" .

Incidentally, the next exercise in (GR) misquotes the so called Barber Paradox and in so doing eliminates the paradox in question; see Problem II/10.

(iv) On pages 127-128 there appears an informal explanation of the concept of "ordered pair" , but there is no acceptable definition thereof. The authors also write:

> Ordered pairs constitute the building blocks from which modern mathematics constructs many important concepts. Among these are relations and functions, and the far reaching concept of a cartesian product upon which they are based. We shall develop these ideas *carefully* in this chapter.

The italics are mine. The first two sentences are admirable, but the third raises hopes soon to be dashed for the following reasons.

The absent definition of ordered pair cannot fail to throw serious doubt on the authors' right to adopt the subsequent so called definition

$$[(a, b) = (c, d)] = [(a = c) \wedge (b = d)] \quad ;$$

note the two roles of the sign = here, and see III.1 below.

On page 134, "relation in $U \times V$ " is defined to mean simply "subset of $U \times V$ " ; "relation" (in the nude, so to speak) is nowhere explained. Yet on page 147 one finds " f is a function" defined to mean

$$\{(f \text{ is a relation}) \wedge [(xfy_1) \wedge (xfy_2) \Rightarrow (y_1 = y_2)]\} \quad .$$

This is plainly not good enough; there is lacking something of the sort

$$(r \text{ is a relation}) \equiv (\exists x)(\exists y)(r \text{ is a relation in } x \times y) \quad .$$

In spite of the omission, it is clear from what the authors do that any relation (in particular, any function) has a certain set as its domain and a certain set as its range.

(v) All the examples of functions listed on page 177 of

(GR) involve the idea of defining a function by a domain and a rule, but there is no explanation of the general principle involved (see IV.2 below).

Nor is it made clear what " f is a function on oojahs" or " f is a function on A " are to mean, though the reader may well infer that it certainly implies that the domain of f is the set of all oojahs or the set A .

However, this inference leads to trouble inasmuch as the thirteenth example (appearing on page 178) speaks of a power set function P defined "on sets A " . In view of what has been written in the last paragraph above, it is a reasonable inference that there is a set of all sets (namely, the domain of P). Likewise with the example which follows in (GR). Compare this with (iii) above.

(vi) Example 10 on page 178 reads:

> The <u>divides function</u> is defined on ordered pairs of integers as follows:
>
> $$(a|b) = \exists_{c\in\{\text{Integers}\}}(ac = b) \quad .$$
>
> The range of <u>divides</u> is {True, False} .
> " 3|6 " is True, " 3|8 " is False

(In the formalism used in this book the formula just displayed would in due course appear as

$$(\underline{a}|\underline{b}) \equiv (\exists\underline{c})(\underline{c} \in Z \wedge \underline{a}\cdot\underline{c} = \underline{b}) \quad ,$$

where $\underline{a}$, $\underline{b}$ and $\underline{c}$ denote distinct letters not appearing in the set (of all integers) denoted by Z .)

This example invites chaos.

To begin with, the first = belongs to the metalanguage, while the second = belongs to (or denotes something belonging to) the formal language.

Next, what is in fact defined is an s-form, not a set; hence neither a function nor a relation.

It would seem that the authors are flirting with idea of a function $(a, b) \rightsquigarrow (a|b)$ with domain $Z \times Z$ and range a set of s-forms. But this is again a gross blunder: the formal concept of "set" involved in that of "function" has no room for so called sets of metamathematical entities (such as s-forms, True, False, for example).

Overlooking this, one would obtain a function with range {True, False} , only after forming the composite of $(a, b) \rightsquigarrow (a|b)$ with some truth-value function.

See also Remark (i) following II.3.8.

I.4.4 <u>Discussion of (A)</u> Turning to (A), Chapter 1, the general comments in I.4.2 may be supplemented by numerous detailed criticisms broadly similar to those in I.4.3. I mention only a few.

(i) The book provides for no distinction between a metalanguage and a formal language. There is no mention of "axiom" , leaving the reader with the impression (which can hardly fail to be puzzling) that after all something emerges from nothing.

(ii) On page 23, lines 9 and 10, one encounters the assertion " $p(x)$ logically implies $q(x)$ " . Now "logically implies" has hitherto been defined by this author in relation to statements only (not conditions, that is, statement forms), and the implied extended definition is nowhere supplied. What is said, taken in conjunction with the general approach, seems to suggest that the condition $p(x) \to q(x)$ has attached to it a truth table, but nothing explicit is said about truth tables of conditions. The problem of finitary truth tabling the predicate calculus looms large here; see the comments in I.4.2 above.

(iii) One of the author's aims must surely be to provide the reader with concept of proof which is as clear as he can make it. Indeed, on page 37 we find written:

> In mathematics, we shall be concerned with theorems and proofs. We have used these words already. Let us now make our terminology precise. Many of our theorems are of the form "If a , b and c , then d " ; in other words, that a certain set of premises imply a certain conclusion. What, then, is a proof of such a theorem? We make the following definition.
>
> Definition 1.10-3 Let P , Q be statements, or combinations of statements. A proof is an argument which shows that $P \Rightarrow Q$. Thus, if the argument $P \vdash Q$ is valid, we say we have proved that P implies Q .

So far, so good (perhaps). In search of the necessary further enlightenment, the reader has to back-track to page 33 to find the following:

> Let $p_1, p_2, \ldots, p_n$ and q be statements. Then, an argument is an assertion that the statements $p_1, \ldots, p_n$, called the premises, lead to the statement q , called the conclusion. We write this as
>
> $$p_1, \ldots, p_n \vdash q \ .$$
>
> We have said that an argument is an assertion; that is, it is itself a statement which may be true or false.
>
> Definition 1.10-1 The argument $p_1, \ldots, p_n \vdash q$ is said to be valid, or true, if
>
> $$p_1 \ \& \ p_2 \ \& \ p_3 \ \& \ \ldots \ \& \ p_n \Rightarrow q \ .$$
>
> Equivalently, it is valid if $(p_1 \ \& \ p_2 \ \& \ \ldots \ \& \ p_n) \to q$ is a tautology.

It is unnecessary and pointless for me to back-track any more, except to say that here "is a tautology" means "has a truth table, all of whose entries are T " ; see Definition 1.7-7 on page 19.

I find it impossible to see how the reader can fail to conclude from the material quoted above that a proof, being an argument, is an assertion and therefore a statement. It is difficult to imagine an impression which is more misleading in the present context. Supposedly, what the author intends to convey is that a proof is a succession of statements, interlinked in such a way as to constitute a valid argument; and that a valid argument from premises $p_1, \ldots, p_n$ to conclusion q is some sort of verification that the statement

$$(p_1 \,\&\, \ldots \,\&\, p_n) \to q$$

is a tautology. This verification must (I again suppose) amount in the last analysis to an agreed sort of computation involving truth tables.

Apart from this, the author makes it appear that one can only prove things like $P \Rightarrow Q$, where P and Q are as stipulated in Definition 1.10-3, quoted above. In this connection, the index in (A) refers the reader to pages 13 ff. thereof for the concept of combinations of statements: but I can find at that point no satisfactory definition. (Presumably, one must have so called atomic statements and connectives &, or, ~, et cetera, out of which combinations of statements are to be built ... but precisely how are they to be built?) Note also that P and Q are to be statements, or combinations of such, not conditions; cf. (ii) above.

The reader may be left with the feeling that sentences which are not of the conditional form $P \Rightarrow Q$ are in principle unprovable. The observation that any statement Q is equivalent to $P \Rightarrow Q$, where P is any true statement, may be thought to present an escape; but it is artificial and does not in any case get to the heart of the matter.

Moreover, almost as soon as "real mathematics" begins, the author is concerned (on page 52) with a proof of

$$(x, y, z \in R \,\&\, x + z = y + z) \to (x = y) \quad .$$

It seems to me that there is trouble here, inasmuch as this is seemingly not of the form $P \to Q$ (or $P \Rightarrow Q$), wherein P and Q are statements or combinations of statements. (The appropriate P and Q are conditions, that is, statement forms; cf. I.4.2 above.) One might try to recast what is to be proved into the form of a statement by universal quantification over x, y, and z: the result is a statement, but it is no longer conditional in form. (And anyway, I cannot find in (A) any sound justification for the said quantification, parallel to (X) in I.3.2 above.) In brief, the author's account seems to be inadequate almost as soon as one gets down to "real mathematics".

It is, in my view at any rate, difficult to take seriously the author's claims (appearing in the Preface and the preamble to Part One on page 1 of (A)) concerning the understanding to be imparted to his readers via his Chapter 1. This chapter amounts to what is (again in my view) a frequently vague, incomplete and misleading account of a portion of logic which is, taken by itself, inadequate to support the mathematics which follows.

I.4.5 Discussion of (DO) I now turn my attention to (DO), which is a beautifully produced book in respect of appearances.

The main trouble here is that the authors attempt next to nothing (much less than (GR) or (A)) to prepare the way for the use of symbols like $\Rightarrow$ and $\forall$, which they sometimes abuse in horrifying fashion. Vague use of $\Rightarrow$ begins on page 4; that of $\forall$ begins on page 59 (or perhaps earlier). They are used confusingly. (For instance, why is $\forall$ used in Example 3 on page 59, but not in Examples 1 and 2? The use of $\Rightarrow$ in 3. on page 283 is totally wrong.) Some of the routine parts of the book are acceptable, without really justifying the rather grandiose title.

I turn to a detailed discussion of a few examples, in the treatment of which the authors seem to have thrown caution to the winds.

(i) In the space of less than one page (55), the authors seek to define "proposition" and "axiom". For instance:

> A proposition in mathematics is a statement (that is, a sentence with a verb) which may be definitely labelled either "true" or "false".

I think it certain that the majority of readers of (DO) will interpret the "either ...... or ......" in the exclusive sense and conclude that a proposition is never both true and false; see (ii)(c) below.

They proceed to say that $x + 1 > 3$ is not a proposition, whereas

$$P(n) : 1 + 2 + \ldots + n = \tfrac{1}{2}n(n + 1)$$

is a proposition. There is no attempt to explain where the implied difference lies. The proposed definition is just too vague to be coherently workable.

On this same page 55, " 2^n is even for $n \geq 1$ " is explained to mean " 2^1 is even, 2^2 is even, et cetera", that is (presumably) " 2^n is even for <u>all</u> natural numbers $n \geq 1$ ". In spite of this, this proposition is labelled " $P(n)$ " which (in this context) suggests that it depends on n, even though it plainly does not. This piece of explanation vouchsafed by the authors has bearing on my comments in (ii).

(ii) I propose to consider at length Exercises 6.2 on page 56, which are prefaced by the general instruction:

> Assume the given proposition true for $n \geq 1$, $n \in J^+$, and hence deduce the information asked. (1)

I can find no definition of J^+, but it seems safe to assume that it denotes the set of natural numbers (denoted in this book by $\mathbb{N}$). In view of the final paragraph in (i) above, I assume also that "for $n \geq 1$, $n \in J^+$ " means "for all $n \in J^+$ such that $n \geq 1$ ". The parenthesised sentence in (2) below lends support to this assumption.

As will be seen, the "information asked" turns out to assume the form of a question; but does a question constitute information? More to the point, how does one deduce a question? At all events, I suppose the authors anticipate that their questions will each be answered by the reader with a "Yes" or a "No" .

One more preliminary seems to be relevant. Preceding (1), the authors write:

> In the following exercise it is not intended that you test the truth of the given proposition. You are merely asked to deduce what <u>would</u> be true <u>if</u> the proposition were true. (In <u>fact</u>, some of <u>the</u> propositions are false: it should make no difference to your answer.) (2)

To make things specific, I shall henceforth concentrate on number 6 of the said exercises, wherein the "given proposition" is (I quote verbatim)

$$P(n) \;:\; 2^n - 1 \text{ is divisible by } 3 \text{ for } n \geq 6 \;; \tag{3}$$

and the question asked is

$$\text{Is } 126 \text{ divisible by } 3 \;? \tag{4}$$

Notice the conflict between (1) and (3) : precisely which proposition is to be assumed?

(a) For reasons already given, I assume that the intended "given proposition" (hypothesis) is

$$\underline{\underline{H}} \equiv 2^n - 1 \text{ is divisible by } 3 \text{ for all } n \in J^+ \text{ such that } n \geq 6 \;. \tag{3'}$$

(If the intended "given proposition" is in fact " $2^n - 1$ is divisible by 3 " , the situation becomes a little more complicated, but my subsequent criticisms would apply in modified form.)

It seems clear that the authors intend that the reader should assume $\underline{\underline{H}}$ and, on that basis, decide that at least (and more probably, precisely) one of sentences

$$\underline{\underline{D}} \equiv 126 \text{ is divisible by } 3 \;,$$

$$\neg\underline{\underline{D}} \equiv 126 \text{ is not divisible by } 3$$

is true, that is, that at least (and more probably precisely) one of $\underline{\underline{D}}$ and $\neg\underline{\underline{D}}$ is deducible from $\underline{\underline{H}}$. (Recall that the authors stipulate that a proposition must be either true or false.)

My readers should here refer to I.3.4(vi) and Problem I/4: deductions are made within a background theory Θ . This theory is (understandably) vague in (DO), but there is every reason to suppose that it includes the elementary arithmetic and algebra already informally handled. (There is in (DO) no hint that, for the purposes of this exercise, their readers should ignore what they have learned earlier.)

(b) The first two sentences of (2) are, presumably, intended to make the valid point that one may legitimately make deductions from false propositions (just as well as from true ones). This point should, indeed, be spelled out even more clearly at an appropriate place and with due care. But, for the reasons to appear in (c) below, this is <u>not</u> an appropriate place and due care is <u>not</u> exercised.

Besides this, the final (parenthesised) sentence in (2) is, in my view, grossly misleading. While there is a sense in which it may be said that the method of deduction does not depend on the status (truth or falsity) of the hypothesis, what is or is not deducible (in a given theory) from that hypothesis may so depend on its status. Thus, suppose that Θ is a background theory, assumed consistent, that $\underline{\underline{H}}$ is the hypothesis, and that $\underline{\underline{C}}$ is a sentence not provable in Θ. Then, if $\underline{\underline{H}}$ is true, $\underline{\underline{C}}$ is not deducible (in Θ) from $\underline{\underline{H}}$; while if $\underline{\underline{H}}$ is false, $\underline{\underline{C}}$ is deducible (in Θ) from $\underline{\underline{H}}$. So, if one were instructed to assume $\underline{\underline{H}}$, and then asked the question "Is $\underline{\underline{C}}$ deducible?", one would have to reply "No" or "Yes" according as $\underline{\underline{H}}$ is true or false.

(c) Returning to the hypothesis $\underline{\underline{H}}$ defined in (3'), the reader of (DO) may be expected to recognise that $\underline{\underline{H}}$ is false and that $\underline{\underline{H}} \Rightarrow \neg\underline{\underline{D}}$ is true (in the background theory). (He would in fact supposedly argue thus: In the first place,

$2^7 - 1$ is not divisible by 3 (arithmetic), which implies $\neg\underline{\underline{H}}$, and so $\underline{\underline{H}}$ is false; in the second place, $\underline{\underline{H}}$ implies

$2^7 - 1$ is divisible by 3, which in turn implies (via arithmetic)

that $126 = (2^7 - 1) - 1$ is not divisible by 3, that is, $\underline{\underline{H}}$ implies $\neg\underline{\underline{D}}$.)

Since $\underline{\underline{H}}$ is false, $\underline{\underline{D}}$ is deducible. (Actually, of course, the reader's knowledge of arithmetic would (without any reference to $\underline{\underline{H}}$) lead him to conclude that $\underline{\underline{D}}$ is true.)

Since $\underline{\underline{H}} \Rightarrow \neg\underline{\underline{D}}$ is true, $\neg\underline{\underline{D}}$ is deducible.

It thus seems that the authors should be prepared for a "Yes and No" answer to (4), and that they should deal with this eventuality. There is no doubt in my mind, that the vast majority of their readers would be unable to cope unaided with this predicament. (Very few students at this stage are happy with the idea that a "mathematical proposition" may be both true and false.) Actually, as has been said, it seems certain that most readers will interpret in the exclusive sense the "or" in the authors' definition (see (i) above) of proposition; to them, a downright inconsistency may seem inescapable.

The root cause of the trouble is that the authors (in effect) instruct the reader to consider the outcome of adjoining (as a new axiom) to a theory, a sentence which is false in that theory. The resulting theory is contradictory and thus embraces a host of counter-intuitive conclusions. Their readers have little other than intuition to guide them, yet the authors offer no hint of what is in store, much less any helpful explanation.

(iii) A close analysis of numbers 4 and 5 in Exercise(s) 16.2 on page 229 will show that again there is a serious lack of care in developing a perfectly laudable attempt to impress upon the reader that, although

If $x < y$ and $a < b$, then $x + a < y + b$

is (when suitably qualified) true, the similar-looking sentence

If $x < y$ and $a < b$, then $x - a < y - b$

is not true.

(iv) On page 67 the discussion is likely to confuse the reader over the distinction between (rough) upper bound, abbreviated to U.B. , and least upper bound, abbreviated to L.U.B. ; see VII.5.1. The authors write as if an U.B. is unique, thereby suggesting that perhaps they mean L.U.B.; yet, if the reader _is_ persuaded this way, he will surely be further puzzled by seeing written

$$\text{So } a_n < 2 \ \forall\, n \in J^+ \Rightarrow \text{U.B.} = 2 \ .$$

Similar remarks apply to lower bound (L.B.) and greatest lower bound (G.L.B.).

On page 79 the authors consider a sequence $n \leadsto na^n$ in the case $0 < a < 1$. The sequence is shown to be eventually decreasing and to have positive terms. One then finds written

But all terms are positive $\Rightarrow$ L.B. = 0

$\Rightarrow na^n$ is a null sequence.

The lack of clarity in this sentence is striking and intolerable. Furthermore, taking what seems to be the most plausible clear interpretation, the first implication sign ($\Rightarrow$) is (unless set theory is contradictory) unjustified. That is to say, if set theory is consistent, neither of the sentence schemas denoted by

($\{a_n\}$ is an eventually decreasing sequence of positive real numbers) $\Rightarrow$ (L.B. of $\{a_n\}$ is equal to 0)

and

($\{a_n\}$ is an eventually decreasing sequence of positive real numbers) $\Rightarrow$ (G.L.B. of $\{a_n\}$ is equal to 0)

is a theorem schema.

(v) Following upon an unsatisfactory preliminary discussion of the concept of limit of a real valued sequence $\{a_n\}$, the authors present on page 70 what is apparently their final precise definition, thus (I quote) :

(You may choose to use either the less formal definition as in (a), or the more stringent definition in (b).)

(a) A sequence $\{a_n\}$ has limit L if $|a_n - L|$ may be made as small as we please by taking n large enough ($n \in J^+$) .

(b) A sequence $\{a_n\}$ has limit L if

for every $\varepsilon > 0$ (however small)

$|a_n - L| < \varepsilon$ for all $n > N$ $(n \in J^+)$.

Something akin to (a) is often presented as an admittedly vague but homely explanation of what is meant by "the sequence $\{a_n\}$ has limit L " , to be clarified in due course by something more precise, along the lines of (b). It is, however, rather dangerous to suggest (as is suggested by the opening parenthesised sentence) that (a) and (b) are equivalent. The wording of (a) is in fact so loose that it might reasonably be interpreted as follows :

A sequence $\{a_n\}$ has limit L if, for every $\varepsilon > 0$ and every $N \in J^+$, $|a_n - L| < \varepsilon$ for some (at least one) $n \in J^+$ such that $n > N$.

With this interpretation of (a), (a) and (b) are not equivalent (unless set theory is contradictory).

Leaving this aside, consider (b) , which the authors seemingly intend to take priority. Here trouble arises through the intervention of N which, for all that is indicated to the contrary, may be interpreted as a free variable, or at least (and more probably) as a variable which is free save for the restriction $N \in J^+$. With either interpretation, it is easily derivable that, if $\{a_n\}$ has limit L , then $a_n = L$ for every $n \in J^+$ such that $n > 0$. This conclusion would (to say the very least) occasion surprise and dismay in conventional mathematics. Compare this with the discussion in I.2.9(ix) above.

(vi) On page 92 it is stated

A series $p_1 + p_2 + p_3 + \cdots + p_n + \cdots$

(of positive terms) is convergent if

$p_{n+1}/p_n < r < 1$ holds eventually.

It is, of course, vital that here r must be choosable "independent of n " , and this should be advertised more clearly.

(vii) The authors' proposed definitions of closure (page 240), associativity (page 241) and identity element (page 242) are again ambiguous to a serious degree.

Thus, for example, in relation to a set G and a function $(a, b) \rightsquigarrow a * b$ with domain $G \times G$, they supposedly define

$$G \text{ is closed under } * \qquad (5)$$

by writing (I quote again)

Mathematically, if a , $b \in G^*$ and $a * b = c$, then the set is closed under the operation $*$ if $c \in G$. (6)

Herein, G* is presumably a misprint for G ; and (again presumably) the set referred to is G . Apart from this, it is at least a reasonable presumption that the intended formal substance of (5) is expressible as

$$(\underline{a} \in G \wedge \underline{b} \in G \wedge \underline{c} = \underline{a} \times \underline{b}) \Rightarrow (\underline{c} \in G \Rightarrow G \text{ is closed under } \times) \quad , \qquad (7)$$

wherein (perhaps) $\underline{a}$, $\underline{b}$ and $\underline{c}$ denote distinct letters not appearing in G or $\times$. Then, however, (5) gains a significance quite different from the conventional one.

A formal version of the conventional understanding of (5) would be

$$G \text{ is closed under } \times \equiv (\forall\underline{a})(\forall\underline{b})((\underline{a} \in G \wedge \underline{b} \in G) \Rightarrow (\underline{a} \times \underline{b} \in G)) \qquad (8)$$

where $\underline{a}$ and $\underline{b}$ denote distinct letters not appearing in G or $\times$; the universal quantifiers appearing in (8) should not be omitted.

In comparing (7) and (8) , one should note the theorem schema

$$(\forall\underline{c})((\underline{\underline{A}} \wedge (\underline{c} = S)) \Rightarrow \underline{\underline{B}}) \Leftrightarrow (\forall\underline{c})((S|\underline{c})\underline{\underline{A}} \Rightarrow (S|\underline{c})\underline{\underline{B}}) \quad ,$$

$\underline{c}$ denoting a letter not appearing in S ; see Problem II/37.

(Incidentally, there is another intentionally totally different concept of " G is closed" which finds use in mathematics; see VIII.4.1 below.)

In none of the preceding criticisms of (DO) is the main issue the lack of fully formal expressions (which would assuredly be inappropriate in this context). The issue and complaint is that the adopted conventionally informal expressions are intolerably vague and ambiguous.

Incomplete though the above list of criticisms is, enough has been said to illustrate the pressing need for care and also the frequency with which that care is lacking.

Although the informal development described in Chapters 1 - 3 of Gleason (1) is similar in many respects to those discussed in I.4.2 - I.4.4 above and in the present subsection, he cannot be justly criticised in the same way. This is because he is at pains to make it quite clear that he is concerned with the informal development, makes no claims to be presenting a formal system, and does not hide the differences; see loc. cit., p.8, lines 10 - 15. (Note, however, that he attaches a less exacting meaning to the term "semiformal" than is done in I.2.9(vi) and I.3.4(i).) Moreover, and what is perhaps even more important, he is not guilty of the gross ambiguities in the informal development which are incorporated in some of the accounts mentioned in I.4.2 - I.4.5. (This happy circumstance is probably the outcome of Gleason's extensive and deep experience of informal mathematics combined with excellent intuition. Nevertheless, formal blunders can still occasionally intrude; see V.5.4 below.)

Certain portions of my own writings give me cause to blush, even though they feature in books making no attempt to expound foundational matters and even though the shortcomings are probably no worse than those to be found in most books on similar topics at a similar level. Thus, many criticisms may fairly be levelled against Section 0.1 of Edwards (1), despite the purely summary-like nature of that Section and the clear statement on page 1 of the book to the effect that the adopted framework is naive set theory. The present book incorporates incidentally an analysis of these shortcomings. Another example (pointed out to me by Dr Jeff Sanders) graces page 174 of the same book, where I write

Suppose that $T = \{t\}$ is a topological space.

This sentence may quite fairly be interpreted in a way which I did not intend and which would then promote confusion. (Concerning $\{t\}$, see II.4.1 below.) A similar instance appears in the first complete paragraph on page 52 of Volume I of Edwards (2). (This shortcoming is eliminated from the forthcoming revised edition of that book.) There are doubtless other similar blunders. Regarding the analysis of the last two shortcomings, it will suffice to say this : the first incorporates a commonly used device intended to indicate briefly a notational convention (that t is to denote a "generic" element of T); the second incorporates a compressed form of wording which is open to misinterpretation on a technical issue (the reader may be led to believe that I am asserting something which I do not intend to assert) which lies outside the technical scope of the present book.

I.4.6 <u>Degrees of formalisation</u> Almost every teacher or author dealing with mathematics will, often without much conscious thought, adopt at least a vestige of what appears to be naive formalism, vaguely based though it usually is. Many of those who purport to be advancing the cause of the "new mathematics" and "better understanding" speak or write with some pride of "mathematical language" as opposed to "everyday language" . The degree and standard of formalisation adopted may, as it often must, vary in deference to the aims and the expected audience. The outcome of a conventional degree and standard is what I repeatedly refer to as informal or conventional mathematics.

If the teacher or author works all the time with skill and care, the outcome can be an overall success, at least for the limited aims and audience in view. Otherwise, the outcome is less happy. The trouble is that even a small degree of formalisation, if upheld to the letter, sooner or later imposes annoying restrictions. What often happens is that the teacher or author rejects the strictures on occasions where they seem to be a nuisance or to conflict with a less formal viewpoint. The outcome is then confusion in the mind of any student or reader who has accepted and absorbed the degree and standard of formalisation initially adopted, unless the teacher or author takes special care to explain the apparent change of rules and standards. Such changes ought to be seen and recognised by the student or reader, even if fully detailed explanations are judged to be out of place. It is intolerable that definitions should be materially altered without notice; or that fallacious reasoning (however intuitively plausible it may appear) should be presented as valid.

The contents of I.4.2-I.4.5 illustrate the outcome of the lack of care on the part of authors.

High school and tertiary mathematics do not consist solely of the formation of definitions and the construction of logical arguments: experiment and conjecture play vital roles. But definitions and logical arguments are certainly major components and, when they <u>are</u> being attempted, great care is essential.

I.5 Summary of the chapter

The following is a list of principal features and concepts.

1. An <u>alphabet</u> of <u>primitive signs</u> of the formal language. Logical signs $\neg$, $\underline{\vee}$, τ , $\square$ and logical letters $\underline{x}$, $\underline{y}$, $\underline{z}$, ... ; one mathematical sign $\underline{\in}$. See I.1.1. These are later supplemented or replaced by semiformal signs $\vee$, $\wedge$, $\Rightarrow$, $\Leftrightarrow$, $\in$, $\exists$, $\forall$; see I.1.7.

2. <u>Strings</u> These are (finite) sequences of primitive signs, in which some appearances of the sign τ may be linked by ties with certain appearances of the sign $\square$. See I.1.2.

3. Two kinds of strings are described. Strings of one kind are termed <u>(formal) sentences</u>; strings of the other kind are termed <u>sets</u>. See I.1.6. At this stage there is no assumed concept of truth of sentences.

4. A <u>theory</u> is specified by selecting (i) at most finitely many

(perhaps none) sentences, called the explicit axioms of that theory; and (ii) finitely many sentence schemas called axiom schemas of that theory. Each instance of an axiom schema of the theory is called an implicit axiom of that theory. See I.2.1-I.2.3.

For the version of set theory discussed in this book, the axiom schemas are S1 - S5 in I.2.2, S6 and S7 in II.1.2, and S8 in II.12.1. The three explicit axioms are listed in II.12.2.

5. Certain finite sequences of sentences, the initial terms of which are axioms (explicit or implicit) of a chosen theory, are termed proofs of (or in) that theory; see I.2.4. Proofs are syntactically defined (no resort to truth tables).

6. Sentences which appear in proofs of a given theory are said to be provable, or to be true, or to be theorems, in that theory; see I.2.4. A sentence is false in a given theory, if and only if its negation is true in that theory.

7. In a given theory, certain sentences may be neither true nor false, that is, may be undecidable in that theory. Certain sentences may be both true and false in that theory. If the latter situation arises, the theory is contradictory or inconsistent: otherwise the theory is consistent; see I.2.7. Set theory is believed to be consistent, but no conclusive "proof" is known.

8. The description of the formal language, and its properties, is effected in a metalanguage, statements in which are termed metastatements. Intuitively true metastatements are metatheorems (examples are the replacement rules in I.1.4 and I.1.8 and the proof methods in I.3.2 and I.3.3). Sometimes it is absolutely vital to distinguish between the metalanguage and the formal language.

Formal sentences are a priori meaningless, but they are in practice endowed with intuitive meanings. This is more than a mere convenience, since it is the basis on which intuitive ideas influence formal developments (see I.1.7). On the other hand, this "translation" has subjective elements which can lead to trouble (see I.2.9, I.3.8 and II.3.9). Experience has shown that complete reliance on intuition can lead to inconsistency, and the role of a background formal theory is to act as a regulator. One has to struggle to preserve a balance.

In this book, there is initially an emphasis on the formalities. Then, gradually, the emphasis shifts towards what is the conventional style.

9. Proof methods The essence of the remainder of this chapter comprises the proof methods and their use. A proof method is a (meta)theorem about theorems in the formal theory; it is not itself a part of the formal theory. It may also be regarded as a rule of mathematical logic. The simplest proof method, which is also quite typical, reads

> If $\underline{\underline{A}}$ and $\underline{\underline{B}}$ denote arbitrary sentences, and if (the sentences denoted by) $\underline{\underline{A}}$ and $\underline{\underline{A}} \Rightarrow \underline{\underline{B}}$ are both theorems, then (the sentence denoted by) $\underline{\underline{B}}$ is also a theorem.

Each proof method is an ultracondensed expression of prohibitively lengthy sections of formal proofs which have a certain pattern in common and which reappear over and over again. In principle, such proof methods could be eliminated. But in practice resort to them is virtually necessary in order to achieve adequate descriptions, within practically possible spans of time and space, of formal proofs.

Chapter II. Elements of Set Theory

Introduction Most readers for whom this book is primarily intended will realise that sets are in some way the formal, theoretical counterparts in mathematics of collections. The formal theory recognises and is governed by this relationship to the extent that, initially at any rate, one makes definitions suggested by one's informal feeling for collections and one looks for theorems likewise suggested by intuitive knowledge. (On the other hand, of course, the proofs of the suspected theorems have to correspond more or less closely with the rules and formalities of the theory; and, as one progresses, intuition comes to be tempered by technical experience and know-how.) Because of this, attempts will sometimes be made to illustrate and motivate theoretical developments by looking at simple illustrative examples expressed in terms of collections, though naturally readers of this book will not be so dependent upon this as will high school students themselves.

It therefore comes about that for illustrative purposes alone, and on a purely informal basis, I will from time to time refer to collections of "basic objects" , collections of such collections, and so on. As basic objects I might as well take the numerals 0, 1, 2, ... (or the numbers they denote), again thought of intuitively; or these numerals could be used as labels for real objects. For many illustrative purposes, it will indeed suffice to think only of finite collections, that is, collections whose members are - or could, given sufficient time and patience, be - explicitly indicated one by one. Thus, in this context, I will employ the usual braces symbol, as in $\{3, 2, 19, 6, 10\}$, to denote the collection whose members are the numerals or collections exhibited between the outermost braces and no other things. Similarly, for example, $\{1, 5, \{1, 3\}, \{2, 4, 6\}\}$ will denote the collection whose members are: 1 , 5 , the collection $\{1, 3\}$, and the collection $\{2, 4, 6\}$. At the same time, I will (still informally) use $\in$ to denote the relationship of an object to a collection of which it is a member. Thus, for example

$$2 \in \{3, 2, 19\} \ ,$$
$$\{1, 3\} \in \{1, 5, \{1, 3\}\} \ ,$$

while

$$4 \notin \{3, 2, 19\} \ ,$$
$$2 \notin \{1, 5, \{1, 3\}\} \ .$$

(The reader may be amused to note that he is probably intuitively at a loss to decide at this stage whether $0 \in 1$ is true or false; the question will, of course, remain unresolved until 0 and 1 have been formally defined as sets. See Chapter V.)

On the matter of the intuitive use of collections, it should always be borne in mind that one has direct and reasonably comprehensive knowledge only of finite collections, and indeed only of collections having a relatively small number of members. In the formal theory there is room (in fact, a need) for both finite and infinite sets. In respect of the latter, one's intuitive feelings are much less reliable.

Again on an informal basis, and for similar purposes, the reader may wish to use Venn diagram type illustrations. (The Introduction to Section II.6 includes a discussion in relation to which such diagrams may be thought to be helpful.) As a rule, such diagrams will not be provided : most readers for whom this book is primarily intended will be capable of supplying them. If the reader chooses to do this, he should devote some thought to the matter and exercise some caution; see the remarks concerning the use of diagrams to be found in the Mise-en-scène in Volume 2.

The reader is reminded that from here on I shall diverge more and more from Bourbaki (1) on many points of definition and detail.

The reader is reminded of what has been said in Chapter I, especially I.3.4(ii), regarding proofs (or, better, descriptions of proofs). These incorporate over and over again English phrases of the form

By (or : using) _________, _________ and _________

and

As a corollary of _________, _________ and _________

and

From _________, _________ and _________ it follows (or : it may be deduced or derived) .

Each such pronouncement refers to a proof or subproof which would involve, not only the metatheorems and theorems referred to in _________, _________ and _________, but usually numerous others in addition. The references actually cited are merely those which are judged to be adequate to indicate a path the reader could construct and follow. On occasions, the reader should take the trouble to fill in the missing details, all the way back to the metatheorems and theorems in Chapter I.

II.1 Equality of sets

II.1.1 *Formal definition of equality* Intuitively, two collections are regarded as equal if and only if every object which belongs to (that is, is a member of) either one also belongs to the other. For example, the collection $\{1, 2, \{1, 2\}\}$ is equal to the collection $\{2, \{1, 2\}, 1\}$, but is not equal to the collection $\{\{1, 2\}, 3\}$. This at once suggests that, if X and Y denote sets, $X = Y$ should be defined to denote, or to be a name for, the formal sentence otherwise denoted by

$$(\forall \underline{x})(\underline{x} \in X \Leftrightarrow \underline{x} \in Y) \quad ,$$

wherein $\underline{x}$ is a letter. However, in view of the remarks in I.3.5(ii) one will take pains to ensure that this last denotes a unique sentence in the formal language, as soon as X and Y are specified sets; that is, that the allowed choice of the letter $\underline{x}$ is immaterial. Thus one will need to restrict the letter $\underline{x}$ appearing above in such a way that the sentence indicated is independent of the permitted choice of $\underline{x}$ (or at the very least that only equivalent sentences arise from the permitted choices of $\underline{x}$). Now the replacement rule (e) in I.1.8 ensures that this will be so, provided $\underline{x}$ is restricted to being a letter which <u>does not appear</u> in X or Y. As will be seen in the remarks below, this restriction upon $\underline{x}$ is desirable for yet other reasons.

Thus I will formally and unconditionally define

$$X = Y \equiv_{\text{def}} (\forall \underline{x})(\underline{x} \in X \Leftrightarrow \underline{x} \in Y) \quad , \tag{1}$$

where X and Y denote arbitrary strings and $\underline{x}$ any letter which does not appear in X or Y. (What is involved here is a definition schema, rather than an individual definition; see I.3.5(ix).)

With this definition, $X = Y$ denotes a string whenever X and Y denote strings; it denotes a sentence if X and Y denote sets.

(I am choosing to define $=$. Sometimes $=$ is taken alongside $\underline{\in}$ as a second primitive mathematical sign. If that is done (and the change to $\in$ is made as in I.1.7(i)), the balance is restored by adopting a corresponding explicit axiom, called the Axiom of Extensionality. This axiom is the sentence denoted by

$$(\forall \underline{x})(\forall \underline{y})((\forall \underline{z})(\underline{z} \in \underline{x} \Leftrightarrow \underline{z} \in \underline{y}) \Leftrightarrow (\underline{x} = \underline{y})) \quad .)$$

The reader is reminded of the metatheorem (see I.3.5(iii)):

If X and Y denote sets, and if $X \equiv Y$,

then $X = Y$ is a theorem.

Implicit use of this metatheorem will be widespread, but rarely if ever is the use made explicit.

Remarks Although many (perhaps most) writers omit reference to the stated restriction on $\underline{x}$ in (1) , it is fairly certain that either the omission is an oversight, or that these writers judge (wrongly, I think, cf. II.4.2(i) and (ii) and II.6.7) that this restriction and others similar to it are always clearly and correctly dictated by intuition and therefore deserve no mention. (This omission seems to be almost characteristic of accounts of informal set theory, even when it is presented in naive axiomatic style; see I.0.6.) If one violates the restriction, one can be led to paradoxes. These are not genuine contradictions in the formal theory (from which definitions may, and ultimately must, be eliminated); rather, they are indicators of errors in the metamathematical commentary on the formal theory. (Compare the remarks in the fifth paragraph of I.2.7 referring to fallacious arguments alleged to establish the inconsistency of the formal theory.) The crux of the trouble here is the fact that one can exhibit (see Problem II/1) letters $\underline{x}$ and sets X and Y (for example, $X \equiv \{\underline{x}\}$ and $Y \equiv \{\underline{x}\} \cup \{\emptyset, \{\emptyset\}\}$) such that the sentence

$$(\forall \underline{x})(\underline{x} \in X \Leftrightarrow \underline{x} \in Y) \Rightarrow (X = Y)$$

is false; a fortiori, the sentence

$$(\forall \underline{x})(\underline{x} \in X \Leftrightarrow \underline{x} \in Y) \Leftrightarrow (X = Y)$$

is false.

See also the discussion in II.1.3(i) below.

It should also be noted that one has not defined the sign $=$ standing in isolation, nor $X =$, nor $= Y$: that is, no recipe has been given which prescribes a formal sentence denoted by $=$, $X =$, or $= Y$. One therefore has to make sure that these combinations of symbols never appear in isolation.

A consequence of (1) and I.3.5(iii) is that one has the theorem schema (cf. I.2.9(v))

$$\text{T!} \quad : \qquad (X = Y) \Leftrightarrow (\forall \underline{x})(\underline{x} \in X \Leftrightarrow \underline{x} \in Y) \ , \qquad (1_1)$$

wherein X and Y denote arbitrary sets and $\underline{x}$ any letter not appearing in X or Y . (Every similar formal difinition has attached to it in like fashion a theorem schema, of which explicit mention will often be omitted.)

In place of $\neg(X = Y)$, one almost invariably writes $X \neq Y$. In other words,

$$X \neq Y \equiv_{def} \neg(X = Y) \quad .$$

In an attempt to avoid ambiguities and facilitate the reading, $(X) = (Y)$ will sometimes be written in place of $X = Y$. The same applies later in connection with $\subseteq$, $\cup$, $\cap$, $\times$, et cetera. (Parentheses used this way are not intended as essential parts of names of sets or sentences - as they are on other occasions, as for example in $\tau_{\underline{x}}(\underline{\underline{A}})$, $(\exists \underline{x})\underline{\underline{A}}$, $f(\underline{x})$, et cetera.)

II.1.2 Two schemas related to equality In connection with equality thus defined set theory Θ_0 adopts two schemas, namely:

$$\text{S6} \qquad (S = T) \Rightarrow ((S|\underline{x})\underline{\underline{A}} \Leftrightarrow (T|\underline{x})\underline{\underline{A}}) \quad ;$$

$$\text{S7} \qquad (\forall \underline{x})(\underline{\underline{A}} \Leftrightarrow \underline{\underline{B}}) \Rightarrow (\tau_{\underline{x}}(\underline{\underline{A}}) = \tau_{\underline{x}}(\underline{\underline{B}})) \quad ;$$

herein $\underline{\underline{A}}$ and $\underline{\underline{B}}$ denote arbitrary sentences, S and T denote arbitrary sets, and $\underline{x}$ denotes an arbitrary letter.

(Strictly speaking, the schemas referred to are the outcomes of eliminating the defined symbols $\Rightarrow$, $\Leftrightarrow$ and $=$ from the above semiformal expressions, thereby leaving only signs of the primitive alphabet together with the set symbols S and T and the sentence symbols $\underline{\underline{A}}$ and $\underline{\underline{B}}$.)

Of these, S6 is intuitively used almost without thought; S7 rarely if ever appears explicitly in conventional mathematical texts and is thus obtrusive by its strangeness. Both schemas are in fact quite vital to formal developments. For instance, when (in IV.1.3) values of functions are defined, one will intuitively expect that $f = g$ will imply $f(\underline{x}) = g(\underline{x})$ for all $\underline{x}$: the proof of this falls back on appeals to both S6 and S7 . Both schemas may also be involved when (as frequently happens in practice) one wishes to interchange two formal definitions on the basis of certain equalities. Such interchanges would not arise in strictly formalised mathematics - at least, not without the justification appearing through explicit use of S6 and/or S7 on every occasion. But they occur frequently in informal mathematics, the justification often being left without comment : "equal" definitions of sets (which are more often than not spoken of as equivalent definitions) are (like equivalent definitions of sentences) treated as if they were identical definitions. (A typical example appears in VI.6.8.)

It may be noted that the theorem schema II.1.1(1_1) is a prototype case of S6 , from which it results by substitution in the sense of I.3.9(ii).

II.1.3 Remarks The substance of II.1.1 and II.1.2 raises numerous points which I discuss in turn.

(i) What has been said in II.1.1 and the Remarks following it, relating to the restriction that, in (1) , $\underline{x}$ should not appear in X or Y , shows that violation of this restriction may lead one to conclude that (for example)

$$\{\underline{x}\} = \{\underline{x}\} \cup \{\emptyset, \{\emptyset\}\}$$

is true, a conclusion which is intuitively "wrong" and unwanted. This conclusion is the outcome of attaching the same name to nonidentical sets, a procedure which is always and everywhere perilous (see I.3.5(ii)); it is ultimately harmless only if the nonidentical sets are in fact equal. The conclusion is not an indicator of inconsistency in formal set theory. In fact, as may be simply verified quite independently, if set theory is consistent, the above sentence is not a theorem of set theory.

This illustrates the imperative need at times for metamathematical restrictions on the choice of letters, which restrictions are rarely (if ever) made explicit in informal mathematics. The claim that intuition will always protect one from the possible ambiguities and paradoxes is one with which few mathematicians, if pressed, would pretend satisfaction.

See also II.2.1 below.

Other ambiguities also arise. Informal texts (for example, Halmos (1), p.2 and Bartle and Ionescu Tulcea (1), p.7) are often not sufficiently specific about the significance of X = Y , and may suggest that it is to be understood to denote

$$\underline{x} \in X \Leftrightarrow \underline{x} \in Y \quad , \qquad (1_2)$$

possibly with (but usually without) the stricture that $\underline{x}$ shall not appear in X or Y . It will usually be an unspoken assumption that $\underline{x}$ is a "variable" (though it is hardly ever explained what this means). (1_2) differs from the right hand number of (1) or (1_1) by omission of the quantifier $(\forall \underline{x})$ (see I.2.9(ix)). Formally, this omission is illegitimate; and this notwithstanding the fact that, in a theory of which $\underline{x}$ is a variable, (1_2) is true if and only if $X = Y$ is true. More specifically, although

$$(\forall \underline{x})(\underline{x} \in X \Leftrightarrow \underline{x} \in Y) \Rightarrow (X = Y)$$

is a theorem of Θ_0 for arbitrary sets X and Y in which the letter $\underline{x}$ does not appear, one can exhibit sets X and Y and a letter $\underline{x}$ not appearing in X or Y, such that

$$(\underline{x} \in X \Leftrightarrow \underline{x} \in Y) \Rightarrow (X = Y) \qquad (1_3)$$

is not a theorem of Θ_0 (assumed to be consistent). Indeed, if (1_3) were a theorem of Θ_0, I.3.2(IX) entails

$$(\exists \underline{x})(\underline{x} \in X \Leftrightarrow \underline{x} \in Y) \Rightarrow (X = Y)$$

would also be a theorem; but then, "choosing $\underline{x} \notin X \cup Y$" (see II.3.10(36)), I.3.2(I) would entail that $X = Y$ is a theorem of Θ_0; if X and Y are chosen so that $X \neq Y$ is true in Θ_0 ($X \equiv \emptyset$, $Y \equiv \{\emptyset\}$ for example; see II.4 and II.5), a metamathematical contradiction emerges.

(ii) Some writers might go about defining $X = Y$ (and other components of the formal language) in an alternative way leading to the same outcome; cf. I.3.5(vi). They would begin with the case in which X and Y are merely distinct letters $\underline{x}$ and $\underline{y}$, agreeing that then

$$\underline{x} = \underline{y} \equiv (\forall \underline{z})(\underline{z} \in \underline{x} \Leftrightarrow \underline{z} \in \underline{y}) \quad ,$$

where $\underline{z}$ is any letter different from $\underline{x}$ and $\underline{y}$ (which proviso would in all probability be taken as being made clear by the typography itself). The replacement rule I.1.8(e) ensures that the choice of $\underline{z}$ is immaterial, subject to this condition. The definition would then be extended by agreeing that, for any two strings X and Y (not necessarily distinct), $X = Y$ shall denote the string which results from the <u>simultaneous</u> replacement, in the string denoted by $\underline{x} = \underline{y}$, of $\underline{x}$ by X and $\underline{y}$ by Y : in symbols,

$$X = Y \equiv (X|\underline{x}, Y|\underline{y})(\underline{x} = \underline{y})$$

$$\equiv (X|\underline{x}, Y|\underline{y})(\forall \underline{z})(\underline{z} \in \underline{x} \Leftrightarrow \underline{z} \in \underline{y}) \quad ,$$

cf. Problem I/17. One has here to reckon with the possibility that $\underline{x}$ and/or $\underline{y}$ may appear in either or both of X and Y. To cover such cases, it is in practice important on occasions to be able to express simultaneous replacements in terms of iterated single replacements of one letter at a time; cf. Problem I/21.

A way in which this can be done has been indicated in Problem I/17. Thus, suppose that A, X and Y denote arbitrary strings and $\underline{x}$ and $\underline{y}$ distinct letters, and that $(X|\underline{x}, Y|\underline{y})A$ denotes the string obtained by simultaneous replacement of $\underline{x}$ by X and $\underline{y}$ by Y wherever they appear in the string denoted by A; if $\underline{x}'$ and $\underline{y}'$ denote distinct letters, each different from both $\underline{x}$ and $\underline{y}$ and neither appearing in A, X or Y, then (see Bourbaki (1), p.17 and §2 of the Appendix below)

$$(X|\underline{x}, Y|\underline{y})A \equiv (X|\underline{x}')(Y|\underline{y}')(\underline{x}'|\underline{x})(\underline{y}'|\underline{y})A \qquad (*)$$

where on the right one has only iterated single replacements.

Thus, taking $A \equiv (\forall \underline{z})(\underline{z} \in \underline{x} \Leftrightarrow \underline{z} \in \underline{y})$, one is led finally to the definition

$$X = Y \equiv (X|\underline{x}')(Y|\underline{y}')(\underline{x}'|\underline{x})(\underline{y}'|\underline{y})(\forall\underline{z})(\underline{z} \in \underline{x} \Leftrightarrow \underline{z} \in \underline{y}) ,$$

where $\underline{x}'$ and $\underline{y}'$ are distinct letters, different from $\underline{x}$, $\underline{y}$, $\underline{z}$ and appearing neither in X nor in Y. Two applications of I.1.8(f) show that the last-written string is identical with

$$(X|\underline{x}')(Y|\underline{y}')(\forall\underline{z})(\underline{z} \in \underline{x}' \Leftrightarrow \underline{z} \in \underline{y}') .$$

If one supposes (as again one may) that $\underline{z}$ appears neither in X nor in Y, two further applications of I.1.8(f) show, since $\underline{x}'$ and $\underline{y}'$ have been assumed not to appear in either X or Y, that one has finally the string

$$(\forall\underline{z})(\underline{z} \in X \Leftrightarrow \underline{z} \in Y) .$$

In view of I.1.8(e), this agrees perfectly with the definition (1). Notice that once again the restriction that $\underline{z}$ appears neither in X nor in Y has been imposed and utilised.

(In spite of this, if one chooses a letter $\underline{z}$ different from $\underline{x}$ and $\underline{y}$ and not appearing in X or Y, one finds that

$$\begin{aligned}(X|\underline{x})(Y|\underline{y})(\underline{x} = \underline{y}) &\equiv (X|\underline{x})(Y|\underline{y})(\forall\underline{z})(\underline{z} \in \underline{x} \Leftrightarrow \underline{z} \in \underline{y}) \\ &\equiv (X|\underline{x})(\forall\underline{z})(\underline{z} \in \underline{x} \Leftrightarrow \underline{z} \in Y) && (\text{I.1.8(f)}) \\ &\equiv (\forall\underline{z})(\underline{z} \in X \Leftrightarrow \underline{z} \in (X|\underline{x})Y) && (\text{I.1.8(f)}) \\ &\equiv X = (X|\underline{x})Y , && (\text{II.1.1(1)})\end{aligned}$$

which may not be identical with $X = Y$ (though it is whenever $\underline{x}$ does not appear in Y). Thus one must take care when contemplating a switch from $(X|\underline{x})(Y|\underline{y})$ ..... to $(X|\underline{x}, Y|\underline{y})$ .... .)

Similar remarks apply to definitions which appear later, where both approaches are used at various times (cf., for example, II.4.1, II.6.1, II.6.2, III.1.1, III.1.2, IV.1.4(vii)).

(iii) The rule (*) makes it possible to verify that

$$(X|\underline{x}, Y|\underline{y})A \equiv (X|\underline{x}', Y|\underline{y}')A'$$

where $A' \equiv (\underline{x}'|\underline{x})(\underline{y}'|\underline{y})A$ and $\underline{x}, \underline{y}$ and $\underline{x}', \underline{y}'$ are any two pairs of distinct letters, $\underline{x}'$ and $\underline{y}'$ not appearing in A. (This is the counterpart of (a) in I.1.4.) This fact is needed to ensure (for example) that $X = Y$, as defined in (ii) above, depends only on the strings X and Y and not on choice of the distinct letters $\underline{x}, \underline{y}$.

In general, when one has by means of a definition introduced an abbreviatory symbol Ω for a certain string B, it is tacitly agreed to use the symbol obtained by writing A in place of $\underline{x}$ wherever the latter occurs in Ω to denote the string $(A|\underline{x})B$; sometimes, indeed, one may for the same purpose use the symbol obtained by writing in place of each occurrence of $\underline{x}$ in Ω, a symbol which abbreviates A. In other words, one may apply abbreviation conventions to previous abbreviations to further compress the notation. Care is often needed; see the end of II.3.7 and the end of III.1.1.

For example, in III.1.2, I make the definition

$$\underline{z} \text{ is an ordered pair } \equiv_{\text{def}} (\exists\underline{x})(\exists\underline{y})(\underline{z} = (\underline{x}, \underline{y})) ,$$

where $\underline{x}$ and $\underline{y}$ denote distinct letters, each different from $\underline{z}$, their choice being otherwise immaterial; from then on it will be a tacit convention that, for an arbitrary string A ,

$$A \text{ is an ordered pair} \equiv (A|\underline{z})(\underline{z} \text{ is an ordered pair})$$

$$\equiv (A|\underline{z})(\exists\underline{x})(\exists\underline{y})(\underline{z} = (\underline{x}, \underline{y})) \ .$$

Since here one may choose $\underline{x}$ and $\underline{y}$ to be letters not appearing in A , I.1.8(f) shows that accordingly

$$A \text{ is an ordered pair} \equiv (\exists\underline{x})(\exists\underline{y})(A|\underline{z})(\underline{z} = (\underline{x}, \underline{y}))$$

$$\equiv (\exists\underline{x})(\exists\underline{y})(A = (\underline{x}, \underline{y})) \ ,$$

the last step because $\underline{x}$ and $\underline{y}$ are supposed each to be different from $\underline{z}$, so that $\underline{z}$ does not appear in (the string denoted by) $(\underline{x}, \underline{y})$.

On occasions this tacit convention can lead to confusions which are then averted by ad hoc devices ... such as the insertion of parentheses to replace $\underline{x}$ by (A) rather than by A . For instance, $\underline{x} \cap \underline{y}$ denotes a string in which the letter $\underline{y}$ appears: if one replaces $\underline{y}$ by the string $A \cup B$, the resulting string would usually be denoted by $\underline{x} \cap (A \cup B)$ rather than by $\underline{x} \cap A \cup B$ (since the latter is open to confusion). Once again it is to be noted that parentheses used in this way have to be distinguished from parentheses added to the semiformal language via formal definitions, as for example in the functional notation $f(\underline{x})$ introduced in IV.1.3 below.

Also note again that the matters just discussed do not refer to the internal working of the formal theory, but rather to descriptions of and reports about the said theory.

(iv) One further piece of notation (hinted at in I.1.6) suggests itself at this stage. In cases where (as in (iii) above) one has a string A and one or two distinct letters $\underline{x}$ and $\underline{y}$ in which one is temporarily especially interested, one sometimes writes $A[\![\underline{x}]\!]$ or $A[\![\underline{x}, \underline{y}]\!]$ in place of A . (This is not meant to imply that $\underline{x}$ or $\underline{x}$ and $\underline{y}$ do in fact appear in A , though this will usually be the case when this notation is contemplated; nor that they are the only letters appearing in A .) If this be done, one will also write $A[\![X]\!]$ and $A[\![X, Y]\!]$ for $(X|\underline{x})A$ and $(X|\underline{x}, Y|\underline{y})A$ respectively. This notation often helps in the intuitive reading and interpretation of sentences.

In practice one sometimes finds $A_{\underline{x}}$ or $A(\underline{x})$ written instead of $A[\![\underline{x}]\!]$, and $A_{\underline{x},\underline{y}}$ or $A(\underline{x}, \underline{y})$ written instead of $A[\![\underline{x}, \underline{y}]\!]$; see, for example, the discussion in V.6.2. As was remarked in I.1.6, this is formally objectionable since these notations clash with the formalises versions of standard notations used in connection with functions and sequences. If A and X denote strings, one should not confuse A⟦X⟧ (defined as above) with A(X) (as defined formally in IV.1.3 below) - and this not even when A denotes the special sort of set termed a function and X denotes a set which is an element of the domain of that function; see IV.1.4(ii).

(v) The introduction of the equality sign may be taken to illustrate the imperative practical need for the use of abbreviations. Thus one may embark on the following procedure designed to lead back from $\underline{x} = \underline{y}$ to the primitive language:

$$(\underline{x} = \underline{y}) \equiv (\forall\underline{z})(\underline{z} \in \underline{x} \Leftrightarrow \underline{z} \in \underline{y})$$

$$\equiv (\forall\underline{z})((\underline{z} \in \underline{x} \Rightarrow \underline{z} \in \underline{y}) \wedge (\underline{z} \in \underline{y} \Rightarrow \underline{z} \in \underline{x}))$$

$$\equiv (\forall\underline{z})\neg(\neg(\underline{z} \in \underline{x} \Rightarrow \underline{z} \in \underline{y}) \vee \neg(\underline{z} \in \underline{y} \Rightarrow \underline{z} \in \underline{x}))$$

$$\equiv (\forall\underline{z}) \neg \underline{\vee} \neg \underline{\vee} \neg (\underline{z} \in \underline{x})(\underline{z} \in \underline{y}) \neg \underline{\vee} \neg (\underline{z} \in \underline{y})(\underline{z} \in \underline{x})$$

$$\equiv (\forall\underline{z}) \neg \underline{\vee} \neg \underline{\vee} \neg \underline{\in} \underline{z} \underline{x} \underline{\in} \underline{z} \underline{y} \neg \underline{\vee} \neg \underline{\in} \underline{z} \underline{y} \underline{\in} \underline{z} \underline{x}$$

$$\equiv \neg (\exists\underline{z}) \neg \neg \underline{\vee} \neg \underline{\vee} \neg \underline{\in} \underline{z} \underline{x} \underline{\in} \underline{z} \underline{y} \neg \underline{\vee} \neg \underline{\in} \underline{z} \underline{y} \underline{\in} \underline{z} \underline{x} \quad ;$$

at this point one has to introduce the definition of $(\exists\underline{z})$ in terms of τ, which leads to

$$\neg (\tau_{\underline{z}}(\ldots\ldots)|\underline{z}) \neg \neg \underline{\vee} \neg \underline{\vee} \neg \underline{\in} \underline{z} \underline{x} \underline{\in} \underline{z} \underline{y} \neg \underline{\vee} \neg \underline{\in} \underline{z} \underline{y} \underline{\in} \underline{z} \underline{x} \quad ,$$

where the ellipsis indicates the string appearing after the last parenthesis. Patience gives out here and I content myself with the remark that the string finally obtained is of length 106, which is already somewhat cumbersome to handle conveniently. But this is only the very beginning and much worse than this happens: see the comments on the Pairing Axiom in II.12.2 and the end of V.2.1.

(vi) The reader should notice the replacement rule for = which ensues from the earlier ones in I.1.4 and I.1.8 coupled with the definition of equality. This rule reads:

(p) If S, X and Y denote arbitrary strings and $\underline{y}$ an arbitrary letter, then

$$(S|\underline{y})(X = Y) \equiv ((S|\underline{y})X = (S|\underline{y})Y) \quad .$$

To verify this, choose a letter $\underline{x}$ different from $\underline{y}$ and not appearing in S, X, Y. Then, by (1),

$$(S|\underline{y})(X = Y) \equiv (S|\underline{y})(\forall \underline{x})(\underline{x} \in X \Leftrightarrow \underline{x} \in Y)$$

$$\equiv (\forall \underline{x})(S|\underline{y})(\underline{x} \in X \Leftrightarrow \underline{x} \in Y) \qquad \text{(I.1.8(f))} ,$$

$$\equiv (\forall \underline{x})((S|\underline{y})(\underline{x} \in X) \Leftrightarrow (S|\underline{y})(\underline{x} \in Y))$$

by the remarks in I.1.4,

$$\equiv (\forall \underline{x})(\underline{x} \in (S|\underline{y})X \Leftrightarrow \underline{x} \in (S|\underline{y})Y)$$

since $\underline{x}$ is different from $\underline{y}$,

$$\equiv ((S|\underline{y})X = (S|\underline{y})Y) \quad ,$$

the last step by reference to (1) since $\underline{x}$ appears neither in $(S|\underline{y})X$ nor in $(S|\underline{y})Y$.

This rule will frequently be used in the sequel, more often than not without explicit mention.

(vii) A subsidiary definition which is universally adopted (usually without explicit mention) reads:

$$(X = Y = Z) \equiv_{def} (X = Y) \wedge (Y = Z) \quad .$$

Similarly,

$$(X = Y = Z = \ldots = V = W) \equiv_{def} (X = Y) \wedge (Y = Z) \wedge \ldots \wedge (V = W) \quad .$$

(viii) Much has been written about the concept of equality as it figures in logic and in set theory; see, for example, Fraenkel et al. (1), pp. 22-30; Kleene (1) and (2) at many places (easily located by reference to his indexes); Mendelson (1), pp. 77, 159, 161.

I remark merely that sometimes equality of X and Y is defined to signify

$$(\forall \underline{z})(X \in \underline{z} \Leftrightarrow Y \in \underline{z}) \quad ,$$

where $\underline{z}$ denotes any letter appearing neither in X nor in Y ; cf. Fraenkel et al., loc.cit., p.28 and Mendelson, loc. cit., p.161. In our formalism, the theorem schema

$$X = Y \Rightarrow (\forall \underline{z})(X \in \underline{z} \Leftrightarrow Y \in \underline{z}) \tag{1_4}$$

is provable on the basis of the schema S6 in II.1.2. Further, the converse theorem schema

$$(\forall \underline{z})(X \in \underline{z} \Leftrightarrow Y \in \underline{z}) \Rightarrow X = Y \tag{1_5}$$

is provable on the basis of the Pairing Axiom appearing in II.12.2 together with the theorem schemas in II.4.1. (Routinely speaking, assume the left hand side of (1_5), omit the quantifier, and then replace $\underline{z}$ by $\{X\}$.) However, it is doubtful whether (1_5) is provable without the use of the Pairing Axiom or something similar. See Problem II/43.

(ix) As has been indicated in II.1.1, Bourbaki's version of formal set theory (in common with most others) is expressed in a language in which $=$ is a primitive sign, related to which there are the schemas S6 and S7 and an additional axiom called the Axiom of Extension (or Extent), labelled A1 .

We have proceeded differently by

(a) defining $=$ in terms of our shorter list of primitive signs

and have followed this up by

(b) postulating the schemas S6 and S7 , expressing properties of $=$.

This difference raises the question of relative consistency. Let us assume Bourbaki's theory to be consistent; see the remarks in I.2.7. Step (a) in our procedure does not in itself influence the consistency or otherwise of Θ_0 ; in fact, our $=$ is a semiformal sign and would never appear in the strictly formal developments, being always eliminated via definition II.1.1(1). (Compared with Bourbaki's theory, (a) may be said to reduce the number of degrees of freedom.) However, step (b) might, in conjunction with (a) , introduce inconsistency (supposedly absent from Bourbaki's theory). Closer examination shows that such is not the case : if Bourbaki's theory is consistent, so too is Θ_0 . (They share the purely logical primitive signs $\vee$ (or $\underline{\vee}$), $\neg$, τ , $\square$ and the schemas and axioms of Θ_0 ; Θ_0 has one primitive specific sign $\in$, while Bourbaki's theory has also $=$ and $\supset$; and Bourbaki's theory has additional axioms A1 and A3 (Bourbaki (1), pp.67, 72).)

A similar situation arises in connection with ordered pairs, in so far as III.1.1 below is adopted in our theory as a definition of ordered pairs whereas Bourbaki's theory has an appropriate primitive sign $\supset$ and an associated axiom (Bourbaki (1), p.72, axiom A3). In this case, however, there is the important difference that we postulate no associated schemas or axioms (that is, there is no stage (b)), and therefore no new question of relative consistency arises.

(x) Concerning conventional informal terminology, it is convenient to indicate here that a sentence of the type

(Under the stated hypotheses) one may write (or : it may be supposed that) $S = T[\![X]\!]$, where $\underline{\underline{A}}[\![X]\!]$

has to be interpreted to mean that

$$(\exists \underline{x})(\underline{\underline{A}} \wedge S = T[\![\underline{x}]\!])$$

is true in whatever theory is being considered at the time (which theory will usually depend upon the said stated hypotheses), $\underline{x}$ denoting a letter not appearing in S and the notation being as explained in (iv) above. The reader will (by implication) be expected to supply a proof within the said theory.

Similar interpretations have to be made of sentences or phrases of the type

..... one may write $S = T[\![X, Y]\!]$, where $\underline{\underline{A}}[\![X, Y]\!]$.

II.1.4 Properties of equality Whatever the style of the adopted definition of equality, one expects certain theorems to follow, namely, those which express the reflexivity, symmetry and transivity of equality. It is necessary to state these properties as sentences in the formal language and to prove that these sentences are theorems (of Θ_0).

The expected theorems are as follows, wherein $\underline{x}$, $\underline{y}$, $\underline{z}$ denote distinct letters:

T! : $$(\forall \underline{x})(\underline{x} = \underline{x}) \quad ; \qquad (2)$$

T! : $$(\forall \underline{x})(\forall \underline{y})((\underline{x} = \underline{y}) \Leftrightarrow (\underline{y} = \underline{x})) \quad ; \qquad (3)$$

T! : $(\forall\underline{x})(\forall\underline{y})(\forall\underline{z})(((\underline{x} = \underline{y}) \wedge (\underline{y} = \underline{z})) \Rightarrow (\underline{x} = \underline{z}))$. (4)

Before considering semiformal proofs of these theorems, I remark that one ay in (2) , (3) and (4) omit all the universal quantifiers. To be sure, his changes the sentences appearing, but it leaves unchanged the claim that these entences are theorems. (This is a consequence of I.3.2(X) and the fact that set neory has no constants, which fact is almost immediately verifiable on inspecting he explicit axioms of set theory listed collectively in II.12.2.)

Another possible variation is that, in place of claiming (for example) hat the sentence in (2) is a theorem, one may (and frequently does) state a orresponding theorem schema which in this instance would read

$$X = X \quad ,$$

herein it is understood that for X one may substitute an arbitrary set. (That no change in content results is now a consequence of I.3.2(II), plus the fact that $\mathcal{S}_0$ has no constants; cf. the end of I.3.4(i) and I.3.8(vi).) Similar variations are possible in the case of (3) and (4) , and in many analogous situations in the sequel. It is these theorem schemas which most frequently appear in informal accounts of set theory. They (and/or (2) - (4) themselves) are used over and over again, usually without comment.

I turn now to a semiformal proof of (2) . By the first clause of I.3.3(j) ,

T : $\underline{y} \in \underline{x} \Leftrightarrow \underline{y} \in \underline{x}$;

hence, by I.3.2(X) ,

T : $(\forall\underline{y})(\underline{y} \in \underline{x} \Leftrightarrow \underline{y} \in \underline{x})$.

Reference to (1) shows that the last written (formal) sentence is identical with $\underline{x} = \underline{x}$ (recall the discussion immediately preceding II.1.1(1)). This completes the proof. □

Now consider the semiformal proof of (3) . By I.3.7(1)(iv) ,

$$T \quad : \qquad ((\underline{z} \in \underline{x}) \Leftrightarrow (\underline{z} \in \underline{y})) \Leftrightarrow ((\underline{z} \in \underline{y}) \Leftrightarrow (\underline{z} \in \underline{x})) \quad .$$

By I.3.3(1) used twice over, together with I.3.3(j) ,

$$T \quad : \qquad (\forall \underline{z})(\underline{z} \in \underline{x} \Leftrightarrow \underline{z} \in \underline{y}) \Leftrightarrow (\forall \underline{z})(\underline{z} \in \underline{y} \Leftrightarrow \underline{z} \in \underline{x}) \quad .$$

(I am minimising the use of parentheses to whatever extent seems compatible with adequate clarity.) Reference to II.1.1(1) shows that this is identical with

$$T \quad : \qquad (\underline{x} = \underline{y}) \Leftrightarrow (\underline{y} = \underline{x}) \quad ,$$

as alleged. □

The semiformal proof of (4) is just like that of II.2.2(3) below. The reader is invited to construct such a proof before looking ahead.

The reader will probably expect to see an analogue of S6 in which the arbitrary sentence $\underline{\underline{A}}$ is replaced by an arbitrary set U . The appropriate theorem schema may now be deduced, namely:

$$T! \quad : \qquad (S = T) \Rightarrow ((S|\underline{x})U = (T|\underline{x})U) \quad , \qquad (5)$$

wherein S , T and U denote arbitrary sets and $\underline{x}$ an arbitrary letter.

<u>Proof of (5)</u> Introduce letters $\underline{y}$ and $\underline{z}$ different from $\underline{x}$ and not appearing in S , T or U . Define $U[\![X]\!] \equiv (X|\underline{x})U$ for arbitrary sets X . Adjoin the axiom $\underline{y} = \underline{z}$ and denote by Θ the theory thus obtained. Apply S6 in II.1.2 with $\underline{z}$ in place of $\underline{x}$, $\underline{\underline{A}}$ taken to be the sentence $U[\![\underline{y}]\!] = U[\![\underline{z}]\!]$ and

S and T taken to be $\underline{y}$ and $\underline{z}$ respectively: the result is

$$T\Theta \quad : \qquad (\underline{y}|\underline{z})\underline{\underline{A}} \Leftrightarrow (\underline{z}|\underline{z})\underline{\underline{A}} \quad ,$$

that is (noting that $\underline{z}$ does not appear in U and making use of I.1.4(a)),

$$T\Theta \quad : \qquad (U[\![\underline{y}]\!] = U[\![\underline{y}]\!]) \Leftrightarrow (U[\![\underline{y}]\!] = U[\![\underline{z}]\!]) \quad .$$

Hence, by (specialisation on $\underline{x}$ in) (2) , I.3.3(j) and I.3.2(I) ,

$$T\Theta \quad : \qquad U[\![\underline{y}]\!] = U[\![\underline{z}]\!] \quad .$$

Hence, by I.3.2(VI) ,

$$T \quad : \qquad (\underline{y} = \underline{z}) \Rightarrow (U[\![\underline{y}]\!] = U[\![\underline{z}]\!]) \quad .$$

So, by I.3.2(II) used twice over,

$$T \quad : \qquad (S|\underline{y})(T|\underline{z})(\underline{y} = \underline{z}) \Rightarrow (S|\underline{y})(T|\underline{z})(U[\![\underline{y}]\!] = U[\![\underline{z}]\!]) \quad .$$

In view of (a) and (b) in I.1.4 and the choice of the letters $\underline{y}$ and $\underline{z}$, this may be verified to be identical with the sentence appearing in (5) , and the proof is complete. □

Similarly one has the theorem schema

$$T! \quad : \qquad (\forall\underline{x})(S = T) \Rightarrow ((U|\underline{x})S = (U|\underline{x})T) \quad , \qquad (5')$$

wherein S , T , U denote arbitrary sets and $\underline{x}$ an arbitrary letter.

<u>Proof of (5')</u> (This proof and that of (5") are both somewhat

condensed. The reader should expand them into semiformal style; cf. the proof in II.1.5.) Adjoin the axiom $(\forall \underline{x})(S = T)$. Then, by (XI) and (I) in I.3.2,

$$(U|\underline{x})(S = T)$$

is true. By (p) in II.1.3(vi), this last is identical with $(U|\underline{x})S = (U|\underline{x})T$. Another appeal to I.3.2(VI) completes the proof. □

Combining the last two theorem schemas, there results

$$\text{T!} \quad : \quad ((S = T) \wedge (\forall \underline{x})(A = B)) \Rightarrow ((S|\underline{x})A = (T|\underline{x})B) \quad , \tag{5''}$$

wherein S , T , A , B denote arbitrary sets and $\underline{x}$ an arbitrary letter.

<u>Proof of (5")</u> Adjoin the axiom $(S = T) \wedge (\forall \underline{x})(A = B)$. In the strengthened theory,

$$(S|\underline{x})A = (T|\underline{x})A$$

and

$$(T|\underline{x})A = (T|\underline{x})B$$

are both true, the first by (5) and the second by (5') . (Use is also being made of I.3.3(j) and I.3.2(I).) Hence, by transitivity of equality, $(S|\underline{x})A = (T|\underline{x})B$ is true and one appeals to I.3.2(VI) to derive (5") . □

Alongside (1) - (5") must be placed a simple but vital theorem schema, namely:

$$\text{T!} \quad : \quad \tau_{\underline{x}}(\underline{x} = X) = X \quad , \tag{6}$$

wherein X denotes an arbitrary set and $\underline{x}$ any letter not appearing in X .

The proof is left to the reader in Problem II/13. See also Problem II/46.

> There is a recurrent irksome point which, I think, is unavoidable. In due course, sentences $X \subseteq Y$ and sets $\{X, Y\}$, $\cup X$, $\cap X$, $P(X)$, (X, Y) , $X \times Y$, Dom X , Ran X , and so on, will be defined. It is natural to expect that $X = X'$ and $Y = Y'$ will imply $X \subseteq Y \Leftrightarrow X' \subseteq Y'$, $\{X, Y\} = \{X', Y'\}$, $\cup X = \cup X'$, and so on; but I see no way of avoiding a check on this separately in each case. In practice this is done by using replacement rules to reduce each such problem to a form in which use of S6, S7 , (5) - (5") and cases already encountered and proved yield what is required. I shall not usually give any details, and may even sometimes omit explicit reference to the result; the reader should fill the gaps (see Problem II.2.3, the end of II.4.1 and IV.1.5(2) for illustrations).
>
> (The mere fact that one has in some way assigned to every set X a set λX , say, is no guarantee that $X = X'$ ensures $\lambda X = \lambda X'$. (Consider the case in which one first makes a choice of construction for every set X , and then defines λX to be the first set appearing in the chosen construction of X .) What ensures the expected result in the above cases is the fact that the assignments involved are expressible within the theory, so to speak.)

II.1.5 Remark I insert here a remark illustrating the opening paragraph of I.3.4(iv). Suppose one is concerned with proving the theorem schema

$$T \quad : \quad (\forall \underline{y})((\underline{y} = X \vee \underline{y} = Y) \Rightarrow \underline{x} \in \underline{y})) \Rightarrow (\underline{x} \in X \wedge \underline{x} \in Y) \quad , \qquad (7)$$

where X and Y denote arbitrary sets and $\underline{x}$ and $\underline{y}$ distinct letters not appearing in X or Y .

One form of semiformal proof might read as follows:

> Adjoin to Θ_0 the explicit axiom
>
> $$(\forall \underline{y})((\underline{y} = X \vee \underline{y} = Y) \Rightarrow \underline{x} \in \underline{y})$$

to obtain a theory Θ'' . By I.3.2(VI) and I.3.3(j) it is enough to prove that

$$T\Theta'' \quad : \quad \underline{x} \in X \quad , \qquad (7_1)$$

$$T\Theta'' \quad : \quad \underline{x} \in Y \quad . \qquad (7_2)$$

Now

$$T\Theta'' \quad : \qquad (\forall \underline{y})((\underline{y} = X \vee \underline{y} = Y) \Rightarrow \underline{x} \in \underline{y}) \quad ,$$

and so by I.3.2(XI), replacing $\underline{y}$ by X ,

$$T\Theta'' \quad : \qquad (X = X \vee X = Y) \Rightarrow \underline{x} \in X \quad . \qquad (7_3)$$

However, $X = X$ is true in Θ_0 (see II.1.4(2)); hence $X = X \vee X = Y$ is true in Θ_0 (by S2 in I.2.2 and I.3.2(I)). Hence, by (7_3) and I.3.2(I), (7_1) follows. The proof of (7_2) is exactly similar.

A more strictly formal presentation might appear as follows:

T :	$X = X$	(II.1.4(2)),	(7_4)
T :	$X = X \vee X = Y$	(S2, I.3.2(I) and (7_4)),	(7_5)
$T\Theta''$:	$X = X \vee X = Y$	((7_5); Θ'' stronger than Θ_0),	(7_6)
$T\Theta''$:	$(\forall \underline{y})((\underline{y} = X \vee \underline{y} = Y) \Rightarrow \underline{x} \in \underline{y})$	(axiom of Θ''),	(7_7)
$T\Theta''$:	$(X\|\underline{y})((\underline{y} = X \vee \underline{y} = Y) \Rightarrow \underline{x} \in \underline{y})$	(I.3.2(XI), (I) and (7_7)),	(7_8)
$T\Theta''$:	$(X = X \vee X = Y) \Rightarrow \underline{x} \in X$	((7_8); $\underline{x}$, $\underline{y}$ do not appear in X, Y),	(7_9)
$T\Theta''$:	$\underline{x} \in X$.	((7_6), (7_9) and I.3.2(I)),	(7_{10})

In a similar way one would arrive at

$$T\Theta'' \quad : \qquad \underline{x} \in Y \quad . \qquad (7_{11})$$

Then

$T\Theta''$: $(\underline{x} \in X) \wedge (\underline{x} \in Y)$ ((7$_{10}$), (7$_{11}$) (7$_{12}$) and I.3.3(j)),

T : $(\forall \underline{y})((\underline{y} = X \vee \underline{y} = Y) \Rightarrow (\underline{x} \in \underline{y}))$ ((7$_{12}$) and I.3.2(VI)),

$\Rightarrow (\underline{x} \in X \wedge \underline{x} \in Y)$

which is (7) .

This second procedure may well be formally preferable, but it is (I suggest) less intuitively obvious when one is granted the proof methods, and less likely to suggest itself to anybody intent upon proving (7) .

II.2 Inclusion and subsets

II.2.1 Formal definition of inclusion A second basic concept relating to the comparison of sets is that of inclusion or containment, denoted by $\subseteq$. (Some writers use $\subset$ in place of my $\subseteq$, but I shall reserve $\subset$ to mean " $\subseteq$ and $\neq$ " .) This concept springs from the intuitive idea of one collection being included or contained in a second (in the sense that every member of the first is a member of the second, without the converse being necessarily the case). For example, the collection $\{1, \{3, 2\}\}$ is contained in the collection $\{\{3, 2\}, 2, 1\}$. Having said this, the discussion appearing in II.1.1 and II.1.3 may be expected to prepare the reader for the following formal definition (schema):

$$X \subseteq Y \equiv_{\text{def}} (\forall \underline{x})(\underline{x} \in X \Rightarrow \underline{x} \in Y) \quad , \tag{1}$$

with the important proviso that $\underline{x}$ is (or denotes) a letter which appears neither in X nor Y . (Regarding this proviso, see also Problem II/1.) In place of $X \subseteq Y$ one often writes $Y \supseteq X$; and the sentence $X \subseteq Y$ is often rendered into English by saying that X is a subset of, or is contained (or included) in, Y

and that Y <u>is a superset of</u>, or <u>contains</u> (or <u>includes</u>), X .

Thus, in the role of a name, " X is a subset of Y " is a name for (that is, denotes) the formal sentence otherwise denoted by $X \subseteq Y$.

The verb "contains" can be interpreted either as "contains as an element" (expressed by $\in$) or as "contains as a subset" (expressed by $\subseteq$) . Confusion here can be fatal. The sentences $X \in Y$ and $X \subseteq Y$ should never be confused (even though they might, for all we know, be equivalent for certain choices of sets X and Y) . As a matter of passing interest, it is deducible from (1), (1_1), V.6.1(2') and V.6.1(8) that, for natural numbers X and Y , $X \in Y$ is equivalent to $X \subset Y$.

As in the case of equality (see II.1.3(i)), the quantifier $(\forall \underline{x})$ should not be omitted from the right side of (1) .

One should note the following replacement rule, analogous to II.1.3(p) :

> If S , X and Y denote arbitrary strings and u an arbitrary letter, then
>
> $$(S|\underline{u})(X \subseteq Y) \equiv (S|\underline{u})X \subseteq (S|\underline{u})Y \quad .$$

It would be possible, by using (*) in II.1.3(ii), to obtain analogous results concerning $(S|\underline{u},\ T|\underline{v})(X \subseteq Y)$, but I shall not record these extensions here or elsewhere.

Alongside (1) goes the formal definition (schema)

$$(X \subset Y) \equiv_{def} (X \subseteq Y) \wedge (X \neq Y) \quad . \qquad (1_1)$$

In place of $X \subset Y$ one often writes $Y \supset X$; and the sentence $X \subset Y$ is usually expressed (or named) in English by saying that X <u>is a proper subset of</u>, or <u>is properly contained</u> (or <u>properly included</u>) <u>in</u>, Y , and that Y <u>is a proper superset of</u>, or <u>properly contains</u> (or <u>properly includes</u>), X . Then, in analogy with the case of $\subseteq$, one has the replacement rule

$$(S|\underline{u})(X \subset Y) \equiv (S|\underline{u})X \subset (S|\underline{u})Y \quad ,$$

et cetera.

II.2.2 Properties of inclusion There is a basic theorem in which the signs $=$ and $\subseteq$ both appear and which serves to relate the two, namely

$$T! \quad : \qquad (\forall \underline{x})(\forall \underline{y})((\underline{x} = \underline{y}) \Leftrightarrow ((\underline{x} \subseteq \underline{y}) \wedge (\underline{y} \subseteq \underline{x})) \quad ; \tag{2}$$

compare this with the Axiom of Extensionality mentioned in II.1.1. The theorem schema version of (2) is

$$T! \quad : \qquad (X = Y) \Leftrightarrow ((X \subseteq Y) \wedge (Y \subseteq X)) \quad , \tag{2'}$$

wherein it is understood that X and Y denote arbitrary sets. Either of (2) or (2') is the basis of many proofs of the equality of certain sets.

In addition one has the theorem schema

$$T! \quad : \qquad ((X \subseteq Y) \wedge (Y \subseteq Z)) \Rightarrow (X \subseteq Z) \tag{3}$$

for arbitrary sets X , Y and Z , expressing transitivity of $\subseteq$. I leave the reader to express this as a genuine theorem.

The reader is invited to construct a (semiformal) proof of (2) ; cf. II.5.8. Meanwhile, I will write out such a proof of (3) . (The reader should, here and at all similar points, recall the remarks in I.3.4(ii) about the more modest versions of semiformal proofs.)

Semiformal proof of (3) Adjoin to Θ_0 the explicit axiom

$$(X \subseteq Y) \wedge (Y \subseteq Z) \tag{4}$$

to obtain a theory Θ . (Properly speaking, (4) is an axiom only when definite sets replace X , Y and Z ; and the proof to follow is really only applicable when this replacement is made. In particular, the proof method I.3.2(VI) should be applied to the theorem from which (3) derives; cf. the discussion in I.3.4(i).) Let $\underline{x}$ denote a letter not appearing in X , Y or Z . By (4) , definition (1) and I.3.2(XI),

$$T\Theta \ : \quad \begin{array}{c} \underline{x} \in X \Rightarrow \underline{x} \in Y \ , \\ \underline{x} \in Y \Rightarrow \underline{x} \in Z \ . \end{array} \tag{5}$$

Adjoin to Θ the explicit axiom

$$\underline{x} \in X$$

to obtain a theory Θ' . Then, by (5) ,

$$T\Theta' \ : \quad \underline{x} \in X \ , \tag{6}$$

$$T\Theta' \ : \quad \underline{x} \in X \Rightarrow \underline{x} \in Y \ , \tag{7}$$

$$T\Theta' \ : \quad \underline{x} \in Y \Rightarrow \underline{x} \in Z \ . \tag{8}$$

By I.3.2(I), (6) and (7) ,

$$T\Theta' \ : \quad \underline{x} \in Y \ . \tag{9}$$

By I.3.2(I), (8) and (9) ,

$$T\Theta' \ : \quad \underline{x} \in Z \ . \tag{10}$$

At this point make an appeal to I.3.2(VI) to conclude that

$$T\Theta \quad : \qquad \underline{x} \in X \Rightarrow \underline{x} \in Z \ . \tag{11}$$

Since Θ_0 has no constants, and since $\underline{x}$ does not appear in (4) , $\underline{x}$ is a variable of Θ . Hence I.3.2(X) and (11) combine to yield

$$T\Theta \quad : \qquad (\forall \underline{x})(\underline{x} \in X \Rightarrow \underline{x} \in Z) \ .$$

Reference to (1) gives now

$$T\Theta \quad : \qquad X \subseteq Z \ . \tag{12}$$

A second appeal to I.3.2(VI) leads from (12) to (3) . This ends the semiformal proof. □

A routine style proof of (3) would be much shorter and might look somewhat as follows: Assume $X \subseteq Y$ and $Y \subseteq Z$. Suppose $\underline{x} \in X$. Then (1) gives $\underline{x} \in Y$ and so (using (1) with X , Y replaced by Y , Z) also $\underline{x} \in Z$. So $\underline{x} \in X$ implies $\underline{x} \in Z$. Another reference to (1) , this time with X , Z in place of X , Y , gives $X \subseteq Z$, Q.E.D. (This shows clearly how appeals to proof methods become obliterated in everyday mathematics.)

II.2.3 Problem Give semiformal proofs of the theorem schemas

$$T! \quad : \qquad (X = X' \wedge Y = Y') \Rightarrow (X \subseteq Y \Leftrightarrow X' \subseteq Y')$$

and

$$T! \qquad (T \in X \wedge X \subseteq Y) \Rightarrow (T \in Y) \ .$$

II.3 The set builder or classifier

Introduction In the introductory remarks to this chapter, I introduced, for illustrative purposes and on a purely informal basis, a use of the braces for indicating collections. I am now about to introduce the use of braces in a formal manner as part of the formal theory. The formal development aims from the outset to cover situations much more general than the informal use; in particular, it handles indifferently finite and infinite sets (in fact the formal theory at this stage makes no distinction between finite and infinite sets). What is more, the strictly formal counterpart of the informal use of braces in expressions like $X_1, X_2, \ldots, X_n$ must (rather annoyingly!) await some further theoretical developments (essentially, the formal introduction of natural numbers; cf. the remarks in II.3.9 and II.6.4).

This is almost certainly a place at which to recall Cantor's definition of the concept of set (given around 1895 at the outset of his final exposition of the work he had done over many years):

> A set is a collection into a whole of definite distinct objects of our intuition or of our thought.

The discovery of certain paradoxes (see, for example, II.3.10) indicated that this definition, which would now be viewed as very informal in nature, indicated the need for more stringency. To cut a long story short, one might say that mathematicians turned toward the view that a set was to be characterised by some property common to the elements of that set and possessed by no other objects. However, this left open the precise meaning of "property" , which is initially as vague as the concept one seeks to define. Again cutting short a lengthy story, it came to be accepted that "property" was to be expressed by a sentence in the formal language; and also that one had to reckon with the possibility that not every property so expressed generates a set comprising precisely those objects which possess the said property. This is the starting point for II.3.1.

> One small point is worth noting. In an understandable attempt to simplify matters as much as possible, one sometimes finds in school texts a "definition" of the sort: "a set is a collection of objects of the same nature" . As has been indicated many times (see, for example, Godement (1), p.43) this "definition" is not very effective, inasmuch as the term "collection" is vague; also, the users of this "definition" subsequently experience no difficulty in forming sets, some of the objects of which are apples, while others are oranges. Clearly, "of the same nature" has to be interpreted in very elastic fashion.

II.3.1 Collectivising sentences One frequently imagines the formation of a collection as the outcome of choosing in advance a sentence (expressing a property) and then collecting together all those objects, and only those, for which the sentence is true (that is, those objects and only those having the said property). The use of the set builder is the mathematical analogue of this process, used repeatedly and universally in the formation of sets. It turns out, however, that in the formal theory one has to take care over the selection of the sentence, if one is to arrive at a set having the intuitively expected characteristic; that is, one has to abandon the hope that to every conceivable property corresponds a set whose elements are precisely the objects possessing that property. One therefore has to begin by describing the sentences which express the allowed properties. Meanwhile, and informally, one may say that $\{\underline{x} : \underline{\underline{A}}\}$ will denote that set E , if there is one, such that $(\forall\underline{x})(\underline{x} \in E \Leftrightarrow \underline{\underline{A}})$ is true, that is, such that the elements of E are precisely those objects $\underline{x}$ for which the condition on $\underline{x}$ expressed by $\underline{\underline{A}}$ is satisfied; the definition of equality ensures that there is (up to equality) at most one such set E ; see II.3.3. The symbol $\{ : \}$ is termed the <u>set builder</u> or <u>classifier</u>; the former term is especially suggestive.

(i) In conventional mathematics, and at a later stage, the set builder is used repeatedly and almost without conscious thought. (The formal justification is in almost all cases provided by the proof method (XII) in II.3.7 below.) It is the tool providing the passage from (certain) properties to the corresponding sets. Its use is frequently the key to framing mathematical properties and questions about them in very simple (almost "atomic") and innocent-looking set theoretical language.

A typical instance relates to the famous Fermat conjecture, to the effect that there exist no positive natural numbers x , y , z , a such that

$$x^{a+2} + y^{a+2} = z^{a+2} .$$

In due course, this conjecture may be expressed as the truth of the simple-looking sentence $S = \emptyset$, where

$$S \equiv \{\underline{a} : \underline{a} \in \dot{N} \wedge (\exists\underline{x})(\exists\underline{y})(\exists\underline{z})(\underline{x} \in \dot{N} \wedge \underline{y} \in \dot{N} \wedge \underline{z} \in \dot{N} \wedge (\underline{x}^{\underline{a}+2} + \underline{y}^{\underline{a}+2} = \underline{z}^{\underline{a}+2}))\} ,$$

$\dot{N}$ denoting the set of positive natural numbers (see V.2.1) and $\underline{x}$, $\underline{y}$, $\underline{z}$, $\underline{a}$ denoting distinct letters.

Another instance relates to another famous unsolved problem of number theory, Goldbach's conjecture; see III.2.3(iv) below.

Again, once one has formally defined " $\underline{x}$ is a prime " (see V.9.2), and then

$$\underline{z} \text{ is a prime pair} \equiv (\exists\underline{x})(\exists\underline{y})((\underline{x} \text{ is a prime}) \wedge (\underline{y} \text{ is a prime}) \wedge (\underline{y} = \underline{x} + 2) \wedge (\underline{z} = (\underline{x}, \underline{y}))) ,$$

where $\underline{x}$, $\underline{y}$, $\underline{z}$ denote distinct letters, the famous unsolved problem to know whether there exist infinitely many prime pairs amounts to knowing whether the set

$$\{\underline{z} : \underline{z} \text{ is a prime pair}\}$$

is infinite in the sense of V.7.

Still more simple-looking instances appear in proofs by induction (dealt with in Chapter V), where one may repeatedly introduce sets of the type $\{\underline{n} : \underline{n} \in N \wedge \underline{\underline{A}}\}$. For instance, one may seek to prove that " $2^n > n$ for every natural number n " by introducing the set

$$E \equiv \{\underline{n} : \underline{n} \in N \wedge 2^{\underline{n}} > \underline{n}\}$$

and seeking to prove by induction that $E = N$. Or again, one may seek to prove the commutative law of addition for natural numbers by handling in a similar way the set

$$E \equiv \{\underline{n} : \underline{n} \in N \wedge (\forall\underline{m})(\underline{m} \in N \Rightarrow \underline{n} + \underline{m} = \underline{m} + \underline{n})\} .$$

However, a lot of ground has to be covered before one is in a position to do this with a formal backing.

(ii) Adopting more formal terms, suppose that $\underline{\underline{A}}$ denotes a sentence. The sentence

$$\text{Coll}_{\underline{x}}\underline{\underline{A}} \equiv_{\text{def}} (\exists\underline{y})(\forall\underline{x})(\underline{x} \in \underline{y} \Leftrightarrow \underline{\underline{A}}) , \qquad (1)$$

wherein $\underline{y}$ denotes any letter different from $\underline{x}$ and not appearing in $\underline{\underline{A}}$, is named or read " $\underline{\underline{A}}$ is collectivising in $\underline{x}$ " ; and if $\mathrm{Coll}_{\underline{x}}\underline{\underline{A}}$ is true, the sentence $\underline{\underline{A}}$ is said to be <u>collectivising in</u> $\underline{x}$. (The choice of $\underline{y}$, subject to the stated conditions, is immaterial, thanks to (e) and (f) in I.1.8.) Not surprisingly, it is the sentences $\underline{\underline{A}}$ which are collectivising in $\underline{x}$ which lead to sets $\{\underline{x} : \underline{\underline{A}}\}$ with the intuitively expected properties. To assert $\mathrm{Coll}_{\underline{x}}\underline{\underline{A}}$ is to assert the existence of a set whose elements are precisely these objects $\underline{x}$ such that $\underline{\underline{A}}$.

The concept of collectivisability is of supreme importance to formal developments. Of the remaining four schemas and axioms of set theory (laid out together in II.12), three are directly concerned to affirm that certain sentences are collectivising in certain letters. The two simplest ones are the Pairing Axiom,

$$(\forall \underline{x})(\forall \underline{y})\mathrm{Coll}_{\underline{z}}((\underline{z} = \underline{x}) \vee (\underline{z} = \underline{y})) \quad ,$$

and the Subset Axiom,

$$(\forall \underline{x})\mathrm{Coll}_{\underline{y}}(\underline{y} \subseteq \underline{x}) \quad .$$

(As will be discussed in II.12, the choice here of distinct letters $\underline{x}$, $\underline{y}$, $\underline{z}$ and $\underline{x}$, $\underline{y}$ does not alter the indicated sentences.)

II.3.2 <u>Formal definition of the set builder</u>; <u>solution sets</u> Although the remarks in II.3.1 convey the suggestion that probably $\{\underline{x} : \underline{\underline{A}}\}$ will have the intuitively expected properties, only when the sentence $\underline{\underline{A}}$ is collectivising in $\underline{x}$, the discussion in I.3.5(iv) prompts one to make the formal definition of $\{\underline{x} : \underline{\underline{A}}\}$ unconditional, that is, free from the hypothesis $\mathrm{Coll}_{\underline{x}}\underline{\underline{A}}$. Accordingly I frame the formal definition schema

$$\{\underline{x} : A\} \equiv_{\mathrm{def}} \tau_{\underline{y}}((\forall \underline{x})(\underline{x} \in \underline{y} \Leftrightarrow A)) \quad , \tag{2}$$

where A denotes an arbitrary string, $\underline{x}$ an arbitrary letter, and $\underline{y}$ any letter different from $\underline{x}$ and not appearing in A. (The choice of $\underline{y}$ subject to these conditions is immaterial by virtue of I.1.4(c) and I.1.8(f). Regarding the formal construction involved, see the end of I.1.8.)

Other authors may use one of the notations { ; } , { | } ,

$\{\ ,\ \}$ in place of $\{\ :\ \}$. Bourbaki (1), p.68 and some other authors write $\mathcal{E}_{\underline{x}}(\underline{\underline{A}})$ in place of our $\{\underline{x} : \underline{\underline{A}}\}$.

Note that $\underline{x}$ does not appear in the strings denoted by $\{\underline{x} : A\}$ and $\mathrm{Coll}_{\underline{x}}A$.

Attached to the formal definition (2) is the theorem schema

$$\text{T!} \quad : \qquad \{\underline{x} : \underline{\underline{A}}\} = \tau_{\underline{y}}((\forall \underline{x})(\underline{x} \in \underline{y} \Leftrightarrow \underline{\underline{A}})) \quad ; \qquad (2')$$

cf. the remarks following II.1.1(1).

At least in those cases in which $\underline{\underline{A}}$ is collectivising in $\underline{x}$, the set denoted by $\{\underline{x} : \underline{\underline{A}}\}$ is termed <u>the set of</u> (<u>all</u>) $\underline{x}$ <u>such that</u> $\underline{\underline{A}}$ or sometimes <u>the solution set</u>, <u>relative to</u> $\underline{x}$, <u>of</u> (the condition expressed by) $\underline{\underline{A}}$; see I.1.9 and I.3.4(viii). In particular, if X , T , U denote sets, and if $\underline{x}$ does not appear in X , the set denoted by $\{\underline{x} : \underline{x} \in X \wedge T = U\}$, later (see II.3.7) also denoted by $\{\underline{x} \in X : T = U\}$, is often termed <u>the solution set in</u> X <u>of the equation</u> $T = U$ <u>relative to</u> $\underline{x}$.

Accordingly, to <u>solve relative to</u> $\underline{x}$, or <u>to find a solution relative to</u> $\underline{x}$ <u>of</u> (the condition expressed by) $\underline{\underline{A}}$, is to exhibit a set S such that S is an element of the solution set, relative to $\underline{x}$, of $\underline{\underline{A}}$; and to <u>solve</u> (et cetera) <u>in</u> X , the equation $T = U$ <u>relative to</u> $\underline{x}$, is to exhibit a set S such that S is an element of the solution set in X of the equation $T = U$ relative to $\underline{x}$.

See also IV.1.7(v).

The developments in the remainder of this section are all rather technical and somewhat forbidding. It is suggested that a first reading should skim over the details (including the semi-formal proofs) and concentrate on taking stock of the informal substance of the theorems and theorem schemas labelled T! . The reader should at the same time seek to grasp the significance of the substance of II.3.7 (the technical details of which appear in §2 of the Appendix), for it is the basis of almost everything to follow and the reason why in practice one very rarely encounters explicit mention of collectivisability. Subsections II.3.9 - II.3.11 are cautionary in nature.

II.3.3 Properties of the set builder By virtue of II.3.2(2) and II.3.1(1) and the definition of the existential quantifier (see I.1.7), together with the replacement rules in I.1.4 and I.1.8,

$$\mathrm{Coll}_{\underline{x}}\underline{\underline{A}} \equiv (\forall \underline{x})(\underline{x} \in \{\underline{x} : \underline{\underline{A}}\} \Leftrightarrow \underline{\underline{A}}) \quad . \tag{3}$$

Also, by (3) , the replacement rules in I.1.4 and (XI) in I.3.2, one has the theorem schema

$$\text{T! :} \qquad \mathrm{Coll}_{\underline{x}}\underline{\underline{A}} \Rightarrow (S \in \{\underline{x} : \underline{\underline{A}}\} \Leftrightarrow (S|\underline{x})\underline{\underline{A}}) \quad , \tag{3'}$$

wherein S denotes an arbitrary set, $\underline{\underline{A}}$ an arbitrary sentence, and $\underline{x}$ an arbitrary letter.

It cannot be over-emphasised that it is only when $\mathrm{Coll}_{\underline{x}}\underline{\underline{A}}$ is available as a theorem, that one can be sure that $\{\underline{x} : \underline{\underline{A}}\}$ lives up to the name "set of all $\underline{x}$ such that $\underline{\underline{A}}$ " , that is, that the conclusion of (3') is true. (Compare with the remarks in I.1.7 concerning the interpretation of the selector.) Without this prerequisite, it may be that one can a priori say nothing useful about $\{\underline{x} : \underline{\underline{A}}\}$. This is why subsequent definitions of certain sets (see, for example, II.4.1, II.6.1 and III.2.2) are prefaced by a discussion centreing on the collectivising nature of the appropriate sentences $\underline{\underline{A}}$. In such cases one could display the definition prior to any such discussion; where the latter appears, is a matter of choice. But it is usually only after such discussion that one feels free to use the definition in the intuitively expected way. See the illustration in II.3.10(iii).

The situation may also be expressed in the following way. Given a sentence $\underline{\underline{A}}$ and a letter $\underline{x}$, it is deducible from II.1.1(1) that there is up to equality at most one set E in which $\underline{x}$ does not appear and for which the sentence

$$(\forall \underline{x})(\underline{x} \in E \Leftrightarrow \underline{\underline{A}}) \tag{4}$$

is true; furthermore, if $\underline{\underline{A}}$ is collectivising in $\underline{x}$, there exists such a set E , namely $\{\underline{x} : \underline{\underline{A}}\}$. (It is not the case that any two sets E and E' making (4) true are identical, merely that they are equal; indeed, if $\underline{\underline{A}}$ and $\underline{\underline{A}}'$ are nonidentical but equivalent sentences which are collectivising in $\underline{x}$, then (3) above and II.3.6(18) below confirm that $E \equiv \{\underline{x} : \underline{\underline{A}}\}$ and $E' \equiv \{\underline{x} : \underline{\underline{A}}'\}$, although nonidentical, both make (4) true.)

The matter is formally expressed by the theorem schema

$$\mathrm{T!} \quad : \qquad \mathrm{Coll}_{\underline{x}}\underline{\underline{A}} \Rightarrow (E = \{\underline{x} : \underline{\underline{A}}\} \Leftrightarrow (\forall \underline{x})(\underline{x} \in E \Leftrightarrow \underline{\underline{A}})) \quad , \tag{5}$$

wherein $\underline{\underline{A}}$ denotes an arbitrary sentence, E an arbitrary set, and $\underline{x}$ any letter not appearing in E .

<u>Semiformal proof of (5)</u> Define $T \equiv \{\underline{x} : \underline{\underline{A}}\}$. Let Θ denote the theory obtained by adjoining to Θ_0 the axiom $\mathrm{Coll}_{\underline{x}}\underline{\underline{A}}$. Then, by (3) ,

$$\mathrm{T}\Theta \quad : \qquad (\forall \underline{x})(\underline{x} \in T \Leftrightarrow \underline{\underline{A}}) \quad . \tag{6}$$

By I.3.2(VI), it suffices to prove that

$$\mathrm{T}\Theta \quad : \qquad E = T \Leftrightarrow (\forall \underline{x})(\underline{x} \in E \Leftrightarrow \underline{\underline{A}}) \quad ;$$

and, by I.3.3(j), it suffices to prove that $\Rightarrow$ and that $\Leftarrow$ are both theorems of Θ .

(i) <u>Proof of $\Rightarrow$</u> Let Θ' denote the theory obtained by adjoining to Θ the axiom $E = T$, so that

$$\mathrm{T}\Theta' \quad : \qquad E = T \tag{7}$$

By I.3.2(VI) again, it suffices to prove that

$$T\Theta' \quad : \qquad (\forall \underline{x})(\underline{x} \in E \Leftrightarrow \underline{\underline{A}}) \quad . \tag{8}$$

Let $\underline{y}$ denote a letter different from $\underline{x}$ and not appearing in $\underline{\underline{A}}$. Since $\underline{x}$ does not appear in T , the replacement rule I.1.8(f) permits one to rewrite (6) in the form

$$T\Theta \quad : \qquad (T|\underline{y})(\forall \underline{x})(\underline{x} \in \underline{y} \Leftrightarrow \underline{\underline{A}}) \quad .$$

Hence, since Θ' is stronger than Θ ,

$$T\Theta' \quad : \qquad (T|\underline{y})(\forall \underline{x})(\underline{x} \in \underline{y} \Leftrightarrow \underline{\underline{A}}) \quad . \tag{9}$$

By (9) , (7) , S6 in II.1.2 and I.3.2(I),

$$T\Theta' \quad : \qquad (E|\underline{y})(\forall \underline{x})(\underline{x} \in \underline{y} \Leftrightarrow \underline{\underline{A}}) \quad . \tag{10}$$

Since $\underline{x}$ does not appear in E , the same replacement rule I.1.8(f) leads from (10) to (8) .

(ii) <u>Proof of $\Leftarrow$</u> Let Θ'' denote the theory obtained by adjoining to Θ the axiom $(\forall \underline{x})(\underline{x} \in E \Leftrightarrow \underline{\underline{A}})$, so that

$$T\Theta'' \quad : \qquad (\forall \underline{x})(\underline{x} \in E \Leftrightarrow \underline{\underline{A}}) \quad . \tag{11}$$

By I.3.2(VI) yet again, it will suffice to prove that

$$T\Theta'' \quad : \qquad E = T \quad . \tag{12}$$

Now, by (6) , (11) and I.3.2(XI),

$$T\Theta'' \quad : \qquad \underline{x} \in T \Leftrightarrow \underline{\underline{A}} \quad , \tag{13}$$

$T\Theta''$: $$\underline{x} \in E \Leftrightarrow \underline{\underline{A}} \quad . \qquad (13')$$

By (13) , (13') and I.3.7(1),

$T\Theta''$: $$\underline{x} \in E \Leftrightarrow \underline{x} \in T \quad . \qquad (14)$$

Now $\underline{x}$ is a variable of Θ'' , and so I.3.2(X) leads from (14) to

$T\Theta''$: $$(\forall \underline{x})(\underline{x} \in E \Leftrightarrow \underline{x} \in T) \quad .$$

Since $\underline{x}$ appears neither in E nor in T , reference to II.1.1(1) leads now to (12) . □

<u>Remark</u> The opening sentence of the proof of (5) , namely :

Define $T \equiv \{\underline{x} : \underline{\underline{A}}\}$,

is a typical instance of a definition which is intended to be purely temporary, applying (in this instance) only throughout this one proof. See I.3.5(viii). The symbol T can be expected to be used elsewhere to denote other sets. Such temporary definitions will appear over and over again, almost always without specific comment. The potential confusion rarely materialises in practice; when it does, it is itself dispelled by a little thought on the part of the reader; and it is in any case always avoidable in obvious though perhaps tedious ways. For instance, instead of using T to denote other sets in other contexts, one might be careful to use T' , T" , ... or T_1 , T_2 , ... , et cetera. Similar remarks apply equally well to temporary definitions assigning names to sentences.

<u>Concerning terminology</u> As has been indicated, $\{\underline{x} : \underline{\underline{A}}\}$ is usually referred to as "the set of all $\underline{x}$ such that $\underline{\underline{A}}$ " . One may also encounter the

phrase "the set of all $\underline{x}$ such that $\underline{\underline{A}}$ is true" .

This last phrase may be used as another name for $\{\underline{x} : \underline{\underline{A}}\}$. As such, it is a mildly unhappy choice because in many cases the sentence $\underline{\underline{A}}$ in question is not true (assuming set theory to be consistent).

> On the other hand, this same phrase might conceivably be intended to mean something quite different, involving an informal concept of "set" (metamathematical rather than mathematical). In the latter case, what seems to be involved is a collection (external to the formal theory) such that the set S is a member of the collection if and only if $\underline{\underline{A}}[\![S]\!] \equiv (S|\underline{x})\underline{\underline{A}}$ is true. Even assuming that $\mathrm{Coll}_{\underline{x}}\underline{\underline{A}}$ is true,
>
> this situation should not be confused with what may be thought to be expressed by the conclusion of (3') . Part of the trouble stems from confusing the formal sentence $\underline{\underline{A}}$ with the metasentence " $\underline{\underline{A}}$ is true" . Such mixing of the metalanguage and the formal language can engender apparent paradoxes; see II.3.9(iv), V.3.2(iii) and V.4.1(ii). In addition there is involved the confusion (cf. I.2.9(i)) between " $P \Leftrightarrow Q$ is a theorem" and " P is a theorem if and only if Q is a theorem" .
>
> Remark It should be noted that (2) is an unconditional definition. A relatively informal and conditional version of (2) might read as follows :
>
> If $\underline{\underline{A}}$ is collectivising in $\underline{x}$, then $\{\underline{x} : \underline{\underline{A}}\}$ is defined to be the set (denoted by)
>
> $$\tau_{\underline{y}}((\forall\underline{x})(\underline{x} \in \underline{y} \Leftrightarrow \underline{\underline{A}})) ,$$
>
> which is the unique set E such that, for all $\underline{x}$, $\underline{x} \in E$ if and only if $\underline{\underline{A}}$.
>
> Most conventional texts lack both the selector and clear-cut concepts of formal letter and of collectivisability, and the anticipated informal definition would most probably read somewhat as follows :
>
> $\{x : \underline{\underline{A}}\}$ is defined to be the unique set E (if there is one) such that (for all x) $x \in E$ if and only if $\underline{\underline{A}}$.

II.3.4 Replacement rules As consequences of (c) , (d) , (e) and (f) in I.1.4 and I.1.8, one has the following rules:

(q) If $\underline{x}$ and $\underline{x}'$ denote letters and A an arbitrary string, and if $\underline{x}'$ does not appear in A , then

$$\{\underline{x} : A\} \equiv \{\underline{x}' : (\underline{x}'|\underline{x})A\}$$

and

$$\mathrm{Coll}_{\underline{x}}A \equiv \mathrm{Coll}_{\underline{x}'}(\underline{x}'|\underline{x})A \quad .$$

(r) If $\underline{x}$ and $\underline{y}$ denote distinct letters and A and S denote arbitrary strings, and if $\underline{x}$ does not appear in S , then

$$(S|\underline{y})\mathrm{Coll}_{\underline{x}}A \equiv \mathrm{Coll}_{\underline{x}}(S|\underline{y})A$$

and

$$(S|\underline{y})\{\underline{x} : A\} \equiv \{\underline{x} : (S|\underline{y})A\} \quad .$$

These rules will often be used without explicit mention in the sequel.

II.3.5 <u>An example</u> As a simple instance of the use of the set builder, if X is a set and $\underline{x}$ a letter not appearing in X , then (as one would hope and expect) the sentence $\underline{x} \in X$ is collectivising in $\underline{x}$ and one has the theorem schema

$$\text{T! :} \qquad X = \{\underline{x} : \underline{x} \in X\} \quad . \tag{15}$$

To prove this in some detail, define $\underline{\underline{A}} \equiv \underline{x} \in X$. By (X) in I.3.2 and (j) in I.3.3,

$$\text{T :} \qquad (\forall\underline{x})(\underline{x} \in X \Leftrightarrow \underline{\underline{A}}) \quad . \tag{16}$$

Assuming $\underline{y}$ to be a letter not appearing in $\underline{\underline{A}}$ (that is, different from $\underline{x}$ and not appearing in X) , I.1.8(f) shows that the sentence featured in (16) is identical with

$$(X|\underline{y})(\forall\underline{x})(\underline{x} \in \underline{y} \Leftrightarrow \underline{\underline{A}}) \quad .$$

Hence, by (16) , I.3.2(I) and S5 in I.2.2 (that is, by proof by exhibition; see I.3.4(viii))

$$\mathrm{T} \ : \qquad (\exists \underline{y})(\forall \underline{x})(\underline{x} \in \underline{y} \Leftrightarrow \underline{\underline{A}}) \ . \tag{17}$$

Reference to II.3.1(1) confirms that (17) implies that $\underline{\underline{A}}$ is collectivising in $\underline{x}$. Hence, if $E \equiv \{\underline{x} : \underline{\underline{A}}\}$, II.3.3(3) gives

$$\mathrm{T} \ : \qquad (\forall \underline{x})(\underline{x} \in E \Leftrightarrow \underline{\underline{A}}) \ ,$$

that is,

$$\mathrm{T} \ : \qquad (\forall \underline{x})(\underline{x} \in E \Leftrightarrow \underline{x} \in X) \ .$$

Since $\underline{x}$ appears neither in X nor in E , reference to II.1.1(1) now shows that $E = X$ is true, and II.1.4(3) now entails (15) . □

II.3.6 A basic property of the set builder It has been seen that $\underline{\underline{A}}$ is collectivising in $\underline{x}$ if and only if there is a set whose elements are precisely those objects $\underline{x}$ such that $\underline{\underline{A}}$, and that then $\{\underline{x} : \underline{\underline{A}}\}$ is the unique such set. This leads one to expect that, if $\underline{\underline{A}}$ and $\underline{\underline{B}}$ are equivalent sentences, then they should be together collectivising in $\underline{x}$ or not; and if they are collectivising in $\underline{x}$, then $\{\underline{x} : \underline{\underline{A}}\}$ and $\{\underline{x} : \underline{\underline{B}}\}$ should be equal sets. That all this comes to pass follows, in view of (I) and (X) in I.3.2 and I.3.3(j), from the theorem schema

$$\mathrm{T!} \ : \quad (\forall \underline{x})(\underline{\underline{A}} \Leftrightarrow \underline{\underline{B}}) \Rightarrow ((\mathrm{Coll}_{\underline{x}}\underline{\underline{A}} \Leftrightarrow \mathrm{Coll}_{\underline{x}}\underline{\underline{B}}) \wedge (\{\underline{x} : \underline{\underline{A}}\} = \{\underline{x} : \underline{\underline{B}}\})) \ . \tag{18}$$

Semiformal proof of (18) Consider the theory Θ obtained by adjoining to Θ_0 the axiom $(\forall \underline{x})(\underline{\underline{A}} \Leftrightarrow \underline{\underline{B}})$. By I.3.2(VI) and I.3.3(j), it will suffice to prove that

$$T\Theta \quad : \qquad \{\underline{x} : \underline{\underline{A}}\} = \{\underline{x} : \underline{\underline{B}}\} \tag{19}$$

and

$$T\Theta \quad : \qquad \mathrm{Coll}_{\underline{x}}\underline{\underline{A}} \Leftrightarrow \mathrm{Coll}_{\underline{x}}\underline{\underline{B}} \quad . \tag{20}$$

As to (19), choose a letter $\underline{y}$ different from $\underline{x}$ and not appearing in $\underline{\underline{A}}$ or $\underline{\underline{B}}$. Notice that $\underline{x}$ and $\underline{y}$ are variables of Θ. In Θ, $\underline{\underline{A}} \Leftrightarrow \underline{\underline{B}}$ is a theorem and so (as follows from the theorem schemas in I.3.7(1)),

$$T\Theta \quad : \qquad ((\underline{x} \in \underline{y}) \Leftrightarrow \underline{\underline{A}}) \Leftrightarrow ((\underline{x} \in \underline{y}) \Leftrightarrow \underline{\underline{B}}) \quad . \tag{21}$$

By I.3.3(l) used twice, and I.3.3(j),

$$T\Theta \quad : \qquad (\forall \underline{x})((\underline{x} \in \underline{y}) \Leftrightarrow \underline{\underline{A}}) \Leftrightarrow (\forall \underline{x})((\underline{x} \in \underline{y}) \Leftrightarrow \underline{\underline{B}}) \quad .$$

By I.3.2(X),

$$T\Theta \quad : \qquad (\forall \underline{y})((\forall \underline{x})((\underline{x} \in \underline{y}) \Leftrightarrow \underline{\underline{A}}) \Leftrightarrow (\forall \underline{x})((\underline{x} \in \underline{y}) \Leftrightarrow \underline{\underline{B}})) \quad .$$

Then, by S7 in II.1.2,

$$T\Theta \quad : \qquad \tau_{\underline{y}}((\forall \underline{x})((\underline{x} \in \underline{y}) \Leftrightarrow \underline{\underline{A}})) = \tau_{\underline{y}}((\forall \underline{x})((\underline{x} \in \underline{y}) \Leftrightarrow \underline{\underline{B}}))$$

which, by II.3.2(2), is identical with (19).

Moreover, since $\underline{y}$ is a variable of Θ, (21) and I.3.3(l) used twice and I.3.3(j) together entail

$$T\Theta \quad : \qquad (\exists \underline{y})((\underline{x} \in \underline{y}) \Leftrightarrow \underline{\underline{A}}) \Leftrightarrow (\exists \underline{y})((\underline{x} \in \underline{y}) \Leftrightarrow \underline{\underline{B}}) \quad ,$$

which, by II.3.1(1), is identical with (20). □

For a sort of "partial converse" to (18), see (18') in II.3.8.

II.3.7 <u>The proof method (XII) and some corollaries</u> The practical use of the set builder relies upon axioms and schemas whose role is to settle once and for all that many sentences are collectivising in certain letters. In the formulation of set theory being discussed in this book, there is one schema S8 and two explicit axioms of this nature, all of which are listed in II.12. One vital metatheorem which ensues from S8 (see Appendix, §2 for the detailed verification) is the following:

(XII) Suppose that $\underline{\underline{A}}$ denotes a sentence, X a set, Θ a theory, $\underline{x}$ a variable (free letter) of Θ not appearing in X. If $\underline{\underline{A}} \Rightarrow (\underline{x} \in X)$ is a theorem of Θ, then $\mathrm{Coll}_{\underline{x}}\underline{\underline{A}}$ is a theorem of Θ. In particular, $\mathrm{Coll}_{\underline{x}}((\underline{x} \in X) \wedge \underline{\underline{A}})$ is a theorem of Θ.

(This is a metatheorem; a stronger theorem schema appears in Problem II.3.14. See also Problems II/33 and II/39.)

The metatheorem (XII) is quite easy to use in practice, much more so than the schema S8 from which it is derived. It is the tacit assumption of this metatheorem which explains why, in everyday mathematics, one hardly ever sees reference to collectivisability. The fact is that almost (but not quite : see II.3.10) always one is seeking to form a set whose elements are those objects which belong to a given set X and which satisfy $\underline{\underline{A}}$ (rather than all objects which satisfy $\underline{\underline{A}}$). In deference to this prevalent feature, one introduces the definition schema

$$\{\underline{x} \in X : A\} \equiv_{\mathrm{def}} \{\underline{x} : (\underline{x} \in X) \wedge A\}$$

for arbitrary strings X and A and an arbitrary letter $\underline{x}$.

With this notation, II.3.3(3) yields the theorem schema

$$T! \quad : \qquad (\forall \underline{x})(\underline{x} \in \{\underline{x} \in X : \underline{A}\} \Leftrightarrow ((\underline{x} \in X) \wedge A)) \ , \tag{22}$$

with the restriction that $\underline{x}$ does not appear in X .

Then, by I.3.2(XI), one deduces the theorem schema

$$T! \quad : \qquad T \in \{\underline{x} \in X : \underline{A}\} \Leftrightarrow (T \in X) \wedge (T|\underline{x})\underline{A} \ , \tag{23}$$

with the restriction that $\underline{x}$ does not appear in X .

In connection with the above formal definition of $\{\underline{x} \in X : A\}$ it should be noted that II.3.4(r) yields the replacement rule

$$(S|\underline{y})\{\underline{x} \in X : A\} \equiv \{\underline{x} \in (S|\underline{y})X : (S|\underline{y})A\}$$

provided $\underline{x}$ and $\underline{y}$ denote distinct letters and $\underline{x}$ does not appear in S . In particular

$$(X|\underline{y})\{\underline{x} \in \underline{y} : A\} \equiv \{\underline{x} \in X : A\} \ ,$$

provided $\underline{x}$ and $\underline{y}$ denote distinct letters, $\underline{x}$ does not appear in X and $\underline{y}$ does not appear in A . (Without these precautions, there is no assurance that $\{\underline{x} \in X : A\}$ is identical with $(X|\underline{y})\{\underline{x} \in \underline{y} : A\}$.)

It is also worth noting that use of (XII) is a major step in proving (see Problem II/33) the theorem schema

$$T! \quad : \qquad \mathrm{Coll}_{\underline{x}}A \Leftrightarrow (\exists \underline{y})(\forall \underline{x})(\underline{A} \Rightarrow \underline{x} \in \underline{y}) \tag{1'}$$

where $\underline{x}$ and $\underline{y}$ denote distinct letters and $\underline{y}$ does not appear in $\underline{A}$. Compare (1') carefully with II.3.1(1).

Informally expressed, (1') signifies that in order that the sentence $\underline{A}$ be collectivising in $\underline{x}$, it is necessary and sufficient that there exist a set E which contains (as an element) every object $\underline{x}$ which satisfies (the condition expressed by) $\underline{\underline{A}}$; the set E may have among its elements objects which do not satisfy $\underline{\underline{A}}$.

II.3.8 Another corollary of (XII) This takes the form of the following theorem schema:

$$T! \quad : \qquad (\mathrm{Coll}_{\underline{x}}\underline{\underline{A}} \wedge (\forall \underline{x})(\underline{\underline{B}} \Rightarrow \underline{\underline{A}})) \Rightarrow (\mathrm{Coll}_{\underline{x}}\underline{\underline{B}} \wedge (\{\underline{x} : \underline{\underline{B}}\} \subseteq \{\underline{x} : \underline{\underline{A}}\})) \quad . \qquad (24)$$

Semiformal proof of (24) Denote by Θ the theory obtained by adjoining to Θ_0 the explicit axiom

$$\mathrm{Coll}_{\underline{x}}\underline{\underline{A}} \wedge (\forall \underline{x})(\underline{\underline{B}} \Rightarrow \underline{\underline{A}}) \quad .$$

By I.3.2(VI) and I.3.3(j), it suffices to prove that

$$T\Theta \quad : \qquad \mathrm{Coll}_{\underline{x}}\underline{\underline{B}} \qquad (25)$$

and

$$T\Theta \quad : \qquad \{\underline{x} : \underline{\underline{B}}\} \subseteq \{\underline{x} : \underline{\underline{A}}\} \quad . \qquad (26)$$

Define $E \equiv \{\underline{x} : \underline{\underline{A}}\}$. By II.3.3(3),

$$T \quad : \qquad \underline{\underline{A}} \Leftrightarrow \underline{x} \in E \quad .$$

By hypothesis,

$$T\Theta \quad : \qquad \underline{\underline{B}} \Rightarrow \underline{\underline{A}} \quad .$$

Hence, by I.3.2(III) and the two theorems just displayed,

$$T_\Theta \quad : \qquad \underline{\underline{B}} \Rightarrow \underline{x} \in E \ . \tag{27}$$

Since $\underline{x}$ does not appear in E, II.3.7(XII) and (27) entail (25). On the other hand, I.3.2(I), (25) and II.3.3(5) entail

$$T_\Theta \quad : \qquad (\forall \underline{x})(\underline{\underline{B}} \Leftrightarrow \underline{x} \in \{\underline{x} : \underline{\underline{B}}\}) \ . \tag{28}$$

By (27), (28), (III) and (X) in I.3.2,

$$T_\Theta \quad : \qquad (\forall \underline{x})(\underline{x} \in \{\underline{x} : \underline{\underline{B}}\} \Rightarrow \underline{x} \in E) \ . \tag{29}$$

Since $\underline{x}$ appears in neither $\{\underline{x} : \underline{\underline{B}}\}$ nor E, II.2.1(1) and (29) lead to (26). □

If one makes use of II.3.6(18), (24) (twice) and II.2.2(2), it is possible (see Problem II/23) to prove the theorem schema

$$T_! \quad : \quad (\mathrm{Coll}_{\underline{x}}\underline{\underline{A}} \wedge \mathrm{Coll}_{\underline{x}}\underline{\underline{B}}) \Rightarrow ((\forall \underline{x})(\underline{\underline{A}} \Leftrightarrow \underline{\underline{B}}) \Leftrightarrow (\{\underline{x} : \underline{\underline{A}}\} = \{\underline{x} : \underline{\underline{B}}\})) \ ; \tag{18'}$$

this is a sort of "partial converse" of (18) in II.3.6.

Also, by making use of (24) and the theorems of II.3.7 it is possible (see Problem II/23 again) to prove the theorem schemas

$$T_! \quad : \quad (\forall \underline{x})(((\underline{x} \in X) \wedge \underline{\underline{B}}) \Rightarrow \underline{\underline{A}}) \Rightarrow (\{\underline{x} \in X : \underline{\underline{B}}\} \subseteq \{\underline{x} \in X : \underline{\underline{A}}\}) \ , \tag{24'}$$

$$T_! \quad : \quad \begin{aligned} &(\forall \underline{x})(((\underline{x} \in X) \wedge \underline{\underline{A}}) \Leftrightarrow ((\underline{x} \in X) \wedge \underline{\underline{B}})) \Leftrightarrow (\{\underline{x} \in X : \underline{\underline{A}}\} \\ &\qquad\qquad\qquad = \{\underline{x} \in X : \underline{\underline{B}}\}) \ , \end{aligned} \tag{18''}$$

where in each case it is supposed that the letter $\underline{x}$ does not appear in X.

In practice, it is (18") and (24') (rather than II.3.6(18) and (24)) which are used over and over again (almost always without explicit mention).

Remarks (i) Some books (for instance Glicksman and Ruderman (1), p.78, Theorem 2) allege a strengthened form of (24') which would in our notation appear as

$$\{\underline{x} \in X : \underline{\underline{B}}\} \subseteq \{\underline{x} \in X : \underline{\underline{A}}\} \text{ if and only if } \underline{\underline{B}} \Rightarrow \underline{\underline{A}} \ .$$

In our system, this is not plausible; and, if Θ_0 is consistent, it is not provable in Θ_0 . In the system inadequately described by Glicksman and Ruderman (see I.4.2), salvage is attempted by regarding X as a "universal set" and by somehow "restricting" the variable x to the so-called "domains" of suitable s-forms (their analogues of our sentences). However, as the authors admit (loc. cit., p.72, lines 13-17), the descriptions they give involve difficulties; they are in fact too vague to carry conviction. The basic trouble goes back to their Theorem 1 (loc. cit., p.76), which transcribes into

$$\underline{x} \in \{\underline{x} \in X : \underline{\underline{A}}\} \Leftrightarrow \underline{\underline{A}} \ ,$$

again neither plausible nor provable (if Θ_0 is consistent). Plausibility hinges upon reading into $\underline{\underline{A}}$ (which might be " $\underline{x} = \emptyset$ " , or " $\underline{x}$ is an ordered pair" , or - to quote loc. cit., p.75 - " $\underline{x}$ is an odd prime") some reference to, or specification of, the appropriate "universal set" X .

(ii) In the same cautionary vein, suppose one were to alter (24) by replacing $(\forall\underline{x})(\underline{\underline{B}} \Rightarrow \underline{\underline{A}})$ by $\underline{\underline{B}} \Rightarrow \underline{\underline{A}}$ and then attempt to prove that the resulting sentence schema is a theorem schema. One would then presumably introduce Θ as the theory obtained by adjoining to Θ_0 the axiom

$$\mathrm{Coll}_{\underline{x}}\underline{\underline{A}} \wedge (\underline{\underline{B}} \Rightarrow \underline{\underline{A}}) \ .$$

But then $\underline{x}$ may (usually will) be a constant of Θ , in which case one cannot appeal to I.3.2(X) in order to derive (24) . In other words, the semiformal proof breaks down.

Not only this : if set theory is consistent, the omission of the quantifier $(\forall\underline{x})$ is definitely illicit (cf. I.2.9(ix)). More precisely,

$$(\mathrm{Coll}_{\underline{x}}\underline{\underline{A}} \wedge \mathrm{Coll}_{\underline{x}}\underline{\underline{B}} \wedge (\underline{\underline{B}} \Rightarrow \underline{\underline{A}})) \Rightarrow (\{\underline{x} : \underline{\underline{B}}\} \subseteq \{\underline{x} : \underline{\underline{A}}\}) \qquad (24_1)$$

is not a theorem schema; the letter $\underline{x}$ and sentences $\underline{\underline{A}}$ and $\underline{\underline{B}}$ may be exhibited such that their substitution in (24_1) generates a sentence which is not a theorem. See Problem II/21.

Similarly (Problem II/21 again),

$$(\forall\underline{x})(\underline{\underline{B}} \Rightarrow \underline{\underline{A}}) \Rightarrow (\{\underline{x} : \underline{\underline{B}}\} \subseteq \{\underline{x} : \underline{\underline{A}}\})$$

is not a theorem schema (unless set theory is contradictory).

(iii) The content of (24) may be very crudely and informally expressed thus : the stronger a sentence, the more likely that it is collectivising, and the smaller the corresponding set. This tempts one to feel that, if $\underline{\underline{A}}$ fails to be collectivising in $\underline{x}$, the failure is due to the "collection" of all objects $\underline{x}$

such that $\underline{A}$ being in some sense too vast to be a set. This vague impression receives some measure of backing when one comes to consider cardinal numbers; see V.7.5(vi) below.

II.3.9 Concerning the use of braces; informal bastard languages again

(i) Rather strangely, the most elementary use of braces in talking about set theory is not directly covered by the set builder. Thus, I have not so far formally introduced symbols like $\{X\}$, $\{X, Y\}$, $\{X, Y, Z\}$, $\{X, Y, Z, W\}$, and so on. Informally these symbols are used to denote the sets whose elements are X , X and Y , X and Y and Z , X and Y and Z and W , ... and so on.

The formal justification for the first two will emerge in II.4; and in II.6 it will be possible to formally define

$$\{X, Y, Z\} \equiv \{X, Y\} \cup \{Z\} ,$$

$$\{X, Y, Z, W\} \equiv \{X, Y, Z\} \cup \{W\} ;$$

and informally one may add "and so on" . Even then, however, one will still be some way from an acceptable formal definition of $\{X_1, X_2, \ldots, X_n\}$ for a general natural number n . Passable informal definitions, useful for illustrations, appear in II.6.4; see also V.5.5 and §3 of the Appendix.

(ii) Even less formally legitimate at this stage is the use of symbols such as $\{1, 3, 5, 7, \ldots\ldots\}$, complete with unexplained " ..... " . When one has formally defined the sentence " $\underline{x}$ is an odd natural number" (that is, when one has clearly specified the string so named) and shown it to be collectivising in $\underline{x}$, the set referred to loosely as $\{1, 3, 5, 7, \ldots\}$ would be that which is properly denoted by $\{\underline{x} : \underline{x}$ is an odd natural number$\}$. (Much has to be done before this stage is reached.) Moreover, this premature use of $\{1, 3, 5, 7, \ldots\}$

introduces a risk: one has no very firm grounds for chiding a student who (cantankerously, perhaps) claims that $128 \in \{1, 3, 5, 7, \ldots\}$ is a theorem. There is, after all, nothing to tell the student that one is not thinking of $\{1, 3, 5, 7, 9, 2, 8, 32, 128, \ldots\}$. Once the set builder has been introduced, there remains little or no excuse for or advantage in retaining a symbol as vague as $\{1, 3, 5, 7, \ldots\}$ in place of (say) $\{\underline{x} : \underline{x}$ is an odd natural number$\}$.

Although the notations $\{X_1, \ldots, X_n\}$ and $\{X_1, X_2, \ldots\}$ and others like them are used abundantly in conventional mathematics, where they are often a great convenience, there is no really pressing <u>need</u> for them at an early stage in this book (other than in illustrative examples and informal motivating discussions). The appropriate formal definitions will be deferred; see II.12.1, V.5.5 and §3 of the Appendix. Until then, neither $\{X_1, \ldots, X_n\}$ nor $\{X_1, X_2, \ldots\}$ has any place in the formalism.

In this connection (as also in connection with the notations $X_1 \cup \ldots\ldots \cup X_n$ and $X_1 \cap \ldots\ldots \cap X_n$ mentioned in II.6.4 and the Introduction to II.7), it is probably fair to say that the desire for these notations at an early stage is largely the result of the traditional informal approach to mathematics, according to which natural numbers present themselves as the most basic and familiar objects of all. The outcome of this is that, whenever one has to consider a nonvoid finite set F , one feels pressured into choosing an appropriate natural number $n \geq 1$ and labelling the elements of F as $X_1, \ldots\ldots, X_n$. Sometimes, indeed, such a labelling is a useful (even essential) device. But it is in principle desirable to regard it as such, rather than as an invariable obligation. (This is especially the case since the labelling is usually not unique; a labelling of F which is helpful for one purpose may be a hindrance for another.) The formal developments are forced to proceed without the labelling device until the natural number concept has been formalised, a stage which is delayed longer than might be expected.

(iii) A modicum of care is also needed in the use of the set builder itself, when one seeks to translate into formal or semiformal terms the definition of a set which is informally expressed.

For instance, does "the set of all even natural numbers" correspond to

$$A \equiv \{\underline{x} : \underline{x} = 2 \cdot \underline{y} \wedge \underline{y} \in N\}$$

or to

$$B \equiv \{\underline{x} : (\exists \underline{y})(\underline{x} = 2\cdot\underline{y} \wedge \underline{y} \in N)\} \quad ?$$

(I am assuming that the set N of natural numbers, the natural number 2 , and the multiplication of natural numbers have been formally defined, and that $\underline{x}$ and $\underline{y}$ denote distinct letters not appearing in N , 2 or $\cdot$.)

Suspicions may be aroused regarding A , simply because of the appearance in (the set denoted by) A of the letter (variable) $\underline{y}$. More to the point, however, use may be made of (XII) in II.3.7, II.3.8(24), II.4.1(1) and II.4.1(5) to deduce that

$$A \subseteq \{\underline{x} : \underline{x} = 2\cdot\underline{y}\} = \{2\cdot\underline{y}\} \quad ,$$

and so that A is a set possessing at most one element, which is intuitively expected not to be the case of "the set of all even natural numbers" . Moreover, $A \neq B$ is a theorem; see Problem V/45.

It is, on the contrary, the set B to which the name "the set of all even natural numbers" is habitually attached, a point which ought to be made uncompromisingly clear as soon as the bastard phrase in quotes is introduced.

Thus, English or bastard phrases (see I.2.9(vi)) intended as names for sets may, even when they are very commonly adopted and used, fail to indicate with all the necessary clarity precisely which sets they are intended to name.

(iv) Further trouble can arise if one inadvertently uses a bastard sentence as if it names some formal sentence, when in fact it does no such thing. This is relevant to the present discussion, because legitimate use of the set builder assumes that what appears between the braces and after the colon should be either a formal sentence, or at least an agreed name for a formal sentence. If this be forgotten, annoying things ensue.

For example, assuming the theory to have been carried to the stage of Chapter V, suppose one were to consider

$$S \equiv \{\underline{n} : \underline{n} \text{ is a natural number nameable by an English sentence of at most fifteen words}\} \quad . \tag{30}$$

One might then argue that S is plainly finite; define (see V.8)

$$\nu \equiv \mathrm{Min}_N(N \setminus S) \quad ,$$

the first (smallest) natural number not belonging to S . Then, on the one hand, ν is a natural number which does not belong to S and so ν is not nameable by an English sentence of at most fifteen words; on the other hand, it would usually be agreed that

the first natural number not nameable by an English sentence of at most fifteen words (31) is a name for ν and is an English sentence of fifteen words. (This is a slight variant of Berry's Paradox.)

Of the flaws in the argument, I need mention but two. (a) In the first place, the alleged definition (30) of S is ineffective, until one has agreed on some formal sentence, say $\underline{\underline{A}}$, for which " $\underline{n}$ is a natural number ..... words" is an agreed name. (b) The feeling that (31) is a name for ν hinges, I suppose, on the fact that the definition of ν ensures that the formal sentence

$$\nu = \mathrm{Min}_N(N \setminus S) \tag{32}$$

is true, and that this formal sentence has a meaning which justifies the name (31) . Apart from the fact that formal sentences are not intended to have meanings, one is still faced by the choice of formal sentence $\underline{\underline{A}}$ in (a) , in which case the "meaning" of (32) (if there is any such) will presumably depend upon $\underline{\underline{A}}$ and there is no guarantee that the "meaning" is indeed expressed by (31) .

There are numerous similar paradoxes, some ancient and some relatively modern. Among them are the Barber Paradox, the Liar Paradox (which exists in two slightly different versions due to Epimenides and Eubulides respectively) and Richard's Paradox; see I.0.7, Problem II/10, Kleene (1), §11 and Kleene (2) , §35. These are all instances of the so-called "semantic (or semantical) paradoxes" . There are still others of a somewhat different nature, termed the "logical (or set theoretical) paradoxes, though a more enlightening term might be "syntactic (or syntactical) paradoxes" . The Russell Paradox discussed in II.3.10 is an instance of this second type of paradox. Again see Kleene loc. cit. As is indicated in Problem II/10, a semantic paradox may sometimes be reformulated as a syntactic one.

In general, the use of an informal bastard language has to proceed with extreme care. As another instance, there are plausible arguments in a bastard language for the conclusion that there are only countably many sets (countable sets are defined in V.7.4). On the other hand, there is a theorem which can be bastardised so as to express that there are already uncountably many subsets of N . This is a crude form of Skolem's Paradox. Expressed in this crude form, there is no real paradox because the bastardised argument cannot be formalised (or, at least, nobody has ever succeeded in showing that it can be formalised); but there is more to be said in terms of models. See V.7.4; Wilder (1), p.237; Kneebone (1), p.304; Stoll (2), p. 453. One learns also from such considerations that there are limitations on the extent to which it is safe to think of sets as collections in the intuitive sense.

Informal mathematical language makes free use of the words "all" , "every" , "any" , "a" , "an" , "some" in ambiguous ways; see, for example, the lengthy discussion in Russell (1), §§60 and 61. Ad hoc attempts to clarify matters are often unsuccessful. To illustrate just one of the various ambiguities, selected from among several others indicated by Russell, consider the following two bastard informal descriptions:

(α) Any element of any element of A belongs to an element of B ; that is, the union of A is contained in the union of B .

(β) Any element of any element of A belongs to some

element of B ; that is, there is an element of B which contains the union of A .

Notice that each description incorporates a second clause, thought to be necessary and intended to clarify the first (primary) clause. The primary clauses of (α) and (β) differ only in the change from "an" to "some" , and there are no clear rules governing the choice in this and similar contexts. Yet the intended meanings, as indicated by the respective second clauses, are quite different; and the distinction made hinges upon the use of technical terms (in this case, the term "union"), a development which is usually the first step towards formalisation. In the formal language the differences are made much clearer (and at no great expense in space). Thus (α) would appear as

$$\cup A \subseteq \cup B$$

or, in more primitive formal language (obtained by eliminating the abbreviatory signs $\cup$ and $\subseteq$ by means of definitions) as

$$(\forall \underline{z})((\exists \underline{x})(\underline{x} \in A \wedge \underline{z} \in \underline{x}) \Rightarrow (\exists \underline{y})(\underline{y} \in B \wedge \underline{z} \in \underline{y})) \quad ,$$

$\underline{x}$, $\underline{y}$ and $\underline{z}$ denoting distinct letters not appearing in A or B . Likewise, (β) would appear as

$$(\exists \underline{y})(\underline{y} \in B \wedge \cup A \subseteq \underline{y})$$

or, in more primitive terms, as

$$(\exists \underline{y})(\underline{y} \in B \wedge (\forall \underline{z})((\exists \underline{x})(\underline{x} \in A \wedge \underline{z} \in \underline{x}) \Rightarrow (\underline{z} \in \underline{y}))) \quad .$$

Compare, also, the intended significance of "any" in the two bastard sentences

(γ) If any element of A is nonvoid, then $\cup A$ is nonvoid

and

(δ) Any natural number x satisfies $2^x > x$

In (γ) the intended meaning of "any" is that of "some" or "at least one" , while in (δ) it is that of "every" . Unambiguous formal versions are

$$(\exists \underline{x})(\underline{x} \in A \wedge \underline{x} \neq \emptyset) \Rightarrow (\cup A \neq \emptyset)$$

and

$$(\forall \underline{x})(\underline{x} \in N \Rightarrow 2^{\underline{x}} > \underline{x})$$

respectively ($\underline{x}$ denoting a letter which, in the first case, does not appear in A).

Russell (loc. cit., Chapter VIII) discusses the aid towards clarification afforded by the introduction of variables, again a step towards formalisation.

The preceding remarks are not intended to place a ban on the (very convenient) use of bastard sentences, merely to indicate that caution is sometimes required. See also the remarks at the end of III.1.2, the discussion of inductive proofs in V.4.1 and V.11.3, and VI.10.

(v) It is necessary to add a further cautionary remark of general significance. The utmost care is needed in comparing rival versions of formal set theory. For example, Suppes (1), pp.34-35 defines the set builder in such a way that in his system

$$\{\underline{x} : \underline{x} = \underline{x}\} = \emptyset$$

is a theorem (albeit a nonintuitive one). In our theory Θ_0, there seems no way of proving either the foregoing sentence or its negation (unless Θ_0 is inconsistent). In the system described in the Appendix to Kelley (1), the set builder is used in yet another way which results in

$$\{\underline{x} : \underline{x} = \underline{x}\} \neq \emptyset$$

being a theorem; and the same theorem ensues in Mendelson (1), p. 170, even though the theories described by Kelley and Mendelson are not exactly the same. (Suppes, Kelley and Mendelson all use 0 in place of $\emptyset$, as I shall do later by means of a traditional definition; see V.2.1.) It is noteworthy that the sentence $\underline{x} = \underline{x}$ is not collectivising in $\underline{x}$ (if Θ_0 is consistent: see II.3.10); and it is fairly safe to say that there is no disagreement over $\{\underline{x} : \underline{\underline{A}}\}$ for sentences $\underline{\underline{A}}$ which are collectivising in $\underline{x}$.

See also II.12.3, Problem II/36, II.1.3, IV.1.2 and Problem IV/2.

(vi) Finally, there is place for a remark about the vagueness of conventional notation. Suppose one has reached a stage at which sets denoted by X , Y , f have been defined; certain letters $\underline{a}$, $\underline{b}$, ... appearing in the sets thus denoted may (usually will) appear in the symbols X , Y , f . If one then sees written a phrase like

Consider the set $\{x \in X : f(x) \in Y\}$

and no further explanation is offered, it is almost certain that the writer feels and expects that the notation "speaks for itself" . Thus, if x does not appear in any of the symbols X , Y , f , this is almost always intended and expected to convey that x denotes a letter which does not appear in any of the sets denoted by X , Y , f . (Of course, x may on occasions appear in one or more of the symbols X , Y , f ; there is no reason why it should not.) In this connection, the reader is again reminded that almost never in conventional texts is any clear typographical distinction made between formal letters (and/or variables) and sets; a symbol may on one occasion denote a letter or a variable and on another a set which is neither a letter nor a variable.

From a formal point of view, such unwritten assumptions and distinctions are sometimes quite vital; nevertheless, the reader is almost invariably left to sort things out by a combination of intuition, experience and study of the context.

II.3.10 Sentences which are not collectivising in $\underline{x}$ Suppose that $\underline{\underline{A}}$ denotes a sentence and $\underline{x}$ a letter. If one sees written

$\underline{\underline{A}}$ is not collectivising in $\underline{x}$,

one may interpret this in one of at least two ways: either

(a) $\neg \mathrm{Coll}_{\underline{x}} \underline{\underline{A}}$ is a theorem

or

(b) $\mathrm{Coll}_{\underline{x}} \underline{\underline{A}}$ is not a theorem .

(Even this ignores a third possible interpretation, namely, that the displayed sentence is intended merely as a name for the formal sentence otherwise denoted by $\neg \mathrm{Coll}_{\underline{x}} \underline{\underline{A}}$.

In view of this, I must begin by stating firmly that here and in the sequel the displayed sentence is to be interpreted as (a) , rather than as (b) . (Notice that (a) entails (b), if Θ_0 is consistent; and that (b) entails (a), if $\mathrm{Coll}_{\underline{x}} \underline{\underline{A}}$ is decidable.)

For comparison, note that in common mathematical parlance " ½ is not a natural number" is almost universally interpreted as " ¬(½ is a natural number) " or " ¬(½ is a natural number) is a theorem" , almost never as " (½ is a natural number) is not a theorem" . At least, if the question of interpretation arose at all .... which it seldom does in common mathematical discussion .... , and if "is a theorem" is taken to mean "is provable" , the considered choice would almost always be for the former interpretation.

With this understanding, it is possible and salutary to display sentences $\underline{\underline{A}}$ and letters $\underline{x}$ such that $\underline{\underline{A}}$ is not collectivising in $\underline{x}$.

(i) An historically important example is contained in

$$\text{T!} \quad : \qquad \neg \text{Coll}_{\underline{x}}(\underline{x} \notin \underline{x}) \quad . \tag{33}$$

Anybody a little acquainted with the history of naive (informal) set theory will be comforted to see (33) . It makes it at once clear that, unless set theory is contradictory, $\text{Coll}_{\underline{x}}(\underline{x} \notin \underline{x})$ is not a theorem. Before the advent of formal set theory, it had been taken for granted that (in effect) $\text{Coll}_{\underline{x}}(\underline{x} \notin \underline{x})$ would be a theorem. But then it ensued (see the proof of (33) below) that the set $W \equiv \{\underline{x} : \underline{x} \notin \underline{x}\}$ would have the properties of Bertrand Russell's notorious "set of all sets which are not members of themselves" ; in particular,

$$W \in W \Leftrightarrow W \notin W$$

would be a theorem, showing (see Problem I/14) that the then-current set theory is contradictory. This conclusion (otherwise predictable on the basis of (33)) was unexpected and shattering; see also (iv) below. However, the present formal theory, embracing (33) as it does, warns us that, if set theory is consistent, there exists no set W for which

$$(\forall \underline{x})(\underline{x} \in W \Leftrightarrow \underline{x} \notin \underline{x})$$

is a theorem ($\underline{x}$ denoting a letter not appearing in W) .

<u>Proof of (33)</u> Define $W \equiv \{\underline{x} : \underline{x} \notin \underline{x}\}$. Let Θ denote the theory obtained by adjoining to Θ_0 the explicit axiom

$$\neg\neg \text{Coll}_{\underline{x}}(\underline{x} \notin \underline{x}) \quad .$$

By I.3.2(I) and I.3.3(k),

$$T\Theta \ : \qquad \mathrm{Coll}_{\underline{x}}(\underline{x} \notin \underline{x}) \ . \qquad (33_1)$$

By II.3.3(3'), I.3.2(I) and (33_1),

$$T\Theta \ : \qquad W \in W \Leftrightarrow W \notin W \ . \qquad (33_2)$$

By (33_2) , I.3.2(I) and I.3.3(j)

$$T\Theta \ : \qquad W \in W \Rightarrow \neg(W \in W) \qquad (33_3)$$

and

$$T\Theta \ : \qquad \neg(W \in W) \Rightarrow W \in W \ . \qquad (33_4)$$

By I.3.3(j) and I.3.2(I),

$$T\Theta \ : \qquad \neg(W \in W) \Rightarrow \neg(W \in W) \qquad (33_5)$$

and

$$T\Theta \ : \qquad W \in W \Rightarrow W \in W \ . \qquad (33_6)$$

By (33_3) , (33_5) and I.3.2(VIII),

$$T\Theta \ : \qquad \neg(W \in W) \ . \qquad (33_7)$$

By (33_4) , (33_6) and I.3.2(VIII),

$$T\Theta \ : \qquad W \in W \qquad (33_8)$$

By (33_7) and (33_8) , Θ is contradictory. So, by I.3.2(VII), (33) follows. □

The step from (33_2) to the conclusion that Θ is contradictory may be verified in other ways; see Problems I/14 and II.3.13.

(ii) Another example is contained in

$$T! \quad : \qquad \neg \mathrm{Coll}_{\underline{x}}(\underline{x} = \underline{x}) \quad ; \tag{34}$$

in informal language, there is no set of all sets; cf. (36) below. This may be deduced from (33) via use of (24) , but I will give a separate proof. In this connection, I will first prove that

$$T \quad : \qquad (\forall \underline{y})(\exists \underline{x})(\underline{x} \notin \underline{y}) \Rightarrow \neg \mathrm{Coll}_{\underline{x}}(\underline{x} = \underline{x}) \quad . \tag{35}$$

<u>Proof of (35)</u> Denote by Θ_1 the theory obtained by adjoining to Θ_0 the axiom $(\forall \underline{y})(\exists \underline{x})(\underline{x} \notin \underline{y})$. By I.3.2(VI), it is enough to prove that

$$T\Theta_1 \quad : \qquad \neg \mathrm{Coll}_{\underline{x}}(\underline{x} = \underline{x}) \quad . \tag{35_1}$$

Denote by Θ_2 the theory obtained by adjoining to Θ_1 the axiom $\neg\neg \mathrm{Coll}_{\underline{x}}(\underline{x} = \underline{x})$; by I.3.2(VII), it is enough to show that Θ_2 is contradictory. Now, by I.3.3(k) and I.3.2(I),

$$T\Theta_2 \quad : \qquad \mathrm{Coll}_{\underline{x}}(\underline{x} = \underline{x}) \quad ,$$

that is,

$$T\Theta_2 \quad : \qquad (\exists \underline{y})(\forall \underline{x})(\underline{x} \in \underline{y} \Leftrightarrow \underline{x} = \underline{x}) \quad ,$$

that is, defining $U \equiv \tau_{\underline{y}}((\forall \underline{x})(\underline{x} \in \underline{y} \Leftrightarrow \underline{x} = \underline{x}))$,

$$T\Theta_2 \quad : \qquad (U|\underline{y})(\forall \underline{x})(\underline{x} \in \underline{y} \Leftrightarrow \underline{x} = \underline{x}) \ .$$

Since $\underline{x}$ does not appear in U , I.1.8(f) leads thence to

$$T\Theta_2 \quad : \qquad (\forall \underline{x})(\underline{x} \in U \Leftrightarrow \underline{x} = \underline{x}) \ .$$

Hence, by (I) and (XI) in I.3.2 and I.3.3(j),

$$T\Theta_2 \quad : \qquad \underline{x} = \underline{x} \Rightarrow \underline{x} \in U \ .$$

So, by II.1.4(2), (I) and (X) in I.3.2 ($\underline{x}$ being a variable of Θ_2),

$$T\Theta_2 \quad : \qquad (\forall \underline{x})(\underline{x} \in U) \ ,$$

that is, by I.1.8(f) again,

$$T\Theta_2 \quad : \qquad (U|\underline{y})(\forall \underline{x})(\underline{x} \in \underline{y}) \ ,$$

and so, by S5 in I.2.2 and I.3.2(I) (that is, proof by exhibition)

$$T\Theta_2 \quad : \qquad (\exists \underline{y})(\forall \underline{x})(\underline{x} \in \underline{y}) \ .$$

By I.3.3(m) and I.3.2(I), therefore,

$$T\Theta_2 \quad : \qquad \neg(\forall \underline{y})(\exists \underline{x})(\underline{x} \notin \underline{y}) \ . \qquad (35_2)$$

But

$$T\Theta_1 \quad : \qquad (\forall \underline{y})(\exists \underline{x})(\underline{x} \notin \underline{y})$$

and so

$$T\Theta_2 \quad : \qquad (\forall \underline{y})(\exists \underline{x})(\underline{x} \notin \underline{y}) \quad . \qquad (35_3)$$

Reference to (35_2) and (35_3) shows that Θ_2 is contradictory. □

By virtue of I.3.2(I), to prove (34) , it now suffices to prove

$$T! \quad : \qquad (\forall \underline{y})(\exists \underline{x})(\underline{x} \notin \underline{y}) \quad ; \qquad (36)$$

which is of some independent interest.

Proof of (36) Define

$$S \equiv \{\underline{x} : \underline{x} \in \underline{y} \wedge \underline{x} \notin \underline{x}\} \quad .$$

In view of (I), (X) in I.3.2 and substitution in S5 in I.2.2, it is enough to prove that

$$T \quad : \qquad S \notin \underline{y} \quad . \qquad (37)$$

(The use of S5 and I.3.2(I) is another instance of proof by exhibition.)

Now by II.3.7(XII), $\mathrm{Coll}_{\underline{x}}(\underline{x} \in \underline{y} \wedge \underline{x} \notin \underline{x})$ is true. Hence, by (5) in II.3.3,

$$T \quad : \qquad (\forall \underline{x})(\underline{x} \in S \Leftrightarrow (\underline{x} \in \underline{y} \wedge \underline{x} \notin \underline{x})) \quad . \qquad (38)$$

By (38) , (XI) and (I) in I.3.2 (and the fact that $\underline{x}$ does not appear in S),

$$T \quad : \qquad S \in S \Leftrightarrow (S \in \underline{y} \wedge S \notin S) \quad . \qquad (39)$$

At this point, a routine proof would continue as follows. Assume $S \in \underline{y}$. Only two cases are possible, namely, <u>either</u> (a) $S \in S$ <u>or</u> (b) $S \notin S$.

In case (a), (39) yields $S \notin S$, a contradiction.

In case (b), (39) and the assumption $S \in \underline{y}$ lead to $S \in S$, and one has again a contradiction. This (allegedly) completes the proof, to this culminating routine element of which I shall return in II.3.11.

(iii) If Θ_0 is consistent, (34) shows that $\text{Coll}_{\underline{x}}(\underline{x} = \underline{x})$ is not a theorem. Even so, there is nothing to prevent one considering the set $V \equiv \{\underline{x} : \underline{x} = \underline{x}\}$. A priori, it may <u>seem</u> sensible to speak of V as "the set of all sets" , and to expect that accordingly

$$(\forall \underline{x})(\underline{x} \in V) \tag{40}$$

is true. However, it may be deduced from (36) that

$$(\exists \underline{x})(\underline{x} \notin V) \tag{41}$$

is true. Thus, if Θ_0 is consistent, (40) cannot be a theorem and the name "set of all the sets" rings hollow.

Similarly, one might expect that

$$V \cap X = X \tag{42}$$

will be true for all sets X . Now since

$$(\underline{x} = \underline{x}) \wedge \underline{\underline{A}} \Leftrightarrow \underline{\underline{A}}$$

is true, and hence

$$(\forall \underline{x})((\underline{x} = \underline{x}) \wedge \underline{\underline{A}} \Leftrightarrow \underline{\underline{A}})$$

is true, it may be deduced from (24) in II.3.8 that

$$\{\underline{x} : (\underline{x} = \underline{x}) \wedge \underline{\underline{A}}\} = \{\underline{x} : \underline{\underline{A}}\}$$

is true for all sentences $\underline{\underline{A}}$ which are collectivising in $\underline{x}$; and it might appear to be deducible from II.7.3 below that (42) is indeed true for all sets X . However, if Θ_0 is consistent, the hypothesis of II.7.3 is not true (precisely because $\underline{x} = \underline{x}$ is not collectivising in $\underline{x}$) and the envisaged deduction collapses. In fact, one may (see Problem II/38) prove

$$\neg(\forall \underline{z})(V \cap \underline{z} = \underline{z}) \quad .$$

Formality and informality are not here reconcilable.

(iv) I return to (i) above. Naive set theory lacks any precise concept of collectivisability, and there is nothing to prevent one forming a set $W \equiv \{\underline{x} : \underline{x} \notin \underline{x}\}$ and _expecting_ it to have the property that

$$(\forall \underline{x})(\underline{x} \in W \Leftrightarrow \underline{x} \notin \underline{x})$$

is true. If this were so, however, then

$$W \in W \Leftrightarrow W \notin W$$

would also be true. This is _Russell's Paradox_ (1902-3), discovered independently by Zermelo. It was one of the prime causes of radical rethinking and, ultimately, attempts at formalisation. See also Problem II/10.

In spite of this, it would be unfair to suggest that Russell's Paradox is akin to certain others in owing its existence solely to the use of too loose a language and/or too vague an intuitive background. In fact, if one were to consider the formal theory Θ obtained by adjoining to Θ_0 the axiom

$$(\exists \underline{y})(\forall \underline{x})(\underline{x} \in \underline{y} \Leftrightarrow \underline{x} \notin \underline{x}) \quad ,$$

where $\underline{x}$ and $\underline{y}$ denote distinct letters, then Russell's Paradox would present itself in the guise of inconsistency of this new formal theory Θ .

It would thus be better to say that Russell's Paradox was suggestive of troubles which, in a naive theory, may be hidden for quite a while and troublesome to trace back to their source when they do emerge, but which emerge more clearly when formalisation is undertaken.

All current versions of formal set theory have (so to speak) taken advantage of the warning provided by Russell's Paradox and incorporate strictures which, in one way or another, block the derivation of the corresponding contradiction. In the case of Θ_0 , there simply is no set with the properties of W . (Of course, other contradictions $\underline{\text{may}}$ be derivable.)

The above examples of sentences which are not collectivising in $\underline{x}$ may be thought to be rather artificial. Less artificial examples arise, especially when one discusses various "mathematical structures" (such as ordered sets, groups, et cetera). See (v) immediately below, Problem II/22 and V.7.5(vi).

(v) Although it is almost universally accepted that "the set of all sets" is, both formally and in all the usual informal mathematical frameworks, a self-contradictory concept, it nevertheless appears to rear its head occasionally in serious discussions.

For example, Gleason (1), p.59 introduces the set of all ordered sets in a way which translates into our formalism as

$$\mathcal{O} \equiv \{(\underline{a}, \underline{b}) : (\underline{b} \subseteq \underline{a} \times \underline{a}) \wedge \underline{\underline{P}} \wedge \underline{\underline{Q}}\} \tag{43}$$

wherein

$$\underline{\underline{P}} \equiv (\forall \underline{x})(\underline{x} \in \underline{a} \Rightarrow (\underline{x}, \underline{x}) \notin \underline{b})$$

$$\underline{\underline{Q}} \equiv (\forall \underline{x})(\forall \underline{y})(\forall \underline{z})((\underline{x} \in \underline{a} \wedge \underline{y} \in \underline{a} \wedge \underline{z} \in \underline{a}$$
$$\wedge (\underline{x}, \underline{y}) \in \underline{b} \wedge (\underline{y}, \underline{z}) \in \underline{b}) \Rightarrow (\underline{x}, \underline{z}) \in \underline{b}) \quad ;$$

herein $\underline{a}$, $\underline{b}$, $\underline{x}$, $\underline{y}$, $\underline{z}$ denote distinct letters and $(\underline{a}, \underline{b})$ and $\underline{a} \times \underline{a}$ are as defined in Chapter III below. (Although Gleason hints that $\mathcal{O}$ is to be regarded as a class, he also says (loc. cit., p.2) that "class" is entirely synonymous with "set".) By saying that an element of $\mathcal{O}$ is called "an $\underline{\text{ordered set}}$" , it is strongly suggested that $\mathcal{O}$ has the expected $\underline{\text{property}}$; that is, that

$$(\underline{a}, \underline{b}) \in \mathcal{O} \Leftrightarrow ((\underline{b} \subseteq \underline{a} \times \underline{a}) \wedge \underline{\underline{P}} \wedge \underline{\underline{Q}}) \tag{44}$$

is a theorem. However, if (44) is a theorem, it is easy to infer that $(a, \emptyset) \in \mathcal{O}$ is also a theorem; hence (see II.6.1 and III.1.2(7)) that $\underline{a} \in \cup\cup\mathcal{O}$ is a theorem; and so finally that $\cup\cup\mathcal{O}$ is the self-contradictory set of all sets. (Although it may savour of "cheating" to consider $(\underline{a}, \emptyset)$, it is not; the empty set is a legitimate choice here. In any event, one could equally consider $(\underline{a}, \subset_{\underline{a}})$, where

$$\subset_{\underline{a}} \equiv \{\underline{z} : (\exists \underline{x})(\exists \underline{y})(\underline{x} \in \underline{a} \wedge \underline{y} \in \underline{a} \wedge \underline{x} \subset \underline{y}$$
$$\wedge \underline{z} = (\underline{x}, \underline{y}))\} \quad ;$$

it will appear later (see especially III.1.4 and III.2.3(vii)) that this set is a relation in $\underline{a}$ having the intuitively expected properties.)

It is fairly safe to assume that all that Gleason does in connection with ordered sets can be purged of all references to the nonexistent set denoted by $\mathcal{O}$. (I have not verified this in detail; I rely on Gleason's intuition to avoid irreparable gaffes!)

(vi) It is worth noting that the proof of II.3.10(36) proves that, if Y is any set, there is at least one subset S of Y which is not an element of Y . The proof in fact describes explicitly such a subset S and makes it plain that, for most sets Y one is likely to think of, this subset S is Y itself.

II.3.11 <u>Comments on the proof of II.3.10(37)</u> The culminating routine phase of the proof of II.3.10(37) plainly masks something more elaborate than is covered directly by the proof methods and theorem schemas in I.3.2 and I.3.3. In addition, the routine style argument is itself ill-expressed and misleading. The alternatives referred to are presumably expressed by the metastatements

(a') $S \in S$ is true

and

(b') $S \notin S$ is true.

It is said that these are the <u>only</u> cases possible. This is debatable inasmuch as there is no apparent guarantee that the formal sentence $S \in S$ is decidable. In any case, what is said is irrelevant to what follows, which proceeds rather on the assumption that <u>at least one</u> of (a') and (b') is realised. But this is objectionable for the same reason; cf. the remarks in I.3.4(vii). (To assert that a certain list of potential situations exhausts the possibilities, is not to say that at least one of the said potential situations is or will be realised.)

The reasoning is, however, reparable in various ways.

One method is to proceed as far as II.3.10(39) and then introduce the theory Θ obtained by adjoining to Θ_0 the axiom $S \in \underline{y}$. It is easy to deduce from II.3.10(39) that

$$S \in S \Leftrightarrow S \notin S \equiv \neg(S \in S)$$

is true in Θ . So, by Problem I/14(ii), Θ is contradictory. Then II.3.10(37) follows by appeal to I.3.2(VII).

One might equally well proceed as far as II.3.10(39), introduce Θ as above, and then Θ' and Θ'' as the theories obtained by adjoining to Θ the axioms $S \in S$ and $S \notin S$ respectively. The routine stage of the argument verifies that both Θ' and Θ'' are contradictory, and it remains only to appeal to Problem I/14(iii) and I.3.2(VII).

Yet another approach is based upon appeal to the following metatheorem, which is an elaboration of both (VII) and (VIII) in I.3.2 (see Problem I/9(ii)) :

> (XIII) Suppose that Θ denotes a theory and $\underline{\underline{A}}$, $\underline{\underline{B}}$ sentences. Denote by Θ' the theory obtained by adjoining to Θ the axioms $\neg\underline{\underline{A}}$ and $\underline{\underline{B}}$, and by Θ'' the theory obtained by adjoining to Θ the axioms $\neg\underline{\underline{A}}$ and $\neg\underline{\underline{B}}$. If both Θ' and Θ'' are contradictory, then $\underline{\underline{A}}$ is a theorem of Θ .

To apply this in the proof of II.3.10(37), one takes $\underline{\underline{A}} \equiv S \notin \underline{y}$ and $\underline{\underline{B}} \equiv S \in S$.

Other applications of (XIII) appear in IV.3.10(iii) and VIII.2.7.

A verification of (XIII) appears in §2 of the Appendix.

II.3.12 Notational convention It will save much tedious repetition if I now make the following blanket convention about notation:

> Throughout the remainder of Volume 1, and unless the contrary is explicitly indicated, if some or all of $\underline{x}$, $\underline{x}'$, $\underline{y}$, $\underline{y}'$, $\underline{z}$, $\underline{z}'$, $\underline{w}$, $\underline{w}'$ and X , X' , Y , Y' , Z , Z' , S , T , A , B , f , g , h appear together in one context, it will be assumed that the letters denoted by $\underline{x}$, $\underline{x}'$, $\underline{y}$, $\underline{y}'$, $\underline{z}$, $\underline{z}'$, $\underline{w}$, $\underline{w}'$ are

(two by two) distinct and do not appear in the strings denoted by X , X' , Y , Y' , Z , Z' , S , T , A , B , f , g , h ; whereas the letters denoted by $\underline{s}$, $\underline{t}$, $\underline{u}$, $\underline{v}$ are distinct but may appear in one or more of the said strings.

(The restrictions imposed by this convention may not always be strict necessities, however.)

Although throughout Chapters II and III, strings and sets will usually be denoted by upper case Roman type (X , Y , Z , et cetera), beginning in II.4.3 and increasingly often thereafter one will find lower case Roman and upper and lower case Greek letters used for this same purpose. Everywhere up to and including Chapter V, however, logical letters will be denoted by underlined lower case Roman letters; thereafter some divergences from this rule will (in accordance with custom) be tolerated.

II.3.13 Problem Prove the theorem schemas

$$(\underline{\underline{A}} \Rightarrow \neg\underline{\underline{A}}) \Leftrightarrow \neg\underline{\underline{A}} \quad ,$$

$$(\neg\underline{\underline{A}} \Rightarrow \underline{\underline{A}}) \Leftrightarrow \underline{\underline{A}} \quad ,$$

$$(\underline{\underline{A}} \Leftrightarrow \neg\underline{\underline{A}}) \Leftrightarrow (\underline{\underline{A}} \wedge \neg\underline{\underline{A}}) \quad .$$

Verify that, if $\underline{\underline{A}} \Leftrightarrow \neg\underline{\underline{A}}$ is true in a theory, then that theory is contradictory. (Compare the proof of (33) in II.3.10(i) and recall Problem I/14.)

II.3.14 Problem Give a semiformal proof of the theorem schema

$$\text{T! :} \qquad (\forall \underline{x})(\underline{\underline{A}} \Rightarrow (\underline{x} \in X)) \Rightarrow \mathrm{Coll}_{\underline{x}}\underline{\underline{A}} \quad ,$$

wherein $\underline{\underline{A}}$ denotes an arbitrary sentence, X an arbitrary set and $\underline{x}$ a letter not appearing in X . (Hint: use I.3.2(VI) and II.3.7(XII), or II.3.8(24).)

II.4 Couples and singletons

I am now about to treat the formal analogue of forming a collection from one or two given objects.

II.4.1 Formal definitions and properties of couples and singletons

Among the explicit axioms of Θ_0 is the so-called Pairing Axiom, which reads

$$(\forall \underline{x})(\forall \underline{y})\mathrm{Coll}_{\underline{z}}((\underline{z} = \underline{x}) \vee (\underline{z} = \underline{y})) \quad ,$$

wherein $\underline{x}$, $\underline{y}$, $\underline{z}$ denote distinct letters. Accordingly,

$$T \;:\qquad (\forall \underline{x})(\forall \underline{y})(\exists \underline{w})(\forall \underline{z})((\underline{z} \in \underline{w}) \Leftrightarrow ((\underline{z} = \underline{x}) \vee (\underline{z} = \underline{y}))) \quad ,$$

wherein $\underline{x}$, $\underline{y}$, $\underline{z}$, $\underline{w}$ denote distinct letters. If X and Y denote sets (recall II.3.12) it accordingly follows from (I) and (XI) in I.3.2 and the replacement rules in I.1.8 that

$$T \;:\qquad (\exists \underline{w})(\forall \underline{z})((\underline{z} \in \underline{w}) \Leftrightarrow ((\underline{z} = X) \vee (\underline{z} = Y))) \quad ,$$

that is,

$$T \;:\qquad \mathrm{Coll}_{\underline{z}}((\underline{z} = X) \vee (\underline{z} = Y)) \quad . \tag{1}$$

I accordingly define

$$\{X, Y\} \equiv_{\mathrm{def}} \{\underline{z} : (\underline{z} = X) \vee (\underline{z} = Y)\} \quad , \tag{2}$$

where $\underline{z}$ denotes any letter not appearing in X or Y . {X, Y} is called the <u>couple</u>, or <u>unordered pair, determined by</u> X <u>and</u> Y . (Do not confuse this with the ordered pair (X, Y) discussed in III.1.) I further define

$$\{X\} \equiv_{\text{def}} \{X, X\} \quad , \tag{3}$$

called the <u>singleton determined by</u> X and usually read simply as "singleton X ". On no account should {X} be confused with X .

The definition schemas (2) and (3) are unconditional and apply to arbitrary strings X and Y ; cf. I.3.5(iv).

There is no compelling reason why {X} should be defined in terms of {X, Y} ; one could equally well treat {X} on its own merits and later define {X, Y} via II.6.2(4).

From (1) and the theorems of II.3, it follows (see Problem II/2) that

$$\text{T!} \quad : \qquad (\forall \underline{z})((\underline{z} \in \{X, Y\}) \Leftrightarrow ((\underline{z} = X) \vee (\underline{z} = Y))) \quad , \tag{4}$$

$$\text{T!} \quad : \qquad \{X\} = \{\underline{z} : \underline{z} = X\} \quad , \tag{5}$$

$$\text{T!} \quad : \qquad (\forall \underline{z})(\underline{z} \in \{X\} \Leftrightarrow \underline{z} = X) \quad , \tag{6}$$

wherein (cf. II.3.12) $\underline{z}$ is any letter not appearing in X or Y in the case of (4) , and any letter not appearing in X in the case of (5) and (6) . In the case of (4) and (6) , nothing is lost because of these restrictions: thanks principally to (XI) , (X) and (I) in I.3.2, one may in (4) and (6) replace $\underline{z}$ by any letter at all.

Put in another way, in place of (4) and (6) one might state the theorem schemas

$$\text{T!} \quad : \qquad Z \in \{X, Y\} \Leftrightarrow (Z = X \vee Z = Y) \quad , \tag{4'}$$

T! : $$Y \in \{X\} \Leftrightarrow Y = X \ . \tag{6'}$$

The reader should deduce (4') and (6') from (4) and (6) respectively by making use of (XI) and (I) in I.3.2; see I.3.4(i).

The following theorem schemas will be needed later:

T! : $$(\forall \underline{u})(\underline{u} \in X \Leftrightarrow \{\underline{u}\} \subseteq X) \ , \tag{7}$$

T! : $$(\forall \underline{u})(\forall \underline{v})((\underline{u} \in X \wedge \underline{v} \in X) \Leftrightarrow \{\underline{u}, \underline{v}\} \subseteq X) \ ; \tag{8}$$

concerning (7) , see Problem II/2.

From (2) , II.3.6(18), and the fact (see (j) in I.3.3) that $\underline{\underline{A}} \vee \underline{\underline{B}} \Leftrightarrow \underline{\underline{B}} \vee \underline{\underline{A}}$ is a theorem, it follows that

T! : $$\{X, Y\} = \{Y, X\} \ , \tag{9}$$

which explains why $\{X, Y\}$ is named the unordered pair determined by X and Y .

It should be noted that use of II.3.4(r) leads to the replacement rules

$$(S|\underline{u})\{X, Y\} \equiv \{(S|\underline{u})X, (S|\underline{u})Y\} \ ,$$

$$(S|\underline{u})\{X\} \equiv \{(S|\underline{u})X\} \ ,$$

and

$$(X|\underline{u}, Y|\underline{v})\{\underline{u}, \underline{v}\} \equiv \{X, Y\} \ ,$$

for arbitrary strings X , Y , S and arbitrary distinct letters $\underline{u}$ and $\underline{v}$.

(On the other hand, one can be sure of the identities

$$(X|\underline{x})(Y|\underline{y})\{\underline{x}, \underline{y}\} \equiv (Y|\underline{y})(X|\underline{x})\{\underline{x}, \underline{y}\} \equiv \{X, Y\}$$

only under certain conditions; those in II.3.12 will suffice (as is seen by applying the first replacement rule above).)

Moreover, one has the theorem schema

$$\text{T!} \quad : \qquad (X = X' \wedge Y = Y') \Rightarrow (\{X\} = \{X'\} \wedge \{X, Y\} = \{X', Y'\}) \quad ;$$

see the closing remarks in II.1.4.

Here, for example, is the tedious verification of the third of the above replacement rules: choose letters $\underline{x}$, $\underline{y}$, $\underline{z}$ so that $\underline{x}$, $\underline{y}$, $\underline{u}$, $\underline{v}$, $\underline{z}$ are distinct and none of $\underline{x}$, $\underline{y}$, $\underline{z}$ appears in either of X or Y . Then

$(X|\underline{u}, Y|\underline{v})\{\underline{u}, \underline{v}\}$

$\equiv (X|\underline{x})(Y|\underline{y})(\underline{x}|\underline{u})(\underline{y}|\underline{v})\{\underline{u}, \underline{v}\}$ ((*) in II.1.3(ii))

$\equiv (X|\underline{x})(Y|\underline{y})(\underline{x}|\underline{u})(\underline{y}|\underline{v})\{\underline{z} : \underline{z} = \underline{u} \vee \underline{z} = \underline{v}\}$ (II.4.1(2))

$\equiv (X|\underline{x})(Y|\underline{y})(\underline{x}|\underline{u})\{\underline{z} : (\underline{y}|\underline{v})\underline{z} = \underline{u} \vee \underline{z} = \underline{v}\}$ (II.3.4(r))

$\equiv (X|\underline{x})(Y|\underline{y})(\underline{x}|\underline{u})\{\underline{z} : \underline{z} = \underline{u} \vee \underline{z} = \underline{y}\}$ (I.1.4 and II.1.3(ii))

$\equiv (X|\underline{x})(Y|\underline{y})\{\underline{z} : (\underline{x}|\underline{u})(\underline{z} = \underline{u} \vee \underline{z} = \underline{y})\}$ (II.3.4(r))

$\equiv (X|\underline{x})(Y|\underline{y})\{\underline{z} : (\underline{z} = \underline{x} \vee \underline{z} = \underline{y}\}$ (I.1.4 and II.1.3(ii))

$\equiv (X|\underline{x})\{\underline{z} : (Y|\underline{y})(\underline{z} = \underline{x} \vee \underline{z} = \underline{y})\}$ (II.3.4(v))

$\equiv (X|\underline{x})\{\underline{z} : (\underline{z} = \underline{x} \vee \underline{z} = Y)\}$ (I.1.4 and II.1.3(ii))

$\equiv \{\underline{z} : (X|\underline{x})(\underline{z} = \underline{x} \vee \underline{z} = Y)\}$ (II.3.4(r))

$\equiv \{\underline{z} : \underline{z} = X \vee \underline{z} = Y\}$ (I.1.4 and (p) in II.1.3(vi))

$\equiv \{X, Y\}$. (II.4.1(2))

Also as an example, here is a proof of the last theorem schema displayed above. Denote by Θ the theory obtained from Θ_0 by adjoining the axiom $X = X' \wedge Y = Y'$. Choose distinct letters $\underline{x}$, $\underline{y}$ not appearing X , Y , X' or Y' . As has been noted,

$$\{X, Y\} \equiv (X|\underline{x})(Y|\underline{y})\{\underline{x}, \underline{y}\} \quad ,$$

$$\{X', Y'\} \equiv (X'|\underline{x})(Y'|\underline{y})\{\underline{x}, \underline{y}\} \quad .$$

Now

$T\Theta$: $Y = Y'$ (axiom of Θ , I.3.3(j) and (I.3.2(I)),

$T\Theta$: $(Y = Y') \Rightarrow ((Y|\underline{y})\{\underline{x}, \underline{y}\} = (Y'|\underline{y})\{\underline{x}, \underline{y}\})$ (II.1.4(5)),

$T\Theta$: $(Y|\underline{y})\{\underline{x}, \underline{y}\} = (Y'|\underline{y})\{\underline{x}, \underline{y}\}$ (I.3.2(I)),

$T\Theta$: $(\forall\underline{x})((Y|\underline{y})\{\underline{x}, \underline{y}\} = (Y'|\underline{y})\{\underline{x}, \underline{y}\})$ (I.3.2(X)),

$T\Theta$: $X = X'$ (axiom of Θ),

$T\Theta$: $(X|\underline{x})(Y|\underline{y})\{\underline{x}, \underline{y}\} = (X'|\underline{x})(Y'|\underline{y})\{\underline{x}, \underline{y}\}$ (I.3.2(I) and S6 in II.1.2)

that is,

$T\Theta$: $\{X, Y\} = \{X', Y'\}$,

and an appeal to I.3.2(VI) ends the proof.

Alternatively: choose a letter $\underline{z}$ not appearing in X , Y , X' or Y' and a letter $\underline{x}$ different from $\underline{z}$. Then

$T\Theta$: $X = X'$ (axiom of Θ , I.3.3(j) and I.3.2(I)),

T : $(X = X') \Rightarrow ((X|\underline{x})(\underline{z} = \underline{x}) \Leftrightarrow (X'|\underline{x})(\underline{z} = \underline{x}))$ (S6 in II.1.2),

$T\Theta$: $(X|\underline{x})(\underline{z} = \underline{x}) \Leftrightarrow (X'|\underline{x})(\underline{z} = \underline{x})$ (I.3.2(I)),

that is,

$T\Theta$: $\underline{z} = X \Leftrightarrow \underline{z} = X'$.

Similarly,

$$T\Theta \quad : \qquad \underline{z} = Y \Leftrightarrow \underline{z} = Y' \quad .$$

So

$$T\Theta \quad : \qquad (\underline{z} = X) \vee (\underline{z} = Y) \Leftrightarrow (\underline{z} = X') \vee (\underline{z} = Y') \qquad (I.3.7(1)),$$

$$T\Theta \quad : \qquad (\forall \underline{z})(\text{———————————}) \qquad (I.3.2(X)).$$

$$T\Theta \quad : \qquad \{\underline{z} : (\underline{z} = X) \vee (\underline{z} = Y)\} = \{\underline{z} : (\underline{z} = X') \vee (\underline{z} = Y')\} \qquad (II.3.6(18), I.3.2(I)),$$

$$T\Theta \quad : \qquad \{X, Y\} = \{X', Y'\} \qquad (II.4.1(2)).$$

II.4.2 Remarks (i) As in the cases of equality (see II.1.1 and II.1.3) and inclusion (see II.2.1), the restriction in II.4.1(2), that the letter $\underline{z}$ shall not appear in X or Y, is important. Likewise for II.4.1(5). Consider the latter by way of example.

Define temporarily $A \equiv \{\emptyset\}$, $\emptyset$ being as defined in II.5 below. An infringement of the metamathematical proviso in question would lead to the expectation that the sentence

$$\{\underline{z} \cap A\} = \{\underline{z} : \underline{z} = \underline{z} \cap A\} \quad ,$$

wherein $\cap$ is as defined in II.7.1(1) and II.7.2(4) below, is a theorem. On the contrary, however, this sentence is false. At a somewhat later stage (if not right now), this falsity would be intuitively clear, since the set on the left has precisely one element (viz. $\underline{z} \cap A$) while the set on the right has every subset of A as an element and so (with the present choice of A) has precisely two elements.

Proceeding more formally, adjoin the said sentence as a new explicit axiom, thus obtaining a theory Θ. Let $\underline{t}$ denote a letter different from $\underline{z}$ and note that $\underline{t}$ is a variable of Θ. Then, by II.1.1(1),

$$T\Theta \quad : \quad (\forall \underline{t})(\underline{t} \in \{\underline{z} \cap A\} \Leftrightarrow \underline{t} \in \{\underline{z} : \underline{z} = \underline{z} \cap A\}) \quad .$$

Now

$$T \quad : \qquad \underline{z} = \underline{z} \cap A \Rightarrow \underline{z} \subset A$$

and

$$T \quad : \qquad \underline{z} \subseteq A \Rightarrow \underline{z} \in P(A) \quad ;$$

see II.8.(8) and II.11.2(2) below. Hence, by (XII) in II.3.7, $\mathrm{Coll}_{\underline{z}}(\underline{z} = \underline{z} \cap A)$ is a theorem. Then, by II.3.3(3'),

$$\underline{t} \in \{\underline{z} : \underline{z} = \underline{z} \cap A\} \Leftrightarrow \underline{t} = \underline{t} \cap A$$

is a theorem. Also, by (6) ,

$$\underline{t} \in \{\underline{z} \cap A\} \Leftrightarrow \underline{t} = \underline{z} \cap A$$

is a theorem. Thus, making several appeals to I.3.2(II),

$T\Theta$: $\underline{t} = \underline{z} \cap A \Leftrightarrow \underline{t} = \underline{t} \cap A$,

$T\Theta$: $\underline{t} \cap A = \underline{z} \cap A$ (replace $\underline{t}$ by $\underline{t} \cap A$ and use II.8(4)),

$T\Theta$: $\emptyset \cap A = \underline{z} \cap A$ (replacing $\underline{t}$ by $\emptyset$),

$T\Theta$: $A \cap A = \underline{z} \cap A$ (replacing $\underline{t}$ by A),

$T\Theta$: $\emptyset \cap A = A \cap A$ (transitivity of equality),

$T\Theta$: $\emptyset = A$ (see (1) and (4) in II.8),

$T\Theta$: $\emptyset = \{\emptyset\}$.

This contradicts the conjunction of (6) and II.5.1(5). Thus Θ is contradictory. Hence, by (VII) in I.3.2,

$$T \quad : \qquad \{\underline{z} \cap A\} \neq \{\underline{z} : \underline{z} = \underline{z} \cap A\} \quad .$$

(ii) In view of (i) immediately above, it is perhaps mildly surprising to find that the analogous restriction on $\underline{z}$ in II.4.1(4) and (6) is not essential. In other words, a theorem (schema) of Θ_0 results if, in (4) and (6) , one writes in place of $\underline{z}$ any letter $\underline{u}$ (possibly appearing in X and/or Y) . This tends to suggest that metamathematical restrictions of this kind are <u>not</u> always clearly and correctly dictated by intuition.

Taking (6) as typical, here is a proof of the theorem schema

$$T! \quad : \qquad (\forall \underline{u})(\underline{u} \in \{X\} \Leftrightarrow \underline{u} = X) \quad , \qquad (10)$$

wherein $\underline{u}$ denotes an arbitrary letter and X an arbitrary set. Choose a letter $\underline{z}$ not appearing in X . Adjoin to Θ_0 the axiom $\underline{u} \in \{X\}$ to obtain a theory Θ_1 , so that

$$T\Theta_1 : \qquad \underline{u} \in \{X\} \quad . \qquad (11)$$

Then

T : $(\underline{u}|\underline{z})(\underline{z} \in \{X\} \Leftrightarrow \underline{z} = X)$ (II.4.1(6), I.3.2(XI), I.3.2(I)), that is (since $\underline{z}$ does not appear in X),

$$T \quad : \quad \underline{u} \in \{X\} \Leftrightarrow \underline{u} = X \quad ,$$

and so

$$T \quad : \quad \underline{u} \in \{X\} \Rightarrow \underline{u} = X \qquad \text{(I.3.3(j), I.3.2(I)),} \qquad (12)$$

$$T\Theta_1 \quad : \quad \underline{u} = X \qquad \text{((11), (12), I.3.2(I)),}$$

$$T \quad : \quad \underline{u} \in \{X\} \Rightarrow \underline{u} = X \qquad \text{(I.3.2(VI))} \quad . \qquad (13)$$

A similar procedure yields

$$T \quad : \quad \underline{u} = X \Rightarrow \underline{u} \in \{X\} \quad . \qquad (14)$$

So

$$T \quad : \quad \underline{u} \in \{X\} \Leftrightarrow \underline{u} = X \qquad \text{((13), (14), I.3.3(j)),}$$

and hence

$$T \quad : \quad (\forall \underline{u})(\underline{u} \in \{X\} \Leftrightarrow \underline{u} = X) \qquad \text{(I.3.2(X)),}$$

as alleged.

(iii) Suppose that $\underline{x}$ and $\underline{y}$ denote distinct letters. From (36) in II.3.10 one may deduce that

$$T! \quad : \quad \neg(\exists \underline{y})(\forall \underline{x})(\underline{x} \in \underline{y}) \quad . \qquad (15)$$

On the other hand, from (6) in II.4.1 one may deduce that

$$T! \quad : \quad (\forall \underline{x})(\underline{x} \in \{\underline{x}\}) \quad . \qquad (16)$$

The first of these might be enunciated informally as:

There exists no set X such that $\underline{x} \in X$ for all $\underline{x}$. (15')

At the same time, however, the second might likewise be enunciated thus:

There exists a set X (namely, the set $\{\underline{x}\}$) such that $\underline{x} \in X$ for all $\underline{x}$. (16')

This confrontation makes clear the occasional potential dangers in the use of bastard informal language, and in particular of phrases like "there exist" . The above clearly illustrates at least two different uses of this phrase. In addition, of course, in (16') one tends to overlook the possibility that $\underline{x}$ may appear in the set denoted by X ; whereas, in (16) , this feature is the explanation of everything. One can avoid the clash by inserting into (15') after the first appearance of " X " the phrase "in which the letter $\underline{x}$ does not appear" . But this addition expresses a metamathematical proviso which is not visibly the translation of anything expressible in the formal language; and the outcome expresses a metatheorem rather than a theorem.

II.4.3 Example I will prove the theorem schemas

$$\text{T!} \quad : \qquad (\forall \underline{x})(\underline{x} \in \{a, b\} \Rightarrow \underline{\underline{A}}) \Leftrightarrow (\underline{\underline{A}}[\![a]\!] \wedge \underline{\underline{A}}[\![b]\!]) \quad , \tag{17}$$

$$\text{T!} \quad : \qquad (\exists \underline{x})(\underline{x} \in \{a, b\} \wedge \underline{\underline{A}}) \Leftrightarrow (\underline{\underline{A}}[\![a]\!] \vee \underline{\underline{A}}[\![b]\!]) \quad , \tag{18}$$

where a and b denote arbitrary sets, $\underline{x}$ a letter not appearing in a or b , and $\underline{\underline{A}}[\![a]\!] \equiv (a|\underline{x})\underline{\underline{A}}$ for an arbitrary string a . (See also II.4.4(iii).)

Actually, I will write out a semiformal proof of (17) and leave the reader to deal likewise with (18) ; see Problem II/3.

Semiformal proof of (17) By I.3.3(j), it is enough to prove that

$$\text{T} \quad : \qquad (\forall \underline{x})(\underline{x} \in \{a, b\} \Rightarrow \underline{\underline{A}}) \Rightarrow (\underline{\underline{A}}[\![a]\!] \wedge \underline{\underline{A}}[\![b]\!]) \tag{17_1}$$

and

$$\text{T} \quad : \qquad (\underline{\underline{A}}[\![a]\!] \wedge \underline{\underline{A}}[\![b]\!]) \Rightarrow (\forall \underline{x})(\underline{x} \in \{a, b\} \Rightarrow \underline{\underline{A}}) \quad . \tag{17_2}$$

Considering (17_1) , let Θ denote the theory obtained by adjoining to Θ_0 the explicit axiom

$$(\forall \underline{x})(\underline{x} \in \{a, b\} \Rightarrow \underline{\underline{A}}) \quad .$$

By I.3.2(X),

$$\text{T}\Theta \quad : \qquad (a|\underline{x})(\underline{x} \in \{a, b\} \Rightarrow \underline{\underline{A}}) \quad .$$

Since $\underline{x}$ appears neither in a or b , the replacement rules show that this is identical with

$$\text{T}\Theta \quad : \qquad a \in \{a, b\} \Rightarrow \underline{\underline{A}}[\![a]\!] \quad . \tag{17_3}$$

On the other hand, by specialisation on $\underline{z}$ followed by substitution in II.4.1(4), it may be deduced that

$$T \quad : \qquad a \in \{a, b\} \quad . \qquad (17_4)$$

By (17_3) , (17_4) and I.3.2(I),

$$T\Theta \quad : \qquad A[\![a]\!] \quad . \qquad (17_5)$$

A similar argument leads to

$$T\Theta \quad : \qquad A[\![b]\!] \quad . \qquad (17_6)$$

By (17_5) , (17_6) and I.3.3(j),

$$T\Theta \quad : \qquad A[\![a]\!] \wedge A[\![b]\!] \quad . \qquad (17_7)$$

By (17_7) and I.3.2(VI), (17_1) follows.

Turning to (17_2) , let Θ' denote the theory obtained by adjoining to Θ_0 the explicit axiom,

$$(A[\![a]\!] \wedge A[\![b]\!]) \wedge \underline{x} \in \{a, b\} \quad .$$

Then, by II.4.1(4),

$$T\Theta' \quad : \qquad (\underline{x} = a) \vee (\underline{x} = b) \quad . \qquad (17_8)$$

Let Θ'' denote the theory obtained by adjoining to Θ' the axiom $\underline{x} = a$. Now

$$T\Theta' \quad : \qquad A[\![a]\!]$$

and so also

$$T\Theta'' \quad : \qquad \underline{\underline{A}}[\![a]\!] \quad . \tag{17_9}$$

Further,

$$T\Theta'' \quad : \qquad \underline{x} = a \quad ,$$

which combines with S6 in II.1.2 to prove that

$$T\Theta'' \quad : \qquad \underline{\underline{A}}[\![\underline{x}]\!] \Leftrightarrow \underline{\underline{A}}[\![a]\!] \quad ,$$

that is, since $\underline{\underline{A}}[\![\underline{x}]\!] \equiv \underline{\underline{A}}$, that

$$T\Theta'' \quad : \qquad \underline{\underline{A}} \Leftrightarrow \underline{\underline{A}}[\![a]\!] \quad . \tag{17_{10}}$$

By (17_9) , (17_{10}) and I.3.2(I),

$$T\Theta'' \quad : \qquad \underline{\underline{A}} \quad . \tag{17_{11}}$$

By I.3.2(VI) and (17_{11}) ,

$$T\Theta' \quad : \qquad (\underline{x} = a) \Rightarrow \underline{\underline{A}} \quad . \tag{17_{12}}$$

A similar argument proves that

$$T\Theta' \quad : \qquad (\underline{x} = b) \Rightarrow \underline{\underline{A}} \quad . \tag{17_{13}}$$

By (17_8) , (17_{12}) , (17_{13}) and I.3.2(VIII),

$$T\Theta' \quad : \qquad \underline{\underline{A}} \quad .$$

Another appeal to I.3.2(VI) entails

$$T \quad : \qquad ((\underline{\underline{A}}[\![a]\!] \wedge \underline{\underline{A}}[\![b]\!]) \wedge \underline{x} \in \{a, b\}) \Rightarrow \underline{\underline{A}} \quad . \qquad (17_{14})$$

So, by I.3.7(2),

$$T \quad : \qquad (\underline{\underline{A}}[\![a]\!] \wedge \underline{\underline{A}}[\![b]\!]) \Rightarrow (\underline{x} \in \{a, b\} \Rightarrow \underline{\underline{A}}) \quad .$$

Hence, by I.3.3(1) and the fact that $\underline{x}$ does not appear in a or b (hence not in $\underline{\underline{A}}[\![a]\!] \wedge \underline{\underline{A}}[\![b]\!])$), (17_2) is derived. □

II.4.4 Remarks (i) The reader may with justice object that the preceding argument does not contain all the steps necessary to make explicit every appeal to the metatheorems in I.3.2 and I.3.3; there are, for instance, appeals to I.3.2(I) which are not explicitly mentioned. See the comments in I.3.4(ii).

(ii) An alternative layout of the proof of (17_1) , which may serve to illustrate the remark made in the first paragraph of I.3.4(iv), is as follows. Suppose Θ is introduced as before. Then

$$T\Theta \quad : \qquad (\forall \underline{x})(\underline{x} \in \{a, b\} \Rightarrow \underline{A}) \qquad \text{(axiom of } \Theta \text{),}$$

$$T\Theta \quad : \qquad (a|\underline{x})(\underline{x} \in \{a, b\} \Rightarrow \underline{A}) \qquad \text{((XI) and (I) in I.3.2),}$$

$$T\Theta \quad : \qquad a \in \{a, b\} \Rightarrow \underline{\underline{A}}[\![a]\!] \qquad \text{(replacement rules),} \qquad (19)$$

$$T \quad : \qquad (\forall \underline{x})(\underline{x} \in \{a, b\} \Leftrightarrow \underline{x} = a \vee \underline{x} = b) \qquad \text{(II.4.1(4)),}$$

$$T \quad : \qquad (a|\underline{x})(\underline{x} \in \{a, b\} \Leftrightarrow \underline{x} = a \vee \underline{x} = b) \quad \text{((XI) and (I) in I.3.2),}$$

$$T \quad : \qquad a \in \{a, b\} \Leftrightarrow a = a \vee a = b \qquad \text{(replacement rules),}$$

$$T \quad : \qquad a = a \vee a = b \Rightarrow a \in \{a, b\} \quad \text{(I.3.3(j) and I.3.2(I)),} \qquad (20)$$

$$T \quad : \qquad (\forall \underline{x})(\underline{x} = \underline{x}) \qquad \text{(II.1.4(2)),}$$

$$T \quad : \qquad (a|\underline{x})(\underline{x} = \underline{x}) \qquad \text{((XI) and (I) in I.3.2),}$$

$$T \quad : \qquad a = a \qquad \text{(replacement rules),}$$

$$T \quad : \qquad a = a \vee a = b \qquad \text{(I.3.3(j) and I.3.2(I)),} \qquad (21)$$

$$T \quad : \qquad a \in \{a, b\} \qquad \text{((20), (21) and I.3.2(I)),}$$

$$T\Theta \quad : \qquad a \in \{a, b\} \qquad (\ \Theta \text{ stronger that } \Theta_0 \ ), \qquad (22)$$

$$T\Theta \quad : \qquad A[\![a]\!] \qquad \text{((19), (22) and I.3.2(I)),} \qquad (23)$$

Similarly

TΘ : $\underline{\underline{A}}[\![b]\!]$, (24)

TΘ : $\underline{\underline{A}}[\![a]\!] \wedge \underline{\underline{A}}[\![b]\!]$ ((23, (24) and I.3.3(j)), (25)

T : $(\forall \underline{x})(\underline{x} \in \{a, b\} \Rightarrow \underline{\underline{A}}) \Rightarrow \underline{\underline{A}}[\![a]\!] \wedge \underline{\underline{A}}[\![b]\!]$ ((25) and I.3.2(VI)).

Although this style may present in their correct order a large number of elementary logical steps, it is not easy to obtain from it an overall picture of what is going on.

(iii) It is possible to generalise (17) and (18) into the theorem schemas

$$T \;:\; (\forall \underline{x})((\underline{x} \in X \cup \{a\}) \Rightarrow \underline{\underline{A}}) \Leftrightarrow (\forall \underline{x})(x \in X \Rightarrow \underline{\underline{A}}) \wedge (a|\underline{x})\underline{\underline{A}} \tag{26}$$

$$T \;:\; (\exists \underline{x})((\underline{x} \in X \cup \{a\}) \wedge \underline{\underline{A}}) \Leftrightarrow (\exists \underline{x})(\underline{x} \in X \wedge \underline{\underline{A}}) \vee (a|\underline{x})\underline{\underline{A}} \tag{27}$$

wherein $\underline{x}$ denotes an arbitrary letter not appearing in X or a .

In view of (17) , (18) , (26) and (27) (note also the cases b = a of (17) and (18)), one may say informally: if F is a finite set,

$$(\forall \underline{x})(\underline{x} \in F \Rightarrow \underline{\underline{A}})$$

is equivalent to an iterated conjunction cf. I.3.7(6)

$$\underline{\underline{A}}[\![a]\!] \wedge \underline{\underline{A}}[\![b]\!] \wedge \ldots\ldots \wedge \underline{\underline{A}}[\![c]\!]$$

where a , b , ... c denote the elements of F and $\underline{\underline{A}}[\![S]\!] \equiv (S|\underline{x})\underline{\underline{A}}$ for arbitrary strings S ; and likewise

$$(\exists \underline{x})(\underline{x} \in F \wedge \underline{\underline{A}})$$

is equivalent to an iterated disjunction

$$\underline{\underline{A}}[\![a]\!] \vee \underline{\underline{A}}[\![b]\!] \vee \ldots\ldots \vee \underline{\underline{A}}[\![c]\!] \quad .$$

Incidentally, it is standard practice to abbreviate

$$(\forall \underline{x})(\underline{x} \in X \Rightarrow \underline{\underline{A}})$$

and

$$(\exists \underline{x})(\underline{x} \in X \wedge \underline{\underline{A}})$$

as $(\forall_X \underline{x})\underline{\underline{A}}$ and $(\exists_X \underline{x})\underline{\underline{A}}$ respectively, and to refer to them as _restricted quantifiers with domain_ X . (Little use of this notation will be made in this book.) As has been noted immediately above, restricted quantifiers with finite domains are equivalent to iterated conjunctions or disjunctions. This is not the case with restricted quantifiers with infinite domains, nor with the original unrestricted quantifiers $(\forall \underline{x})$ and $(\exists \underline{x})$.

A variable $\underline{x}$ referred to in a restricted quantifier $(\forall_X \underline{x})$ or $(\exists_X \underline{x})$ is often spoken of as a _restricted variable with domain_ X (or sometimes : _with range_ X). These concepts and terminology form no fundamental part of the formalism discussed in this book. In this formalism, a variable is unrestricted. (If it had a domain, that domain would be the "set of all sets" ; as has been seen in II.3.10, there is no such set.) The above use of the terms "domain" and "range" in relation to restricted variables must not be confused with their use explained in III.2.2.

See also the Foreword to Volume 2.

II.5 The empty set

II.5.1 _Formal definition of_ $\emptyset$ Informally, the empty set is the set with no elements. The problem is to show that there is indeed a set with this property, and that there is only one (up to equality of sets). It is here understood that to assert that a set X has no elements signifies that

$$(\forall \underline{x})(\underline{x} \notin X)$$

is true, $\underline{x}$ denoting a letter not appearing in X . It is a consequence of II.1.1(1) that this requirement does determine X uniquely (up to equality). To deal with matters more formally, I proceed as follows.

According to (XII) in II.3.7, if $\underline{t}$ and $\underline{u}$ denote distinct letters, the sentence

$$\underline{t} \in \underline{u} \wedge \underline{t} \notin \underline{u}$$

is collectivising in $\underline{t}$. Define

$$S \equiv \{\underline{t} : \underline{t} \in \underline{u} \wedge \underline{t} \notin \underline{u}\} \quad .$$

Then, by II.3.3(3),

$$T \quad : \qquad (\forall \underline{t})(\underline{t} \in S \Leftrightarrow \underline{t} \in \underline{u} \wedge \underline{t} \notin \underline{u}) \quad . \tag{1}$$

From this it follows that

$$T \quad : \qquad (\forall \underline{t})(\underline{t} \notin S) \quad ; \tag{2}$$

the reader should (Problem II/40) provide a semiformal proof of (2).

This exhibits a set S having no elements. However, $\underline{u}$ appears in S and different choices of $\underline{u}$ lead to non-identical sets S . Since a formally satisfactory definition of a set demands the specification of a string, and since it is expected, and will later be a convenience (though not a necessity) to have arranged, that no letters appear in the empty set, we proceed to use the selector τ to make a "choice" , in the following way.

Choose a letter $\underline{s}$ different from $\underline{u}$. Since $\underline{t}$ does not appear in S , I.1.8(f) shows that

$$(S|\underline{s})(\forall\underline{t})(\underline{t} \notin \underline{s}) \equiv (\forall\underline{t})(\underline{t} \notin S) \quad .$$

Hence (2) , I.3.2(I) and S5 in I.2.2 entail

T : $$(\exists\underline{s})(\forall\underline{t})(\underline{t} \notin \underline{s}) \quad . \tag{3}$$

(This is an instance of proof by exhibition, as described in I.3.4(viii).)

At this point define

$$\emptyset \equiv_{\text{def}} \tau_{\underline{s}}((\forall\underline{t})(\underline{t} \notin \underline{s})) \equiv \tau \neg \neg \neg \underline{\in} \tau \neg \neg \underline{\in} \square \square \square \quad ; \tag{4}$$

$\emptyset$ (or, better, the set thus denoted) is called the <u>empty set</u> or <u>void set</u>; see the first of the two constructions described in I.1.8. Then I.1.8(f) shows that (3) is identical with

T! : $$(\forall\underline{t})(\underline{t} \notin \emptyset) \tag{5}$$

so that $\emptyset$ has the desired property. Observe that no letters appear in (the set denoted by) $\emptyset$.

Other popular notations are 0 and Λ in place of $\emptyset$.

II.5.2 <u>Properties of $\emptyset$</u> It is not difficult to prove (see Problem II/4) that

T! : $$(\forall\underline{s})(\underline{s} \in \emptyset \Leftrightarrow \underline{s} \neq \underline{s}) \quad . \tag{6}$$

It follows from II.3.6(18) and (6) that the sentence $\underline{s} \neq \underline{s}$ is collectivising in $\underline{s}$ and that $\{\underline{s} : \underline{s} \neq \underline{s}\} = \{\underline{s} : \underline{s} \in \emptyset\}$; in view of II.3.5 and II.1.4(4) it ensues that

T! : $$\emptyset = \{\underline{s} : \underline{s} \neq \underline{s}\} \quad . \tag{7}$$

Other corollaries are ($\underline{s}$ and $\underline{t}$ denoting any two distinct letters)

T! : $(\forall \underline{s})(\emptyset \subseteq \underline{s})$, (8)

T! : $\underline{t} = \emptyset \Leftrightarrow (\forall \underline{s})(\underline{s} \notin \underline{t})$, (9)

(9) being the formal statement of the fact that II.5.1(5) characterises the empty set uniquely.

<u>Semiformal proof of (8)</u>

T : $\underline{t} \notin \emptyset$ (II.1.5(5) and I.3.2(XI)),

that is,

T : $\neg(\underline{t} \in \emptyset)$,

T : $\neg(\underline{t} \in \emptyset) \vee (\underline{t} \in \underline{s})$ (S2 in I.2.2 and I.3.2(I)),

that is,

T : $\underline{t} \in \emptyset \Rightarrow \underline{t} \in \underline{s}$,

T : $(\forall \underline{t})(\underline{t} \in \emptyset \Rightarrow \underline{t} \in \underline{s})$ (I.3.2(X)),

T : $\emptyset \subseteq \underline{s}$ (II.2.1(1)),

T : $(\forall \underline{s})(\emptyset \subseteq \underline{s})$ (I.3.2(X)).

(This illustrates one way of exhibiting semiformal proofs; it is not always the style which makes things easiest to comprehend, especially if the proof method

I.3.2(VI) is involved one or more times. See I.3.4(iv).) □

Semiformal proof of (9) By I.1.3(j) it suffices to prove

$$T \quad : \quad \underline{t} = \emptyset \Rightarrow (\forall \underline{s})(\underline{s} \notin \underline{t}) \tag{10}$$

and

$$T \quad : \quad (\forall \underline{s})(\underline{s} \notin \underline{t}) \Rightarrow \underline{t} = \emptyset \ . \tag{11}$$

Regarding (10) , let Θ denote the theory obtained by adjoining to Θ_0 the axiom $\underline{t} = \emptyset$, so that

$$T\Theta \quad : \quad \underline{t} = \emptyset \ . \tag{12}$$

By (12) , S6 in II.1.2 and I.1.8(f),

$$T\Theta \quad : \quad (\forall \underline{s})(\underline{s} \notin \underline{t}) \equiv (\underline{t}|\underline{u})(\forall \underline{s})(\underline{s} \notin \underline{u}) \Leftrightarrow (\emptyset|\underline{u})(\forall \underline{s})(s \notin \underline{u}) \equiv (\forall \underline{s})(\underline{s} \notin \emptyset) \ . \tag{13}$$

By (13) and II.5.1(5) (together with I.3.3(j) and I.3.2(I)),

$$T\Theta \quad : \quad (\forall \underline{s})(\underline{s} \notin \underline{t}) \ .$$

From this (10) is derived by appeal to I.3.2(VI).

Concerning (11) , let Θ' denote the theory obtained from Θ_0 by adjoining $(\forall \underline{s})(\underline{s} \notin \underline{t})$ as an axiom. By I.3.2(VI), it suffices to prove

$$T\Theta' : \quad \underline{t} = \emptyset \ . \tag{14}$$

Now, by (XI) and (I) in I.3.2,

$$T\Theta' : \quad \underline{s} \notin \underline{t}$$

that is, $\underline{s} \in \underline{t}$ is false in Θ' , hence implies any sentence in Θ' (see I.2.7); in particular

$$T\Theta' \quad : \qquad \underline{s} \in \underline{t} \Rightarrow \underline{s} \in \emptyset \ . \tag{15}$$

Since $\underline{s}$ is a variable of Θ' , I.3.2(X) and (15) entail

$$T\Theta' \quad : \qquad (\forall \underline{s})(\underline{s} \in \underline{t} \Rightarrow \underline{s} \in \emptyset)$$

that is (see II.2.1(1)),

$$T\Theta' \quad : \qquad \underline{t} \subseteq \emptyset \ . \tag{16}$$

By (16) , (8) , II.2.2(1') and I.3.3(j), (14) follows and completes the proof. □

The following theorem schema should also be noted:

$$T! \quad : \qquad (\forall \underline{t})(\neg A) \Leftrightarrow (\mathrm{Coll}_{\underline{t}}\underline{\underline{A}} \wedge \{\underline{t} : \underline{\underline{A}}\} = \emptyset) \ , \tag{17}$$

whence it follows that $\emptyset = \{\underline{t} : \underline{\underline{A}}\}$ is true whenever $\underline{\underline{A}}$ is a false sentence (see I.3.2(X)).

<u>Semiformal proof of (17)</u> (i) Adjoin to Θ_0 the axiom $(\forall \underline{t})(\neg \underline{\underline{A}})$ to obtain a theory Θ . Then

$$T\Theta \quad : \qquad \neg \underline{\underline{A}}$$

and hence (S2 in I.2.2 and I.3.2(I))

$$T\Theta \quad : \qquad \underline{\underline{A}} \Rightarrow (t \in \emptyset) \ . \tag{18}$$

At the same time, by II.5.1(5),

$$T \quad : \qquad \neg(\underline{t} \in \emptyset)$$

and so (S2 in I.2.2 and I.3.2(I) again)

$$T\Theta \quad : \qquad (\underline{t} \in \emptyset) \Rightarrow \underline{\underline{A}} \quad . \qquad (19)$$

By (18) , (19) and I.3.3(j),

$$T\Theta \quad : \qquad \underline{\underline{A}} \Leftrightarrow (\underline{t} \in \emptyset) \quad .$$

Hence, since $\underline{t}$ is a variable (free letter) of Θ , (XII) in II.3.7 entails

$$T\Theta \quad : \qquad \mathrm{Coll}_{\underline{t}}\underline{\underline{A}} \qquad (20)$$

and then, using II.3.3(5),

$$T\Theta \quad : \qquad \{\underline{t} : \underline{\underline{A}}\} = \emptyset \quad . \qquad (21)$$

By (20) , (21) and I.3.3(j),

$$T\Theta \quad : \qquad \mathrm{Coll}_{\underline{t}}\underline{\underline{A}} \wedge \{\underline{t} : \underline{\underline{A}}\} = \emptyset \quad . \qquad (22)$$

So, by I.3.2(VI), one derives

$$T \quad : \qquad (\forall \underline{t})(\neg \underline{\underline{A}}) \Rightarrow (\mathrm{Coll}_{\underline{t}}\underline{\underline{A}} \wedge \{\underline{t} : \underline{\underline{A}}\} = \emptyset) \quad . \qquad (23)$$

(ii) Adjoin to Θ_0 the axiom $\mathrm{Coll}_{\underline{t}}\underline{\underline{A}} \wedge \{\underline{t} : \underline{\underline{A}}\} = \emptyset$ to obtain a theory Θ' . By S6 in II.1.2, II.3.3(3) and I.3.3(j),

$T\Theta'$: $$(\forall \underline{t})(\underline{t} \in \emptyset \Leftrightarrow \underline{\underline{A}}) \quad .$$

Hence, by S5 in I.2.2,

$T\Theta'$: $$\underline{t} \in \emptyset \Leftrightarrow \underline{\underline{A}}$$

and so, by I.3.3(k),

$T\Theta'$: $$\underline{t} \notin \emptyset \Leftrightarrow \neg\underline{\underline{A}} \quad .$$

Hence, by II.5.1(5),

$T\Theta'$ $$\neg\underline{\underline{A}} \quad . \tag{24}$$

Since $\underline{t}$ is a variable (free letter) of Θ' , (24) and I.3.2(X) entail

$T\Theta'$: $$(\forall \underline{t})(\neg\underline{\underline{A}}) \quad . \tag{25}$$

By (25) and I.3.2(VI),

T : $$(\mathrm{Coll}_{\underline{t}}\underline{\underline{A}} \wedge \{\underline{t} : \underline{\underline{A}}\} = \emptyset) \Rightarrow (\forall \underline{t})(\neg\underline{\underline{A}}) \quad . \tag{26}$$

By (23) , (26) and I.3.3(j), (17) follows. □

II.5.3 Alternative definition of the empty set It might seem more intuitive to base the definition of the empty set on equation II.5.2(7). This is formally quite acceptable.

By II.1.4(2) and I.3.3(j), the following are theorems:

$$\neg(\underline{s} \neq \underline{s}) \quad ,$$

$$\neg(\underline{s} \neq \underline{s}) \vee (\underline{s} \in \underline{t}) \quad ,$$

that is,

$$(\underline{s} \neq \underline{s}) \Rightarrow (\underline{s} \in \underline{t}) \quad .$$

Hence, by (XII) in II.3.7, $\mathrm{Coll}_{\underline{s}}(\underline{s} \neq \underline{s})$ is a theorem.

At this point one might define

$$\emptyset_1 \equiv \{\underline{s} : \underline{s} \neq \underline{s}\} \quad ,$$

in which case II.3.3(3) implies

$$\mathrm{T} \quad : \qquad (\forall \underline{s})(\underline{s} \in \emptyset_1 \Leftrightarrow \underline{s} \neq \underline{s}) \quad . \tag{27}$$

From (27) and II.1.4(2), it may be deduced that

$$\mathrm{T} \quad : \qquad (\forall \underline{s})(\underline{s} \notin \emptyset_1) \quad ;$$

and then II.5.2(9) can be used to prove that $\emptyset = \emptyset_1$.

Note, however, that $\emptyset$ and $\emptyset_1$ are <u>not</u> identical sets. (The string denoted by $\emptyset_1$ is very much longer than that denoted by $\emptyset$.) In spite of this, the two definitions of the empty set are formally equivalent because $\emptyset = \emptyset_1$ is a theorem.

II.5.4 <u>Remarks</u> It is necessary to guard against over-hasty passage from theorems to theorem schemas; cf. I.3.9(ii). For instance, consider the theorem expressed by II.5.2(9). One might rush to conclude that the sentence schema

$$E = \emptyset \Leftrightarrow (\forall \underline{s})(\underline{s} \notin E) \quad , \tag{28}$$

wherein E denotes an arbitrary set, is a theorem schema. Actually, if Θ_0 is consistent, (28) is <u>not</u> a theorem schema of Θ_0 .

Proceeding more circumspectly, I.3.2(II) certifies passage

from II.5.2(9) to

$$\mathrm{T} \quad : \qquad (\mathrm{E}|\underline{t})(\underline{t} = \emptyset) \Leftrightarrow (\mathrm{E}|\underline{t})(\forall \underline{s})(\underline{s} \notin \underline{t}) \quad ;$$

but passage from this to (28) is allowed by the replacement rules, only on the assumption that $\underline{s}$ does not appear in E . With this restriction, (28) becomes a theorem schema.

In a similar way one has the theorem schema

$$\mathrm{T} \quad : \qquad X \neq \emptyset \Leftrightarrow (\exists \underline{x})(\underline{x} \in X) \quad , \tag{29}$$

where now the convention in II.3.12 is in action. (The proof uses I.3.3(j), (k) , (m) and II.5.2(9); see Problem II/4.)

II.5.5 Another comparison of semiformal and routine proofs Consider the theorem schema

$$\mathrm{T!} \quad : \qquad (X \neq \emptyset \wedge X \subseteq Y) \Rightarrow (Y \neq \emptyset) \tag{30}$$

wherein X and Y denote arbitrary sets.

A semiformal proof might proceed as follows. Adjoin to Θ_0 the axiom $X \neq \emptyset \wedge X \subseteq Y$ to obtain a theory Θ . Then

$$\mathrm{T}\Theta \quad : \qquad X \neq \emptyset$$

and so, by II.5.4(29) and I.3.2(I),

$$\mathrm{T}\Theta \quad : \qquad (\exists \underline{x})(\underline{x} \in X) \quad ; \tag{31}$$

here and in what follows $\underline{x}$ denotes a letter not appearing in X or Y . Also, by I.3.2(I) and I.3.3(j),

$$\mathrm{T}\Theta \quad : \qquad X \subseteq Y \quad ,$$

that is, by II.2.1(1),

$$\mathrm{T}\Theta \quad : \qquad (\forall \underline{x})(\underline{x} \in X \Rightarrow \underline{x} \in Y) \quad .$$

Hence, by I.3.2(XI),

$$T\Theta \ : \qquad \underline{x} \in X \Rightarrow \underline{x} \in Y \ . \qquad (32)$$

Now, by S5 in I.2.2, $\underline{x} \in Y \Rightarrow (\exists\underline{x})(\underline{x} \in Y)$ is true in Θ_0, hence true in Θ. So, by (32) and I.3.2(III),

$$T\Theta \ : \qquad \underline{x} \in X \Rightarrow (\exists\underline{x})(\underline{x} \in Y) \ . \qquad (33)$$

Since $\underline{x}$ is a variable of Θ and does not appear in $(\exists\underline{x})(\underline{x} \in Y)$, (31), (33), I.3.3(1) and I.3.2(I) imply that

$$T\Theta \ : \qquad (\exists\underline{x})(\underline{x} \in Y) \ . \qquad (34)$$

By (34), II.5.4(29) and I.3.2(I),

$$T\Theta \ : \qquad Y \neq \emptyset \ . \qquad (35)$$

Finally, an appeal to I.3.2(VI) leads from (35) to (30).

An alternative semiformal proof might proceed as far as (31) and then make the definition

$$E \equiv \tau_{\underline{x}}(\underline{x} \in X) \ ,$$

so that (31) reads

$$T\Theta \ : \qquad E \in X \ . \qquad (36)$$

Then proceed as before to

T_Θ : $$(\forall \underline{x})(\underline{x} \in X \Rightarrow \underline{x} \in Y)$$

and invoke I.3.2(XI) to derive

T_Θ : $$(E|\underline{x})(\underline{x} \in X \Rightarrow \underline{x} \in Y)$$

that is,

T_Θ : $$E \in X \Rightarrow E \in Y \quad . \qquad (37)$$

By (36) , (37) and I.3.2(I),

T_Θ : $$E \in Y \quad ;$$

hence, by S5 in I.2.2,

T_Θ : $$(\exists \underline{x})(\underline{x} \in Y) \quad , \qquad (34)$$

at which point one ends the proof as before.

A routine proof might appear as follows:

Since $X \neq \emptyset$, one can choose $\underline{x}$ so that $\underline{x} \in X$;
then, since $X \subseteq Y$, $\underline{x} \in Y$ too; so $Y \neq \emptyset$.

This is a mixture of what is formally well-founded with intuitive ideas which are not clearly of that category. It is not clear, for instance, what formal counterpart there is to "choosing" . Such a routine proof is in fact usually based on a mental picture of sets being collections of objects in such a way that $X \subseteq Y$ is translatable into "every object belonging to X also belongs to Y ". One is thus virtually assuming that the formal theory has been (or can be) "modelled" in concrete terms, and so that meaning has been attached to the formal theory. All this is foreign to the formal theory and raises its own peculiar doubts which are inherited by routine proofs. Faith in routine proofs is largely the outcome of common experience.

Apart from this, further gaps are opened up by the omission of explicit references to proof methods, which are regarded as given "rules of logic" too familiar to merit special mention.

Some texts on informal set theory might rely implicitly on such intuitive meanings and simply say:

It is clear that if $X \neq \emptyset$ and if $X \subseteq Y$, then $Y \neq \emptyset$

with no further argument whatsoever.

II.5.6 Problem Give a semiformal proof of the following theorem schema

$$T! \quad : \qquad (X = \emptyset) \wedge Y \subseteq X) \Rightarrow (Y = \emptyset) \quad . \qquad (35)$$

II.5.7 Further informal discussion of set formation
It may be instructive to turn aside momentarily to consider one particular question of set formation.

Suppose one begins with a definite set, say the empty set $\emptyset$, and imagines repeated application of singleton formation as defined in II.4.1. This results in the generation of the sets

$$\emptyset \ , \ \{\emptyset\} \ , \ \{\{\emptyset\}\} \ , \ \{\{\{\emptyset\}\}\} \ , \ \ldots\ldots \ . \qquad (36)$$

How might one tackle "bare-handed" the problem of producing a set S such that each of the sets (36) is an element of S ? Supposedly, one would first seek a property common to all the sets (36) . In view of the remarks at the outset of II.3, this seemingly amounts to producing a sentence $\underline{A} \equiv \underline{A}[\![\underline{x}]\!]$ of the formal language such that

$$\underline{A}[\![\emptyset]\!] \ , \ \underline{A}[\![\{\emptyset\}]\!] \ , \ \underline{A}[\![\{\{\emptyset\}\}]\!] \ , \ \underline{A}[\![\{\{\{\emptyset\}\}\}]\!] \ , \ \ldots\ldots \ .$$

are all true. However, as yet one has in sight no device for handling an unending list of proofs. Perhaps the nearest possibly-attainable aim is to produce $\underline{A}[\![\underline{x}]\!]$ such that

$$\underline{A}[\![\emptyset]\!] \wedge (\forall \underline{x})(\underline{A}[\![\underline{x}]\!] \Rightarrow \underline{A}[\![\{\underline{x}\}]\!])$$

is true. If this could be done, and if it could then be proved that $\underline{\underline{A}}$ is collectivising in $\underline{x}$, the set

$$S \equiv \{\underline{x} : \underline{\underline{A}}\}$$

would be such that

$$\emptyset \in S \wedge (\forall \underline{x})(\underline{x} \in S \Rightarrow \{\underline{x}\} \in S)$$

is true; and this last is a sentence of the formal language which expresses, as closely as seems possible at the moment, the initial demand (that $\emptyset \in S$, and $\{\emptyset\} \in S$, and $\{\{\emptyset\}\} \in S$, ... be all true).

The fact is, however, that there is no known way of producing such a sentence $\underline{\underline{A}}$, without appeal to axioms or schemas beyond those listed or mentioned so far. In fact, what seems to be needed is something like the Axiom of Infinity listed in II.12.2 below. Even then, the solution is by no means immediate; see Chapter V, especially V.2.1 and Remarks (i) and (ii) following V.5.3. At that stage we cite a theorem which guarantees the existence of a function u with domain N (the set of natural numbers) such that

$$u(0) = \emptyset \quad , \quad u(\underline{n} + 1) = \{u(\underline{n})\}$$

for all $\underline{n} \in N$. The set $S \equiv \text{Ran } u$ is then a formal counterpart of what one seeks.

II.5.8 Problem Suppose that $\underline{x}$ and $\underline{y}$ denote distinct letters. From II.2.2(2) it follows that the sentence

$$(\underline{x} = \underline{y}) \Rightarrow (\underline{x} \subset \underline{y}) \tag{37}$$

is true. Verify that the converse sentence, namely

$$(\underline{x} \subset \underline{y}) \Rightarrow (\underline{x} = \underline{y}) \tag{38}$$

is undecidable (that is, neither true nor false). (The consistency of set theory is to be assumed.)

(Note the choice of the term "Verify" ; see I.3.4(i).)

II.6 Unions

Introduction Informally, the union of a collection of sets is the set whose elements are precisely those objects which are contained (as an element) in at least one set which is a member of the collection. (The simultaneous use of the terms "collection" and "set" is merely an informal device which may be helpful to the reader's comprehension.) Thus, the union of $\{\{1, 2\}, \{2, 3, 4\}\}$ is equal to $\{1, 2, 3, 4\}$; and the union of $\{\{1\}, \{3\}, \{5\}, \ldots\}$ is equal to $\{1, 3, 5, \ldots.\}$.

Frequently, the union of a collection of sets is spoken of as the union of the sets which are members of that collection. This is reflected in the notation; see II.6.2(3) and the discussion in II.6.4.

Most readers of this book will be familiar (if only through the use of Venn diagrams) with unions of collections of sets in a plane, in cases where the collection has only a small finite number of members. If the collection consists of sets X and Y , or of sets X and Y and Z , the union is usually denoted by $X \cup Y$ or by $X \cup Y \cup Z$ respectively. There are few surprises here. But the situation changes when one considers unions of infinite collections of sets. This comes about most frequently when one has in some way assigned a set X_n to every positive natural number n ; the union of the collection whose members are the sets $X_1, X_2, \ldots, X_n, \ldots.$ is then usually denoted by

$$X_1 \cup X_2 \cup \ldots \ \ldots \cup X_n \cup \ldots$$

or

$$\cup_{n=1}^{\infty} X_n ;$$

cf. the end of II.6.4 below.

In the case of infinite unions $\cup_{n=1}^{\infty} X_n$ surprises come about even in the somewhat simpler case in which all the X_n are sets of real numbers, the latter pictured as points on a line Suppose that to each pair of real numbers a, b one assigns the (bounded) "open interval" $]a, b[$ (see VI.6.11), defined to be the set of real numbers x such that $a < x$ and $x < b$; and also the (bounded) "closed interval" $[a, b]$, defined to be the set of all real numbers x such that $a \leq x$ and $x \leq b$. Thus $]a, b[= \emptyset$ unless $a < b$; $[a, b] = \emptyset$ unless $a \leq b$; $[a, a] = \{a\}$. Now the set R of all real numbers is not equal to the union of any finite collection of bounded intervals (open or closed). But it is equal to the union of certain infinite collections of bounded open (or closed) intervals; for example

$$R = \cup_{n=1}^{\infty}]-n, n[= \cup_{n=1}^{\infty}]-n, 2^n] .$$

This exhibits a striking difference between the finite and the infinite.

Reverting to sets in a plane, referred as usual to axes Oxy , consider the "open square" of points (x, y) in the plane such that $x, y \in]0. 1[$. Define an "open disc" to be the set of points in the plane which are at a distance less than a positive real number (depending on the disc in question and called its radius) from a given point (again depending on the disc and called its centre). Pictures make it seem evident that the said open square is not equal to the union of any finite collection of such open discs. (Can you formulate a more convincing proof?) On the other hand, the same open square is equal to the union of various infinite collections of open discs. For instance, one may label as $(x_n, y_n)(n = 1, 2, 3, \ldots)$ all points of the square, both of whose coordinates are rational numbers belonging to]0, 1[; for each n , there is a largest open disc D_n with centre (x_n, y_n) and contained (as a subset) in the square; and the square can be proved to be equal to the union $D_1 \cup D_2 \cup .. \cup D_n \cup \ldots$.

The reader is strongly recommended to cultivate some familiarity with examples of this sort, treated at first quite informally, and then from a more formal point of view. In doing this, he should consider proofs as well. Many assertions which seem evident from diagrams are surprisingly difficult to prove in any satisfactory way; and some are refutable (see Problem XV/13).

If one seeks to deal formally with $X_1 \cup X_2 \cup \ldots \cup X_n \cup \ldots.$, one faces problems very similar to those mentioned in II.3.9(ii) and II.5.7 above. At the very least, one has to face the problem of the existence of a set S such that $X_1 \subseteq S$, $X_2 \subseteq S$, ..., $X_n \subseteq S$, ... are all true; and, as will emerge in II.11, $X_n \subseteq S \Leftrightarrow \{X_n\} \in P(S)$, where $P(S)$ is the set of all subsets of S .

II.6.1 Formal definition of unions I now return to formalities.

All versions of set theory have an axiom or a schema which justifies the formal counterparts within the theory of the process of forming unions, as informally described above. In the theory being described in this book, for example, the schema S8 enunciated in II.12.1 ensures that, if X is a set, and if x and $\underline{y}$ denote distinct letters not appearing in X , then the sentence

$$(\exists \underline{y})(\underline{y} \in X \wedge \underline{x} \in \underline{y})$$

(which is independent of the choice of the letter $\underline{y}$, different from the letter $\underline{x}$ and not appearing in X) is collectivising in $\underline{x}$; recall the conventions adopted in II.3.12. (This conclusion may be derived by substitution (see I.3.9(ii)) in the metatheorem (XIV) in II.12.1, itself easily derived from S8 , on interchanging $\underline{x}$ and $\underline{y}$ and substituting the sentence $\underline{x} \in \underline{y}$ for $\underline{\underline{A}}$. The derivation appears below.) Hence, if one defines the <u>union of</u> X by

$$\cup X \equiv_{def} \{\underline{x} : (\exists \underline{y})(\underline{y} \in X \wedge \underline{x} \in \underline{y})\} \quad , \tag{1}$$

where $\underline{x}$ and $\underline{y}$ denote distinct letters not appearing in X , then II.3.3(3) implies that this set has the expected property expressed by the theorem schema

$$T! \quad : \qquad (\forall \underline{x})(\underline{x} \in \cup X \Leftrightarrow (\exists \underline{y})(\underline{y} \in X \wedge \underline{x} \in \underline{y})) \quad ; \tag{2}$$

the conventions in II.3.12 are being assumed. (It is important that the replacement rules be used to verify that, in (1) , the choice of distinct letters $\underline{x}$ and $\underline{y}$ not appearing in X is immaterial.)

From (2) and I.3.2(XI) it follows that

$$T! \quad : \qquad T \in \cup X \Leftrightarrow (\exists \underline{y})(\underline{y} \in X \wedge T \in \underline{y}) \quad , \tag{2'}$$

wherein $\underline{y}$ denotes any letter not appearing in X or T . In view of (2') and I.3.2(X) one sees that in (2) one need not assume that $\underline{x}$ does not appear in X (cf. II.3.12).

Here is the proof that the sentence

$$(\exists \underline{y})(\underline{y} \in X \wedge \underline{x} \in \underline{y})$$

is collectivising in $\underline{x}$. By II.3.5, $\text{Coll}_{\underline{x}}(\underline{x} \in \underline{y})$ is a theorem, and so by I.3.2(X), so too is $(\forall\underline{y})\text{Coll}_{\underline{x}}(\underline{x} \in \underline{y})$. I now call upon the theorem schema (XIV) appearing in II.12.1 to deduce that

$$T \quad : \qquad (\forall\underline{w})\text{Coll}_{\underline{x}}(\exists\underline{y})(\underline{y} \in \underline{w} \wedge \underline{x} \in \underline{y}) \quad ,$$

$\underline{w}$ denoting any letter different from $\underline{x}$ and $\underline{y}$. By (I) and (XI) in I.3.2, therefore

$$T \quad : \qquad (X|\underline{w})\text{Coll}_{\underline{x}}(\exists\underline{y})(\underline{y} \in \underline{w} \wedge \underline{x} \in \underline{y}) \quad .$$

Using II.3.4(r), the last written formal sentence is identical with

$$\text{Coll}_{\underline{x}}(X|\underline{w})(\exists\underline{y})(\underline{y} \in \underline{w} \wedge \underline{x} \in \underline{y}) \quad ;$$

and, by I.1.8(f), this is identical with

$$\text{Coll}_{\underline{x}}(\exists\underline{y})(\underline{y} \in X \wedge \underline{x} \in \underline{y}) \quad ,$$

as required.

Sometimes the notations $\bigcup_{\underline{y}\in X} \underline{y}$ and $\bigcup\{\underline{y} : \underline{y} \in X\}$ are found to be intuitively more acceptable than $\bigcup X$; see §3 of the Appendix.

Note the replacement rule

$$(S|\underline{u})(\bigcup X) \equiv \bigcup(S|\underline{u})X \quad ,$$

an instance of which reads

$$(X|\underline{u})(\bigcup\underline{u}) \equiv \bigcup X \quad .$$

Note also the theorem schema

$$T! \quad : \qquad (X = Y) \Rightarrow (\bigcup X = \bigcup Y) \quad .$$

(In verifying the above replacement rule, for example, one may apply (1) , wherein the distinct letters $\underline{x}$ and $\underline{y}$ are chosen to be different from $\underline{u}$

and not appearing in S or X . Then

$$(S|\underline{u})UX \equiv (S|\underline{u})\{\underline{x} : (\exists \underline{y})(\underline{y} \in X \wedge \underline{x} \in \underline{y})\}$$

$$\equiv \{\underline{x} : (S|\underline{u})(\exists \underline{y})(\underline{y} \in X \wedge \underline{x} \in \underline{y})\} \qquad \text{by II.3.4(r)}$$

$$\equiv \{\underline{x} : (\exists \underline{y})(S|\underline{u})(\underline{y} \in X \wedge \underline{x} \in \underline{y})\} \qquad \text{by I.1.8(f)}$$

$$\equiv \{\underline{x} : (\exists \underline{y})(\underline{y} \in (S|\underline{u})X \wedge \underline{x} \in \underline{y})\}$$

$$\equiv U(S|\underline{u})X \quad ,$$

the last step since $\underline{x}$ and $\underline{y}$ do not appear in $(S|\underline{u})X$.)

Remark It is of considerable practical interest to note that, once set theory has got under way, the most frequently occurring case of UX is that in which one knows of a set E such that $\underline{y} \in X \Rightarrow \underline{y} \subseteq E$, where $\underline{y}$ is a letter not appearing in E , is a theorem. In this situation, the collectivising nature of $(\exists \underline{y})(\underline{x} \in \underline{y} \wedge \underline{y} \in X)$ can be derived easily from II.3.7(XII). Thus, choose a letter $\underline{x}'$ different from $\underline{x}$ and $\underline{y}$; by hypothesis coupled with (XI) and (I) in I.3.2, $(\underline{x}'|\underline{x})(\underline{y} \in X \Rightarrow \underline{y} \subseteq E)$ is a theorem, that is, $\underline{y} \in X \Rightarrow \underline{y} \subseteq E'$ is a theorem, where $E' \equiv (\underline{x}'|\underline{x})E$. Note that neither $\underline{x}$ nor $\underline{y}$ appears in E' . Thus, by Problems I/1 and II.2.3,

$$\underline{x} \in \underline{y} \wedge \underline{y} \in X \Rightarrow \underline{x} \in \underline{y} \wedge \underline{y} \subseteq E'$$

and

$$\underline{x} \in \underline{y} \wedge \underline{y} \subseteq E' \Rightarrow \underline{x} \in E'$$

are theorems; hence

$$\underline{x} \in \underline{y} \wedge \underline{y} \in X \Rightarrow \underline{x} \in E'$$

is a theorem. So, since $\underline{y}$ does not appear in E' , one derives via I.3.3(1) that

$$(\exists \underline{y})(\underline{x} \in \underline{y} \wedge \underline{y} \in X) \Rightarrow \underline{x} \in E'$$

is a theorem. Since $\underline{x}$ does not appear in E' , II.3.7(XII) affirms that

$$\mathrm{Coll}_{\underline{x}}(\exists \underline{y})(\underline{x} \in \underline{y} \wedge \underline{y} \in X)$$

is a theorem, as alleged.

Unfortunately, this case is not adequate to cover everything; in other words, one may encounter cases where one cannot guarantee a priori that every set belonging to X is a subset of some fixed set.

II.6.2 A special case It is usual to define

$$X \cup Y \equiv_{\mathrm{def}} \cup\{X, Y\} \quad , \tag{3}$$

in which case one has the theorem schemas (recall II.3.12 and the Problem II/5)

T! : $$\{X, Y\} = \{X\} \cup \{Y\} \quad , \tag{4}$$

T! : $$\mathrm{Coll}_{\underline{x}}(\underline{x} \in X \vee \underline{x} \in Y) \wedge (X \cup Y = \{\underline{x} : \underline{x} \in X \vee \underline{x} \in Y\}) \quad . \tag{5}$$

In elementary work, the second clause of (5) may be presented and adopted as the definition of $X \cup Y$.

Semiformal proof of (5) By II.4.3(11),

T : $$(\exists \underline{y})(\underline{y} \in \{X, Y\} \wedge \underline{x} \in \underline{y}) \Leftrightarrow (\underline{x} \in X \vee \underline{x} \in Y) \quad .$$

Hence, by I.3.2(X),

$$T \quad : \qquad (\forall \underline{x})((\exists \underline{y})(\underline{y} \in \{X, Y\} \wedge \underline{x} \in \underline{y}) \Leftrightarrow (\underline{x} \in X \vee \underline{x} \in Y)) \quad .$$

By II.3.6(18), therefore,

$$T \quad : \qquad \begin{array}{l} (\mathrm{Coll}_{\underline{x}}(\exists \underline{y})(\underline{y} \in \{X, Y\} \wedge \underline{x} \in \underline{y}) \Leftrightarrow \mathrm{Coll}_{\underline{x}}(\underline{x} \in X \vee \underline{x} \in Y)) \\ \wedge \ \{\underline{x} : (\exists \underline{y})(\underline{y} \in \{X, Y\} \wedge \underline{x} \in \underline{y})\} = \{\underline{x} : \underline{x} \in X \vee \underline{x} \in Y\} \quad . \end{array} \qquad (5_1)$$

Since, as has been established in II.6.1,

$$T \quad : \qquad \mathrm{Coll}_{\underline{x}}(\exists \underline{y})(\underline{y} \in \{X, Y\} \wedge \underline{x} \in \underline{y}) \quad ,$$

and since, by II.6.1(1),

$$\cup\{X, Y\} \equiv \{\underline{x} : (\exists \underline{y})(\underline{y} \in \{X, Y\} \wedge \underline{x} \in \underline{y})\} \quad ,$$

(5) follows from (5_1) and the metatheorems in I.3.2 and I.3.3. □

It can be verified (compare the end of II.4.1 and recall II.3.12) that

$$(S|\underline{u})(X \cup Y) \equiv (S|\underline{u})X \cup (S|\underline{u})Y \quad ,$$

and that

$$(X|\underline{u}, Y|\underline{v})(\underline{u} \cup \underline{v}) \equiv X \cup Y \equiv (X|\underline{x})(Y|\underline{y})(\underline{x} \cup \underline{y})$$

$$\equiv (Y|\underline{y})(X|\underline{x})(\underline{x} \cup \underline{y}) \quad .$$

Thus, for example (displaying on the right the location of earlier replacement rules being utilised), II.6.2(3) yields

$$(X|\underline{u},\ Y|\underline{v})(\underline{u} \cup \underline{v}) \equiv (X|\underline{u},\ Y|\underline{v})\mathrm{U}\{\underline{u},\ \underline{v}\}$$

$$\equiv (X|\underline{x})(Y|\underline{y})(\underline{x}|\underline{u})(\underline{y}|\underline{v})\mathrm{U}\{\underline{u},\ \underline{v}\} \quad , \text{((*) in II.1.3(ii))}$$

$\underline{x}$ and $\underline{y}$ denoting distinct letters different from $\underline{u}$ and $\underline{v}$ and not appearing in X or Y ,

$$\equiv (X|\underline{x})(Y|\underline{y})(\underline{x}|\underline{u})\mathrm{U}(\underline{y}|\underline{v})\{\underline{u},\ \underline{v}\} \qquad \text{(II.6.1)}$$

$$\equiv (X|\underline{x})(Y|\underline{y})(\underline{x}|\underline{u})\mathrm{U}\{\underline{u},\ \underline{y}\} \qquad \text{(II.4.1)}$$

$$\equiv (X|\underline{x})(Y|\underline{y})\mathrm{U}(\underline{x}|\underline{u})\{\underline{u},\ \underline{y}\} \qquad \text{(II.6.1)}$$

$$\equiv (X|\underline{x})(Y|\underline{y})\mathrm{U}\{\underline{x},\ \underline{y}\} \qquad \text{(II.4.1)}$$

$$\equiv (X|\underline{x})\mathrm{U}(Y|\underline{y})\{\underline{x},\ \underline{y}\} \qquad \text{(II.6.1)}$$

$$\equiv (X|\underline{x})\mathrm{U}\{\underline{x},\ Y\} \qquad \text{(II.4.1)}$$

$$\equiv \mathrm{U}(X|\underline{x})\{\underline{x},\ Y\} \qquad \text{(II.6.1)}$$

$$\equiv \mathrm{U}\{X,\ Y\} \qquad \text{(II.4.1)}$$

$$\equiv X \cup Y \quad ,$$

the last step by II.6.2(3) again. Similarly, using II.6.2(3) twice,

$$(X|\underline{x})(Y|\underline{y})(\underline{x} \cup \underline{y}) \equiv (X|\underline{x})(Y|\underline{y})\mathrm{U}\{\underline{x},\ \underline{y}\}$$

$$\equiv (X|\underline{x})\mathrm{U}(Y|\underline{y})\{\underline{x},\ \underline{y}\} \qquad \text{(II.6.1)}$$

$$\equiv \mathrm{U}(X|\underline{x})(Y|\underline{y})\{\underline{x},\ \underline{y}\} \qquad \text{(II.6.1)}$$

$$\equiv \mathrm{U}\{X,\ Y\} \qquad \text{(II.4.1)}$$

$$\equiv X \cup Y \quad ;$$

remember here the assumption that $\underline{x}$ and $\underline{y}$ do not appear in X or Y . Likewise, under this assumption,

$$(Y|\underline{y})(X|\underline{x})(\underline{x} \cup \underline{y}) \equiv X \cup Y \quad .$$

The reader should also prove the theorem schema

T! : $$(X = X' \wedge Y = Y') \Rightarrow (X \cup Y = X' \cup Y') \quad .$$

II.6.3 <u>Some general properties</u> I begin with the theorem schema

T! : $$(\forall\underline{u})((\underline{u} \in X \wedge X \in Y) \Rightarrow \underline{u} \in \mathrm{U}Y) \quad ; \qquad (6)$$

or, what is metamathematically equivalent, the theorem schema

$T_!$: $$(S \in X \wedge X \in Y) \Rightarrow S \in \cup Y \quad ; \qquad (6')$$

in (6) , X and Y denote arbitrary sets and $\underline{u}$ an arbitrary letter; while in (6') , X , Y and S denote arbitrary sets. In relation to (6) and (6'), see II.6.7 below.

Semiformal proof of (6) By (I), (XI) and (X) in I.3.2, it may and will be assumed that $\underline{u}$ is a letter $\underline{x}$ which does not appear in X or Y . Adjoin to Θ_0 the explicit axiom

$$\underline{x} \in X \wedge X \in Y$$

to obtain a theory Θ .

Since $\underline{y}$ does not appear in Y (see II.3.12),

$$\underline{x} \in X \wedge X \in Y \equiv (X|\underline{y})(\underline{x} \in \underline{y} \wedge \underline{y} \in Y)$$

and hence, by S5 in I.2.2 and I.3.2(I),

T_Θ : $$(\exists \underline{y})(\underline{x} \in \underline{y} \wedge \underline{y} \in Y) \quad .$$

Hence, by II.6.1(2) (with Y in place of X),

T_Θ : $$\underline{x} \in \cup Y \quad .$$

So, by I.3.2(VI),

T_Θ : $$(\underline{x} \in X \wedge X \in Y) \Rightarrow \underline{x} \in \cup Y \quad .$$

Now use I.3.2(X) to derive (6) . □

It is important to know that, if $\underline{\underline{A}}$ and $\underline{\underline{B}}$ are collectivising in $\underline{u}$, so too is $\underline{\underline{A}} \vee \underline{\underline{B}}$, and that then

$$\{\underline{u} : \underline{\underline{A}} \vee \underline{\underline{B}}\} = \{\underline{u} : \underline{\underline{A}}\} \cup \{\underline{u} : \underline{\underline{B}}\} \quad .$$

This is derivable from the following theorem schema

$$\mathrm{T}! \quad : \quad (\mathrm{Coll}_{\underline{u}}\underline{\underline{A}} \wedge \mathrm{Coll}_{\underline{u}}\underline{\underline{B}}) \Rightarrow (\mathrm{Coll}_{\underline{u}}(\underline{\underline{A}} \vee \underline{\underline{B}}) \wedge \{\underline{u} : \underline{\underline{A}} \vee \underline{\underline{B}}\} = \{\underline{u} : \underline{\underline{A}}\} \cup \{\underline{u} : \underline{\underline{B}}\}) \quad . \qquad (7)$$

<u>Semiformal proof of (7)</u> Let Θ denote the theory obtained by adjoining to Θ_0 the axiom $\mathrm{Coll}_{\underline{u}}\underline{\underline{A}} \wedge \mathrm{Coll}_{\underline{u}}\underline{\underline{B}}$. Let S, T denote $\{\underline{u} : \underline{\underline{A}}\}$ and $\{\underline{u} : \underline{\underline{B}}\}$ respectively. Then, by II.3.3(3),

$$\mathrm{T}\Theta \quad : \quad \underline{\underline{A}} \Leftrightarrow \underline{u} \in S \quad ,$$

$$\mathrm{T}\Theta \quad : \quad \underline{\underline{B}} \Leftrightarrow \underline{u} \in T \quad .$$

Hence, by I.3.6(7),

$$\mathrm{T}\Theta \quad : \quad \underline{\underline{A}} \vee \underline{\underline{B}} \Leftrightarrow (\underline{u} \in S) \vee (\underline{u} \in T) \quad .$$

So, by II.6.2(5),

$$\mathrm{T}\Theta \quad : \quad \underline{\underline{A}} \vee \underline{\underline{B}} \Leftrightarrow \underline{u} \in S \cup T \quad . \qquad (8)$$

By I.3.2(X),

$$\mathrm{T}\Theta \quad : \quad (\forall \underline{u})(\underline{\underline{A}} \vee \underline{\underline{B}} \Rightarrow \underline{u} \in S \cup T) \quad .$$

Since $\underline{u}$ appears in neither S nor T, hence not in $S \cup T$ either, it follows from (XII) in II.3.7 that $\underline{\underline{A}} \vee \underline{\underline{B}}$ is collectivising in $\underline{u}$. Further, by II.3.3(3),

$$T\Theta \quad : \qquad \underline{u} \in \{\underline{u} : \underline{\underline{A}} \vee \underline{\underline{B}}\} \Leftrightarrow \underline{\underline{A}} \vee \underline{\underline{B}} \ . \qquad (9)$$

So by I.3.2(III), (8) and (9) ,

$$T\Theta \quad : \qquad \underline{u} \in \{\underline{u} : \underline{\underline{A}} \vee \underline{\underline{B}}\} \Leftrightarrow \underline{u} \in S \cup T \ ;$$

and, by I.3.2(X),

$$T\Theta \quad : \qquad (\forall \underline{u})(\underline{u} \in \{\underline{u} : \underline{\underline{A}} \vee \underline{\underline{B}}\} \Leftrightarrow \underline{u} \in S \cup T) \ .$$

Reference to II.1.1(1) now shows that

$$T\Theta \quad : \qquad \{\underline{u} : \underline{\underline{A}} \vee \underline{\underline{B}}\} = S \cup T \ ,$$

as alleged. The proof is completed by appeal to I.3.2(VI). □

II.6.4 Examples and discussion Taking up the suggestion made in II.3.9(i), I now formally define

$$\{a, b, c\} \equiv_{def} \{a, b\} \cup \{c\} \ , \qquad (10)$$

$$\{a, b, c, d\} \equiv_{def} \{a, b, c\} \cup \{d\} \ , \qquad (11)$$

wherein a, b, c, d denote strings. (The reader should prepare himself for the use of lower case letters to denote strings. It is not intended in (10) and (11) that the letters a, b, c (or the strings they denote) are necessarily different; thus, for example, (10) is intended to cover also the definition $\{a, b, a\} \equiv \{a, b\} \cup \{a\}$.)

There are very many simple theorem schemas relating to the definitions (10) and (11) . For example,

T! : $\{a, b, c\} = \{a, c, b\}$, (12)

T! : $\{a, b, c, d\} = \{a, b\} \cup \{c, d\}$. (13)

In view of later developments (see the next paragraph but one), there is little point in formulating all the expected properties and giving semiformal proofs.

In line with II.6.2(3) one might define

$$X \cup Y \cup Z \equiv_{def} \cup\{X, Y, Z\} \quad , \tag{14}$$

$$X \cup Y \cup Z \cup W \equiv_{def} \cup\{X, Y, Z, W\} \quad , \tag{15}$$

and prove that, in line with II.6.2(4)

T! : $\{a, b, c\} = \{a\} \cup \{b\} \cup \{c\}$, (16)

T! : $(a, b, c, d) = \{a\} \cup \{b\} \cup \{c\} \cup \{d\}$, (17)

T! : $X \cup Y \cup Z = (X \cup Y) \cup Z$, (18)

T! : $X \cup Y \cup Z \cup W = (X \cup Y \cup Z) \cup W$. (19)

Moreover,

T! : $X \cup Y \cup Z = \{\underline{x} : \underline{x} \in X \vee \underline{x} \in Y \vee \underline{x} \in Z\}$, (20)

with a corresponding theorem schema relating to (19) .

The definitions (10) and (11) immediately suggest defining $\{a_1, \ldots, a_n\}$ for natural numbers n and sets $a_1, \ldots, a_n$ by "recurrence" :

$$\{a_1, \ldots, a_{n+1}\} \equiv \{a_1, \ldots, a_n\} \cup \{a_{n+1}\} \quad ;$$

and (14) and (15) suggest a similar definition by recurrence of unions $X_1 \cup \ldots \cup X_n$. However, one cannot formalise these definitions at present since, for one thing, one has as yet no formal definition of "natural number" . Of course, the definitions may be regarded informally for the purposes of illustrations. The formal theory can, as indeed it must, proceed without any premature "definitions" of this sort; see II.3.9(ii). The matter will be reconsidered in V.5.5.

The fact is that (as has been said in II.3.9(ii)), although $\{X_1, X_2, \ldots, X_n, \ldots\}$ and $X_1 \cup X_2 \cup \ldots \cup X_n \cup \ldots$. or $\bigcup_{n=1}^{\infty} X_n$ and similar notations are very convenient and are used abundantly in conventional texts, they are for us rather a luxury. So too are the concepts denoted by $\bigcup_{\underline{x} \in X} T[\![\underline{x}]\!]$ or $\bigcup\{T[\![\underline{x}]\!] : \underline{x} \in X\}$ and $\bigcup_{\underline{x} \in X} f(\underline{x})$ or $\bigcup\{f(\underline{x}) : \underline{x} \in X\}$, the last two denoting the union of the family $\underline{x} \rightsquigarrow f(\underline{x})$ with domain X (see IV.11). Because they are luxuries, the appropriate formal definitions will be delayed and relegated to §3 of the Appendix. Suffice it to say here that the definition of $\bigcup_{\underline{x} \in X} T[\![\underline{x}]\!]$ will be such that

$$\underline{y} \in \bigcup_{\underline{x} \in X} T[\![\underline{x}]\!] \Leftrightarrow (\exists \underline{x})(\underline{x} \in X \wedge \underline{y} \in T[\![\underline{x}]\!]) \tag{21}$$

is a theorem schema, where $\underline{x}$ and $\underline{y}$ denote distinct letters, $\underline{x}$ not appearing in X and $\underline{y}$ not appearing in X or T . More informally, $\underline{y} \in \bigcup_{\underline{x} \in X} T[\![\underline{x}]\!]$ if and only if there exists $\underline{x} \in X$ such that $\underline{y} \in T[\![\underline{x}]\!]$ (that is, if and only if $\underline{y} \in T[\![\underline{x}]\!]$ for at least one $\underline{x} \in X$).

In the special case $T \equiv \underline{x}$ it appears that

$$\bigcup_{\underline{x} \in X} \underline{x} \equiv \bigcup X \quad ;$$

conventionally, $\bigcup_{\underline{x} \in X} \underline{x}$ is often denoted by $\bigcup_{Y \in X} Y$ or $\bigcup_{A \in X} A$, et cetera (the notation is rather ill regulated).

II.6.5 <u>Problem</u> Write out semiformal proofs of II.6.4(12) and II.6.4(18).

II.6.6 Problem Give semiformal proofs of the theorems

T! : $$(\forall \underline{x})((\underline{x} \cup \emptyset) = \underline{x}) \quad ,$$

T! : $$(\forall \underline{x})(\forall \underline{y})((\underline{x} \cup \underline{y}) = (\underline{y} \cup \underline{x})) \quad .$$

II.6.7 Problem In connection with II.6.3(6) and (6') , and by comparison with II.7.1(3) and (3') , exhibit a letter $\underline{u}$ and sets X and Y such that

T : $$\neg(\forall \underline{u})(\underline{u} \in \cup Y \Rightarrow (\underline{u} \in X \wedge X \in Y)) \quad .$$

Remarks This illustrates the point that appropriate metamathematical restrictions on formal letters are not always intuitively evident; see the Remarks following II.1.1.

(Hints : Let $\underline{u}$ denote an arbitrary letter and define $X \equiv \emptyset$ and $Y \equiv \{\{\underline{u}\}\}$. Argue by contradiction, making use of II.4.2(11) and II.8(1).)

II.6.8 Problem Prove the theorem schema

T! : $$X \in Y \Rightarrow X \subseteq \cup Y \quad .$$

II.7 Intersections

Introduction As in the case of unions, I begin with some informal discussion.

Intuitively, the intersection of a collection of sets is the set whose elements are precisely those objects which are elements of every set which is a member of the given collection of sets, that is, precisely those objects which are common to all sets of the collection. Very often this intersection is spoken of as the intersection of the sets belonging to the collection (rather than of the collection itself); and this is reflected in the notation which is frequently used (see II.7.4 below).

Thus the intersection of $\{\{1, 2\}, \{2, 3, 4\}\}$ is equal to $\{2\}$; that of $\{\{1, 2, 3\}, \{4, 3, 2\}\}$ is equal to $\{2, 3\}$; and that of $\{\{1, 2\}, \{4, 3\}\}$ is equal to $\emptyset$.

There is one curious feature which distinguishes the treatment of intersections from that of unions. This appears as soon as one comes to consider the intersection of a collection having no members. (Intuitively it may seem odd to contemplate this situation, but the formal viewpoint makes it desirable (to say the least) to face it.) In this case, the above prescription would lead one to expect that no object can fail to be an element of the intersection (since to fail to do this, there would have to be at least one member of the collection of which it is not an element). The said intersection would thus appear to be a set, of which every object is a member. In view of the remarks in II.3.10 about "the set of all sets" , it is clear that formal developments will require special care at this point.

By analogy with the case of unions, if m is a positive natural number, and if to every positive natural number $n \le m$ (or to every positive natural number n) one has assigned a set X_n , the intersection of the collections $\{X_1, \dots, X_m\}$ and $\{X_1, \dots, X_m, \dots\}$ is usually denoted by

$$X_1 \cap \dots \cap X_m \quad \text{or} \quad \cap_{n=1}^{m} X_n$$

and

$$X_1 \cap \dots \cap X_m \cap \dots \quad \text{or} \quad \cap_{n=1}^{\infty} X_n$$

respectively. The case of finite intersections will be reconsidered in V.5.5.

As in the case of unions (see the Introduction to II.6), the consideration of intersections of infinite collections of sets of the type $\cap_{n=1}^{\infty} X_n$ can produce mild surprises. Consider, for instance, the case of nonempty collections of sets of real numbers. A bounded closed interval $[a, b]$, where $a \leq b$, is not equal to the intersection of any finite collection of bounded open intervals; but it is equal to the intersection $\bigcap_{n=1}^{\infty}]a - \frac{1}{n}, b + \frac{1}{n}[$ of an infinite collection of bounded open intervals. A bounded open interval $]a, b[$, where $a < b$, is not equal to the union of any finite collection of bounded closed intervals; but it is equal to the union $\bigcup_{n=1}^{\infty} [a + \frac{1}{n}, b - \frac{1}{n}]$ of an infinite collection of bounded closed intervals.

These observations prompt an interesting problem of classification. Denote by G_0 the collection of sets (of real numbers) which are equal to unions of the form $\bigcup_{n=1}^{\infty}]a_n, b_n[$, where a_n and b_n are real numbers. With G_0 as starting point, define collections $G_1, G_2, \ldots.$ in the following way:

> If ν is an even natural number, $G_{\nu+1}$ is the collection of all sets of the form $\bigcap_{n=1}^{\infty} A_n$, where, for each positive natural number n , A_n belongs to the collection $G_0 \cup \ldots \cup G_\nu$; if ν is an odd natural number, $G_{\nu+1}$ is the collection of all sets of the form $\bigcup_{n=1}^{\infty} A_n$, where, for each positive natural number n , A_n belongs to the collection $G_0 \cup \ldots \cup G_\nu$.

(Regarding the phrase "the set of all sets of the form .... , where .... " , see I.3.5(vi) above and II.12.1 below.) It is then evident that $G_0 \subseteq G_1 \subseteq G_2 \subseteq \ldots\ldots.$, and it is an intriguing (and far from trivial) problem to determine the first, if any, of the collections G_ν to which a given set (of real numbers) belongs. It turns out, for example, that the set of irrational numbers belongs to G_1 but not to G_0 , and likewise with the set of transcendental numbers; also that the set of rational numbers belongs to G_2 but not to G_1 , and likewise with the set of algebraic numbers. It can be proved (though not very easily) that, for any natural number ν , there are sets of real numbers which belong to $G_{\nu+1}$ but not to G_ν ; and even that there are sets of real numbers which belong to no G_ν at all. It may in fact come as a surprise that there is so much variety among sets of real numbers.

Two points should be stressed. The first is that the classification of sets of real numbers suggested above transcends the use of diagrams, which are far too coarse to reflect the differences involved. (Try to draw a diagram which usefully represents the set of rational numbers.) The second is that the facts stated above are quite difficult to prove. The proofs involve, among other things, a famous theorem of Baire (see, for example, Gleason (1), p.263) to the effect that, if a union $\bigcup_{n=1}^{\infty} F_n$, in which, for every positive natural number n, F_n is the complement (relative to the set of real numbers) of a set belonging to G_0, contains all real numbers, then at least one of the F_n contains a nonvoid bounded open interval. Compare with XV.2.3.

II.7.1 <u>Formal definition of intersections</u> Reverting to formalities, I introduce the definition schema

$$\cap Y \equiv_{def} \{\underline{x} : (\forall \underline{y})(\underline{y} \in Y \Rightarrow \underline{x} \in \underline{y})\} , \tag{1}$$

wherein Y denotes an arbitrary string and $\underline{x}$ and $\underline{y}$ denote distinct letters not appearing in Y (their choice is otherwise immaterial), and name $\cap Y$ the <u>intersection of</u> Y. If Y is a set, $\cap Y$ is a set; but one must not be too hasty in assuming that $\cap Y$ invariably has the intuitively expected property (that is, the property that the elements of $\cap Y$ are precisely the sets which are elements of every element of Y). It has already been predicted that something singular is to be expected in case $Y = \emptyset$, and it is not difficult to guess that the trouble is linked with doubts as to whether the sentence

$$(\forall \underline{y})(\underline{y} \in Y \Rightarrow \underline{x} \in \underline{y})$$

is collectivising in $\underline{x}$.

The first and principal step in further discussion is the claim:

T! : $$Y \neq \emptyset \Rightarrow \mathrm{Coll}_{\underline{x}}(\forall \underline{y})(\underline{y} \in Y \Rightarrow \underline{x} \in \underline{y}) \tag{2}$$

<u>Semiformal proof of (2)</u> It follows from II.5.2(9), I.3.3(k) and the replacement rules that (cf. II.5.4(29))

T : $$Y \neq \emptyset \Rightarrow (\exists \underline{y})(\underline{y} \in Y) \quad .$$

Hence, by (III) and (IX) in I.3.2, to prove (2) it will suffice to prove

T : $$\underline{y} \in Y \Rightarrow \mathrm{Coll}_{\underline{x}}(\forall \underline{y})(\underline{y} \in Y \Rightarrow \underline{x} \in \underline{y}) \quad .$$

By I.3.2(VI), if Θ denotes the theory obtained from Θ_0 by adjoining $\underline{y} \in Y$ as an axiom, it is enough to prove that

TΘ : $$\mathrm{Coll}_{\underline{x}}(\forall \underline{y})(\underline{y} \in Y \Rightarrow \underline{x} \in \underline{y}) \quad .$$

Further, by I.3.2(I) and (XII) in II.3.7, this last is a corollary of

TΘ : $$(\forall \underline{y})(\underline{y} \in Y \Rightarrow \underline{x} \in \underline{y}) \Rightarrow \underline{x} \in \underline{y} \quad . \tag{2_1}$$

If Θ' denotes the theory obtained from Θ by adjoining the axiom

$$(\forall \underline{y})(\underline{y} \in Y \Rightarrow \underline{x} \in \underline{y}) \quad ,$$

then, by (XI) and (I) in I.3.2,

$$\underline{y} \in Y \Rightarrow \underline{x} \in \underline{y}$$

is true in Θ' . Since also $\underline{y} \in Y$ is true in Θ , hence true in Θ' , I.3.2(I) shows that $\underline{x} \in \underline{y}$ is true in Θ' . By I.3.2(VI), this proves (2_1) and the proof is complete. □

Using (2) , II.3.3(3) and I.3.2(III) it now follows that

$$\mathrm{T!} \; : \quad Y \neq \emptyset \Rightarrow (\forall \underline{x})(\underline{x} \in \cap Y \Leftrightarrow (\forall \underline{y})(\underline{y} \in Y \Rightarrow \underline{x} \in \underline{y})) \; ; \tag{3}$$

and this in turn leads (cf. II.6.3(6) and (6')) to the theorem schema

$$\mathrm{T!} \; : \quad (Y \neq \emptyset) \Rightarrow (S \in \cap Y \Leftrightarrow (\forall \underline{y})(\underline{y} \in Y \Rightarrow S \in \underline{y})) \; . \tag{3'}$$

wherein Y and S denote arbitrary sets and $\underline{y}$ a letter not appearing in Y or S . It is on the theorem schemas (3) and (3') that virtually all use of the concept of intersection is based.

As in the case of unions, the notations $\cap_{\underline{y} \in Y} \underline{y}$ or $\cap\{\underline{y} : \underline{y} \in Y\}$ are sometimes preferred to $\cap Y$; see the end of II.6.1 above. Moreover, there are sets denoted by

$$\cap_{\underline{x} \in X} T[\![\underline{x}]\!] \qquad \text{or} \qquad \cap\{T[\![\underline{x}]\!] : \underline{x} \in X\} \; ,$$

analogous to those relating to unions mentioned at the end of II.6.4 (see again §3 of the Appendix).

It can be verified that

$$(S|\underline{u})(\cap X) \equiv \cap(S|\underline{u})X$$

and, in particular, that

$$(X|\underline{u})(\cap\underline{u}) \equiv \cap X \; .$$

To do this, choose distinct letters $\underline{x}$ and $\underline{y}$, both different from $\underline{u}$ and neither appearing in X or S . Then, by II.6.1(1),

$$\begin{aligned} (S|\underline{u})\cap X &\equiv (S|\underline{u})\{\underline{x} : (\forall \underline{y})(\underline{y} \in X \Rightarrow \underline{x} \in \underline{y})\} \\ &\equiv \{\underline{x} : (S|\underline{u})(\forall \underline{y})(\underline{y} \in X \Rightarrow \underline{x} \in \underline{y})\} \qquad \text{(II.3.4)} \end{aligned}$$

$$\equiv \{\underline{x} : (\forall \underline{y})(S|\underline{u})(\underline{y} \in X \Rightarrow \underline{x} \in \underline{y})\} \qquad \text{(I.1.8(f))}$$
$$\equiv \{\underline{x} : (\forall \underline{y})(\underline{y} \in (S|\underline{u})X \Rightarrow \underline{x} \in \underline{y})\}$$
$$\equiv \cap(S|\underline{u})X \qquad \text{(II.6.1(1) again).}$$

Moreover one can prove the theorem schema

T! : $$(X = Y) \Rightarrow (\cap X = \cap Y) \quad .$$

II.7.2 <u>A special case</u> By analogy with II.6.2(3), it is usual to define formally

$$X \cap Y \equiv_{def} \cap\{X, Y\} \quad , \tag{4}$$

in which case one has the theorem schema (recall II.3.12 and compare with II.6.2(4))

T! : $$\mathrm{Coll}_{\underline{x}}(\underline{x} \in X \wedge \underline{x} \in Y) \wedge (X \cap Y = \{\underline{x} : \underline{x} \in X \wedge \underline{x} \in Y\}) \quad . \tag{5}$$

In elementary work, the second clause of (5) is often presented and adopted as a definition of $X \cap Y$.

The proof of (5) is similar to that of II.6.2(5) , except that use is made of II.4.3(10) rather than II.4.3(11).

One may verify the replacement rules (recall II.3.12 and compare with II.6.2)

$$(S|\underline{u})(X \cap Y) \equiv (S|\underline{u})X \cap (S|\underline{u})Y \quad ,$$

$$(X|\underline{u}, Y|\underline{v})(\underline{u} \cap \underline{v}) \equiv X \cap Y \quad ,$$

and

$$(X|\underline{x})(Y|\underline{y})(\underline{x} \cap \underline{y}) \equiv X \cap Y \equiv (Y|\underline{y})(X|\underline{x})(\underline{x} \cap \underline{y}) \quad .$$

The reader should prove the theorem schema

$$\text{T!} \quad : \qquad (X = X' \wedge Y = Y') \Rightarrow (X \cap Y = X' \cap Y') \quad .$$

A further standard definition schema reads

$$X \ \text{intersects} \ Y \ \equiv_{\text{def}} \ X \cap Y \neq \emptyset \quad ,$$

linked to which (see II.8(2) below) is the theorem schema

$$\text{T!} \quad : \qquad (X \ \text{intersects} \ Y) \Leftrightarrow (Y \ \text{intersects} \ X) \quad .$$

II.7.3 <u>Some general properties</u> It can be proved that if $\underline{\underline{A}}$ and $\underline{\underline{B}}$ are collectivising in $\underline{u}$, then so too is $\underline{\underline{A}} \wedge \underline{\underline{B}}$, in which case

$$\{\underline{u} : \underline{\underline{A}} \wedge \underline{\underline{B}}\} = \{\underline{u} : \underline{\underline{A}}\} \cap \{\underline{u} : \underline{\underline{B}}\} \quad .$$

More precisely, one has the theorem schema

$$\text{T!} \quad : \quad (\text{Coll}_{\underline{u}}\underline{\underline{A}} \wedge \text{Coll}_{\underline{u}}\underline{\underline{B}}) \Rightarrow (\text{Coll}_{\underline{u}}(\underline{\underline{A}} \wedge \underline{\underline{B}}) \wedge (\{\underline{u} : \underline{\underline{A}} \wedge \underline{\underline{B}}\} = \{\underline{u} : \underline{\underline{A}}\} \cap \{\underline{u} : \underline{\underline{B}}\})) \ ; \tag{6}$$

see Problem II/6.

Incidentally, informal texts often use the notation $\{\underline{u} : \underline{\underline{A}}, \underline{\underline{B}}\}$ in lieu of $\{\underline{u} : \underline{\underline{A}} \wedge \underline{\underline{B}}\}$. Since this achieves negligible economy, the explanation is presumably that the comma is thought to look less forbidding than the conjunction sign $\wedge$.

II.7.4 <u>Examples and notation</u> Parallel with (14) and (15) in II.6.4, one defines

$$X \cap Y \cap Z \equiv_{def} \cap\{X, Y, Z\} \quad , \tag{7}$$

$$X \cap Y \cap Z \cap W \equiv_{def} \cap\{X, Y, Z, W\} \quad , \tag{8}$$

these are also suggested by II.7.2(4), of course. One can prove that

$$T! \quad : \qquad X \cap Y \cap X = (X \cap Y) \cap Z \quad , \tag{9}$$

$$T! \quad : \qquad X \cap Y \cap Z \cap W = (X \cap Y \cap Z) \cap W \quad , \tag{10}$$

$$T! \quad : \qquad X \cap Y \cap Z = \{\underline{x} : \underline{x} \in X \wedge \underline{x} \in Y \wedge \underline{x} \in Z\} \quad . \tag{11}$$

Remarks analogous to those at the end of II.6.4 apply to the notations $\cap_{\underline{x} \in X} T[\![\underline{x}]\!]$ or $\cap\{T[\![\underline{x}]\!] : \underline{x} \in X\}$ and $\cap_{\underline{x} \in X} f(\underline{x})$ or $\cap\{f(\underline{x}) : \underline{x} \in X\}$, covering intersections of families; see again §3 of the Appendix. The formal definitions will secure the theorem schema

$$(X \neq \emptyset) \Rightarrow (\underline{y} \in \cap_{\underline{x} \in X} T[\![\underline{x}]\!] \Leftrightarrow (\forall \underline{x})(\underline{x} \in X \Rightarrow \underline{y} \in T[\![\underline{x}]\!])) \quad ,$$

where $\underline{x}$ and $\underline{y}$ denote distinct letters, $\underline{x}$ does not appear in X and $\underline{y}$ does not appear in X or T . More informally: if $X \neq \emptyset$, then $\underline{y} \in \cap_{\underline{x} \in X} T[\![\underline{x}]\!]$ if and only if $\underline{y} \in T[\![\underline{x}]\!]$ for all $\underline{x} \in X$.

In case $T \equiv \underline{x}$,

$$\cap_{\underline{x} \in X} \underline{x} \equiv \cap X \quad ,$$

which suggests the conventional (but ill-regulated) notation $\cap_{Y \in X} Y$ or $\cap_{A \in X} A$, et cetera, in place of $\cap X$.

II.7.5 Problem (i) Give a semiformal proof of the theorem schema

T! : $$\mathrm{Coll}_{\underline{u}} A \Rightarrow \mathrm{Coll}_{\underline{u}} (\underline{\underline{A}} \wedge \underline{\underline{B}}) \quad . \tag{12}$$

(ii) Give a semiformal proof of II.7.4(11) above.

(iii) Prove the theorem schema

T! : $$(X \in Y) \Rightarrow (\cap Y \subseteq X) \quad . \tag{13}$$

II.7.6 Problem Give semiformal proofs of the theorems

T! : $$(\forall \underline{x})((\underline{x} \cap \emptyset) = \emptyset) \quad , \tag{14}$$

T! : $$(\forall \underline{x})(\forall \underline{y})((\underline{x} \cap \underline{y}) = (\underline{y} \cap \underline{x})) \quad . \tag{15}$$

II.7.7 Duality between sentences and sets At this point it may be remarked that the theorems enunciated in II.3.8, coupled with II.6.3(7) and II.7.3(6), may suggest a species of duality between

$$\text{sentences}, \quad \Rightarrow \quad , \quad \vee \quad , \quad \wedge$$

and

$$\text{sets}, \quad \subseteq \quad , \quad \cup \quad , \quad \cap \quad .$$

This duality, such as it is, is often suggestive and useful. It is, however, limited in various ways and has to be treated with some caution.

To begin with, the intuitively natural correspondence between sentences and sets is hampered and restricted by the circumstance that sentences may fail to be collectivising. Moreover, although $\underline{\underline{A}} \Rightarrow \underline{\underline{B}}$ denotes a sentence whenever $\underline{\underline{A}}$ and $\underline{\underline{B}}$ denote sentences, $A \subseteq B$ denotes a sentence (rather than a set) whenever A and B denote sets.

A more specific and practical danger is exemplified by the feature that, although

$$(\underline{\underline{A}} \Rightarrow (\underline{\underline{B}} \vee \underline{\underline{C}})) \Rightarrow ((\underline{\underline{A}} \Rightarrow \underline{\underline{B}}) \vee (\underline{\underline{A}} \Rightarrow \underline{\underline{C}}))$$

is (see Problem I/34) a theorem schema, it is easy to exhibit sets A, B and C such that

$$(A \subseteq B \cup C) \Rightarrow ((A \subseteq B) \vee (A \subseteq C))$$

is not (unless Θ_0 is contradictory) a theorem; this despite the fact that the said duality might lead one from the first to the second. Likewise with the sentences (or sentence schemas)

$$((\underline{\underline{A}} \wedge \underline{\underline{B}}) \Rightarrow \underline{\underline{C}}) \Rightarrow ((\underline{\underline{A}} \Rightarrow \underline{\underline{C}}) \vee (\underline{\underline{B}} \Rightarrow \underline{\underline{C}}))$$

and

$$(A \cap B \subseteq C) \Rightarrow ((A \subseteq C) \vee (B \subseteq C)) \quad ;$$

again see Problem I/34.

II.8 Further basic properties of unions and intersections

II.8.1 Commutativity, associativity and distributivity of finite unions and intersections. According to II.6.2(5) and II.7.2(5)

T! : $$X \cup Y = \{\underline{x} : \underline{x} \in X \vee \underline{x} \in Y\} \quad ,$$

T! : $$X \cap Y = \{\underline{x} : \underline{x} \in X \wedge \underline{x} \in Y\} \quad ,$$

which (as has been indicated) might be adopted as definitions of $X \cup Y$ and $X \cap Y$.

There are a number of basic theorem schemas, largely based on the above, which need to be recorded. Among them are the following (in which I am using a compressed form of statement):

T! : $\emptyset \subseteq X \ , \ \cup\{X\} = X = \cap\{X\} \ , \ X \cup \emptyset = X \ , \ X \cap \emptyset = \emptyset \ ;$ (1)

T! : $X \cup Y = Y \cup X \ , \ X \cap Y = Y \cap X$ (commutativity); (2)

T! : $X \cup (Y \cup Z) = (X \cup Y) \cup Z \ , \ X \cap (Y \cap Z) = (X \cap Y) \cap Z$ (associativity); (3)

T! : $X \cup X = X \ , \ X \cap X = X$ (idempotence); (4)

T! : $X \cap (Y \cup Z) = (X \cap Y) \cup (X \cap Z) \ ,$ $X \cup (Y \cap Z) = (X \cup Y) \cap (X \cup Z) \ ,$ (distributivity); (5)

T! : $X \subseteq Y \Leftrightarrow X \cup Y = Y \Leftrightarrow X \cap Y = X \ ;$ (6)

T! : $X \subseteq X \cup Y \ , \ Y \subseteq X \cup Y \ ;$ (7)

T! : $X \cap Y \subseteq X \ , \ X \cap Y \subseteq Y \ .$ (8)

Semiformal proofs of (1) - (8) would be very long; they are left as exercises for the courageous reader (cf. II.7.6). As a hint, consider the first part of (5) . Arguing in routine fashion one would say: $\underline{x} \in X \cap (Y \cup Z)$ implies $\underline{x} \in X$ and $\underline{x} \in (Y \cup Z)$, which in turn implies $\underline{x} \in X$ and $(\underline{x} \in Y$ or $\underline{x} \in Z)$. One would then use the theorem schema of logic:

$$\underline{\underline{A}} \wedge (\underline{\underline{B}} \vee \underline{\underline{C}}) \Leftrightarrow (\underline{\underline{A}} \wedge \underline{\underline{B}}) \vee (\underline{\underline{A}} \wedge \underline{\underline{C}}) \ ,$$

which is listed in I.3.3(j). This would lead to $(\underline{x} \in X$ and $\underline{x} \in Y)$ or

$(\underline{x} \in X$ and $\underline{x} \in Z)$, that is, to $(\underline{x} \in X \cap Y)$ or $(\underline{x} \in X \cap Z)$, so finally to $\underline{x} \in (X \cap Y) \cup (X \cap Z)$. It would then be necessary to formulate all this in semi-formal fashion. See also Problem II/7.

Associativity of $\cup$ (first clause of (3)) ensures that, were one to ignore II.6.4(14), there would be no material ambiguity involved in writing $X \cup Y \cup Z$ in place of $X \cup (Y \cup Z)$ or $(X \cup Y) \cup Z$, et cetera. Similarly with $\cap$ in place of $\cup$ (second clause of (3)).

In connection with (1) - (8) above (and (2) and (3) in II.10.2 below), the use of diagrams in an illustrative, suggestive role is customary and acceptable; see also II.10.4 below. However, every reader should construct more formal, diagram-free proofs at least once in his career. See also the remarks about the use of diagrams in the Foreword to Volume 2.

II.8.2 <u>Remarks</u> There are numerous theorem schemas involving general unions and intersections of sets and of families of sets. Examples are

$$\cup(X \cup Y) = (\cup X) \cup (\cup Y)$$

and

$$\cup(X \cap Y) \subseteq (\cup X) \cap (\cup Y) \quad .$$

None of these will play any essential role in the rest of this book and further discussion will be omitted. Interested readers should consult Bourbaki (1), pp. 90-110.

II.9 <u>Disjoint sets</u>

To begin with a simple instance, two sets are said to be disjoint if their intersection is equal to $\emptyset$, that is, if they do not intersect (see II.7.2). Thus, if X denotes any set, then X and $\emptyset$ are disjoint.

A seemingly more general concept is covered by the following formal definition (schema)

$$Z \text{ is disjoint} \equiv_{def} (\forall \underline{x})(\forall \underline{y})((\underline{x} \in Z \wedge \underline{y} \in Z \wedge \underline{x} \neq \underline{y}) \Rightarrow (\underline{x} \cap \underline{y} = \emptyset)) ,$$

that is, Z is disjoint if and only if every two non-equal sets belonging to Z are disjoint. If X and Y denote unequal sets, X and Y are disjoint according to the first definition if and only if $\{X, Y\}$ is disjoint according to the second; but X and X are disjoint in the first sense if and only if $X = \emptyset$, while $\{X, X\}$ is in any case disjoint in the second sense.

The general concept rarely occurs in elementary work. An instance where it _does_ arise emerges in III.2.8.

As usual, the reader should note the replacement rule

$$(S|\underline{u})(Z \text{ is disjoint}) \equiv (S|\underline{u})Z \text{ is disjoint}$$

and prove the theorem schema

$$T! \quad : \quad (Z = Z') \Rightarrow ((Z \text{ is disjoint}) \Leftrightarrow (Z' \text{ is disjoint})) .$$

Great care must be taken to distinguish between disjoint sets and disjoint families; see IV.11.4.

II.10 Relative complements

II.10.1 _Formal definition of relative complements_ The formal definition here is

$$X \setminus Y \equiv_{def} \{\underline{x} \in X : \underline{x} \notin Y\} . \qquad (1)$$

If X and Y denote sets, $X \setminus Y$ denotes a set, termed the _complement of_ Y _relative to_ X or the _relative complement of_ Y _in_ X. Alternative notations are in common use: $C_X Y$, $X - Y$, $X \sim Y$ are examples. In practice, one often has in any one context a fixed reference set E (a temporary "universe of discourse") , and then it is convenient to use in place of $E \setminus X$ a handier notation making no explicit reference to E : common choices are X' , $\sim X$, $\overline{X}$. (In the formal theory as here presented, there is no room for an "absolute complement" , because there is no "set of all sets" ; see II.3.10.)

Note the replacement rules (recall II.3.12)

$$(S|\underline{u})(X \setminus Y) \equiv (S|\underline{u})X \setminus (S|\underline{u})Y$$

and

$$(X|\underline{u}, Y|\underline{v})(\underline{u} \setminus \underline{v}) \equiv X \setminus Y \equiv (X|\underline{x})(Y|\underline{y})(\underline{x} \setminus \underline{y})$$

$$\equiv (Y|\underline{y})(X|\underline{x})(\underline{x} \setminus \underline{y}) \quad ,$$

and the theorem schema

T! : $$(X = X' \wedge Y = Y') \Rightarrow (X \setminus Y = X' \setminus Y') \quad .$$

From II.3.7(23) it follows that $X \setminus Y$ has the intuitively expected property expressed in the theorem schema

T! : $$(T \in X \setminus Y) \Leftrightarrow (T \in X \wedge T \notin Y) \quad . \qquad (1_1)$$

II.10.2 Properties of relative complements Typical and basic theorem schemas are the following:

$$X \setminus X = \emptyset \quad ,$$

$$X \setminus \emptyset = X \quad ,$$

$$X \cap (X \setminus Y) = X \setminus Y \quad ,$$

$$Y \cap (X \setminus Y) = \emptyset$$

$$(X \setminus Y) \cup Y = X \cup Y \quad ,$$

$$(X \cup Y) \setminus Z = (X \setminus Z) \cup (Y \setminus Z) \quad ,$$

$$(X \cap Y) \setminus Z = (X \setminus Z) \cap Y = (Y \setminus Z) \cap X = (X \setminus Z) \cap (Y \setminus Z) \ ,$$

T! : $$(X \setminus Y) \setminus Z = X \setminus (Y \cup Z) \quad , \qquad (2)$$

$$X \setminus (Y \setminus Z) = (X \setminus Y) \cup (X \cap Z) \quad ,$$

$$X \setminus (Y \cup Z) = (X \setminus Y) \cap (X \setminus Z) \quad ,$$

$$X \setminus (Y \cap Z) = (X \setminus Y) \cup (X \setminus Z) \quad ,$$

$$X \cap (Y \setminus Z) = (X \cap Y) \setminus (X \cap Z) \quad ,$$

$$(X \cap Y = \emptyset) \Leftrightarrow (X \setminus Y = X) \Leftrightarrow (Y \setminus X = Y) \quad ,$$

$$X = (X \setminus Y) \cup (X \cap Y) \quad ,$$

$$X \cup Y = (X \cap Y) \cup (X \setminus Y) \cup (Y \setminus X) \quad .$$

The unions appearing on the right of the last two theorem schemas are instances of what are usually termed "disjoint unions" , that is, unions of disjoint sets (see II.9 above), namely, the sets

$$\{X \setminus Y, X \cap Y\}$$

and

$$\{X \cap Y, X \setminus Y, Y \setminus X\}$$

respectively.

The reader should at the very least construct routine proofs of (2) . Semiformal proofs would, as usual, be longer; see II.10.4.

For relative complements, as explained above, one has consequently the following theorem schemas, often referred to as <u>de Morgan's laws</u>:

$$\emptyset' = E \quad , \quad E' = \emptyset \quad ,$$

$$\text{T! :} \qquad X \subseteq E \Rightarrow \begin{cases} (X')' = X \quad , \quad X \cap X' = \emptyset \quad , \\ X \cup X' = E \quad , \\ (X \cup Y)' = X' \cap Y' \quad , \\ (X \cap Y)' = X' \cup Y' \quad ; \end{cases} \tag{3}$$

see Problem II/8.

The equality $X \setminus X = \emptyset$ in (2) is frequently made the basis of yet another definition of the empty set.

II.10.3 <u>Connections with the set builder</u> A fundamental result is incorporated in the next theorem schema:

$$\text{T! :} \qquad \{\underline{x} \in X : \neg\underline{\underline{A}}\} = X \setminus \{\underline{x} \in X : \underline{\underline{A}}\} \quad . \tag{4}$$

<u>Semiformal proof of (4)</u> Note first that the appropriate "collectivising" theorems are provable as a result of the substance of II.3.7. Let S and T denote the sets appearing on the left and right sides of (4) respectively. By II.3.3(3) and II.10.1(1_1) ,

$$\text{T :} \qquad \underline{x} \in S \Leftrightarrow (\underline{x} \in X) \wedge (\neg\underline{\underline{A}}) \quad ,$$

$$\text{T :} \qquad \underline{x} \in T \Leftrightarrow (\underline{x} \in X) \wedge (\underline{x} \in \{\underline{x} \notin X : \underline{\underline{A}}\}) \quad .$$

So, by II.3.6(18) and I.3.2(X), it will suffice to prove that

$$T \quad : \qquad (\underline{x} \in X) \wedge (\neg\underline{\underline{A}}) \Leftrightarrow (\underline{x} \in X) \wedge (\underline{x} \notin \{\underline{x} \in X : \underline{\underline{A}}\}) \quad .$$

However, by II.3.3(3) once more,

$$T \quad : \qquad \underline{x} \in \{\underline{x} \in X : \underline{\underline{A}}\} \Leftrightarrow (\underline{x} \in X) \wedge \underline{\underline{A}} \quad ,$$

and so, making use of I.3.3(k),

$$T \quad : \qquad \underline{x} \notin \{\underline{x} \in X : \underline{\underline{A}}\} \Leftrightarrow (\underline{x} \notin X) \vee (\neg\underline{\underline{A}}) \quad .$$

Thus it will suffice to prove that

$$T \quad : \qquad (\underline{x} \in X) \wedge (\neg\underline{\underline{A}}) \Leftrightarrow (\underline{x} \in X) \wedge ((\underline{x} \notin X) \vee (\neg\underline{\underline{A}})) \quad .$$

I claim that in fact one has the more general theorem schema

$$T \quad : \qquad (\underline{\underline{C}} \wedge \underline{\underline{B}}) \Leftrightarrow (\underline{\underline{C}} \wedge (\neg\underline{\underline{C}} \vee \underline{\underline{B}})) \quad . \qquad (4_1)$$

As for this, I will leave to the reader the proof of $\Rightarrow$ and will indicate that of $\Leftarrow$. Adjoin to Θ_0 the axiom $\underline{\underline{C}} \wedge (\neg\underline{\underline{C}} \vee \underline{\underline{B}})$ to obtain a theory Θ ; by I.3.2(VI), it is enough to prove that $\underline{\underline{C}} \wedge \underline{\underline{B}}$ is a theorem of Θ . It is clear from (j) in I.3.3 that $\underline{\underline{C}}$ is a theorem of Θ , and so (also from (j)) that it remains only to show that $\underline{\underline{B}}$ is a theorem of Θ . I do this by contradiction. Adjoin $\neg\underline{\underline{B}}$ to Θ to obtain a theory Θ' ; by I.3.2(VII) it is sufficient to show that Θ' is contradictory. Now $\underline{\underline{C}}$ and $\neg\underline{\underline{B}}$ are true in Θ' . Hence, by (j), $\underline{\underline{C}} \wedge \neg\underline{\underline{B}}$ is true in Θ' . However, (j) and (k) show that $\underline{\underline{C}} \wedge \neg\underline{\underline{B}} \Leftrightarrow \neg(\neg\underline{\underline{C}} \vee \underline{\underline{B}})$ is true in Θ_0 , a fortiori in Θ' , and so by I.3.2(I), $\neg(\neg\underline{\underline{C}} \vee \underline{\underline{B}})$ is true in Θ' . At the same time, (j) and the definition of Θ

show that $\neg\underline{\underline{C}} \vee \underline{\underline{B}}$ is true in Θ , a fortiori true in Θ' . Thus both $\neg\underline{\underline{C}} \vee \underline{\underline{B}}$ and $\neg(\neg\underline{\underline{C}} \vee \underline{\underline{B}})$ are true in Θ' , showing that Θ' is indeed contradictory. □

II.10.4 Discussion of two theorem schemas I will discuss in some detail the proofs of the first, and a special case of the ninth, results in II.10.2(2), namely,

$$T \quad : \qquad X \setminus X = \emptyset \tag{5}$$

and

$$T \quad : \qquad X \setminus (X \setminus Y) = X \cap Y \ . \tag{6}$$

A routine proof of (5) might do no more than remark that either set denotes a set with no elements. More formally, one might call upon II.10.1(1_1), II.5.2(7) and II.3.6(18) to reduce the problem to proving the theorem schema

$$(\forall\underline{x})((\underline{x} \in X \wedge \underline{x} \notin X) \Leftrightarrow (\underline{x} \neq \underline{x})) \ , \tag{5_1}$$

wherein X denotes an arbitrary set and $\underline{x}$ any letter not appearing in X . This is easily done by using II.1.4(2) and the proof methods of I.3. (Recall that any sentence is a theorem of any contradictory theory; see I.2.7.)

A routine proof of (6) might run as follows. The elements of $X \setminus (X \setminus Y)$ are those objects which belong to X and do not belong to $X \setminus Y$, that is, those which belong to X and which either do not belong to X or belong to X and to Y ; and those are precisely the objects which belong to X and to Y , that is, which belong to $X \cap Y$. The argument might well be illustrated by reference to a Venn diagram:

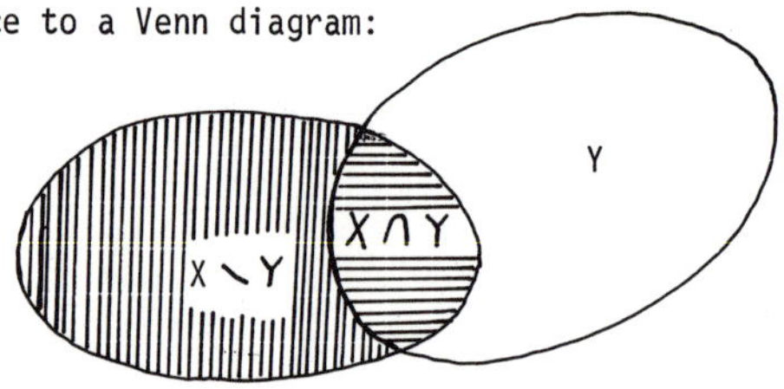

A purist would, of course, reject any appeal to such a diagram; he might even feel uneasy about the argument given. It is indeed not immediately clear that the argument is translatable into formal terms. The nagging doubt arises from the fact that the argument treats sets as if they were collections - which is true almost, but not quite, all of the time.

The way out lies in attempting a semiformal proof. This might be based on a theorem schema (a "rule of logic") derivable from the metatheorems and theorem schemas in I.3, namely,

$$\mathrm{T} \quad : \qquad (\underline{\underline{A}} \wedge \underline{\underline{B}}) \Leftrightarrow (\underline{\underline{A}} \wedge \neg(\underline{\underline{A}} \wedge \neg\underline{\underline{B}})) \quad ;$$

see Problem I/1. One applies this with $\underline{\underline{A}}$ and $\underline{\underline{B}}$ taken to be $\underline{x} \in X$ and $\underline{x} \in Y$ respectively, wherein $\underline{x}$ denotes a letter not appearing in X or Y. One will then derive

$$\mathrm{T} \quad : \qquad X \cap Y = \{\underline{x} : \underline{\underline{A}} \wedge \underline{\underline{B}}\} \qquad \text{(II.7.3 and II.3.5(15))},$$

$$\mathrm{T} \quad : \qquad \{\underline{x} : \underline{\underline{A}} \wedge \underline{\underline{B}}\} = \{\underline{x} : \underline{\underline{A}} \wedge \neg(\underline{\underline{A}} \wedge \neg\underline{\underline{B}})\} \qquad \text{(II.3.6(18))},$$

$$= \{\underline{x} \in X : \neg(\underline{\underline{A}} \wedge \neg\underline{\underline{B}})\} \quad \text{(definition in II.3.7)}$$

(note here the implicit use of the metatheorem following (1) in II.1.1 and also cf. II.1.4(4))

$$= X \setminus \{\underline{x} \in X : \underline{\underline{A}} \wedge \neg\underline{\underline{B}}\} \qquad \text{(II.10.3(4))}.$$

Since $\underline{\underline{A}}$ denotes $\underline{x} \in X$, I.3.3(k) and II.3.6(18) imply that

$$\mathrm{T} \quad : \qquad \{\underline{x} \in X : \underline{\underline{A}} \wedge \neg\underline{\underline{B}}\} = \{\underline{x} \in X : \neg\underline{\underline{B}}\} \quad .$$

So

$$T : \quad X \cap Y = X \setminus \{\underline{x} \in X : \neg \underline{\underline{B}}\}$$

$$= X \setminus \{\underline{x} \in X : \underline{x} \notin Y\}$$

$$= X \setminus (X \setminus Y) \qquad \text{(II.10.1(1))}$$

and (6) is proven.

> Remarks (i) The preceding comments about diagrams are not intended as a ban on their use. Diagrams have their uses, just as has intuition. But they are not intended as a surrogate for formal reasoning. See the Foreword to Volume 2.
>
> (ii) As an aside, consider (5_1) . Metamathematically, its content is the same as that of the theorem schema
>
> $$((Y \in X) \wedge (Y \notin X)) \Leftrightarrow (Y \neq Y) \quad . \qquad (5_2)$$
>
> One's intuition may leave one in some doubt about (5_2) (mine does). It is at such points that a need may be felt for a formal background as a reference point. (See the end of I.1.7.)

II.10.5 An example Prove that if Y and Z are subsets of X , then $Y \subseteq Z$ and $X \setminus Z \subseteq X \setminus Y$ are equivalent.

One is here being asked to prove the theorem schema

$$T! : \quad (Y \subseteq X \wedge Z \subseteq X) \Leftrightarrow (Y \subseteq Z \Leftrightarrow X \setminus Z \subseteq X \setminus Y) \quad . \qquad (7)$$

I will present a semiformal proof in several steps.

As a first step I will prove (what is of some independent interest)

$$T : \quad (Y \subseteq Z) \Rightarrow (X \setminus Z \subseteq X \setminus Y) \quad . \qquad (8)$$

To do this, adjoin to Θ_0 the axiom $Y \subseteq Z$ to obtain a theory Θ . Let $\underline{x}$ denote a letter not appearing in X , Y or Z . By II.10.1(1_1),

T : $\underline{x} \in X \setminus Z \Rightarrow \underline{x} \in X \wedge \underline{x} \notin Z$. (9)

On the other hand,

TΘ : $Y \subseteq Z$

that is,

TΘ : $(\forall \underline{x})(\underline{x} \in Y \Rightarrow \underline{x} \in Z)$ (II.2.1(1))

so

TΘ : $\underline{x} \in Y \Rightarrow \underline{x} \in Z$ (I.3.2(XI))

so

TΘ : $\underline{x} \notin Z \Rightarrow \underline{x} \notin Y$ (I.3.3(k)) . (10)

By (9) , (10) , I.3.2(I) and I.3.3(j),

TΘ : $\underline{x} \in X \setminus Z \Rightarrow \underline{x} \in X \wedge \underline{x} \notin Y$,

hence

TΘ : $\underline{x} \in X \setminus Z \Rightarrow \underline{x} \in X \setminus Y$ (II.3 and II.10.1(1))

TΘ : $(\forall \underline{x})(\underline{x} \in X \setminus Z \Rightarrow \underline{x} \in X \setminus Y)$ (I.3.2(X))

TΘ : $X \setminus Z \subseteq X \setminus Y$. (II.2.1(1)) (11)

By (11) and I.3.2(VI), (8) follows.

From (8) one derives (via (X) and (XI) in I.3.2)

$$T \quad : \quad (X \setminus Z \subseteq X \setminus Y) \Rightarrow (X \setminus (X \setminus Y) \subseteq X \setminus (X \setminus Z)) \ . \tag{12}$$

So, by II.10.4(6) (and S6 in II.1.2),

$$T \quad : \quad (X \setminus Z \subseteq X \setminus Y) \Rightarrow (X \cap Y \subseteq X \cap Z) \ . \tag{13}$$

Using II.8(6), (13) is easily seen to imply

$$T \quad : \quad (Y \subseteq X \wedge Z \subseteq X \wedge X \setminus Z \subseteq X \setminus Y) \Rightarrow (Y \subseteq X \wedge Z \subseteq X \wedge Y \subseteq Z) \ . \tag{14}$$

By (14) and I.3.2(VI) or I.3.7(2),

$$T \quad : \quad (Y \subseteq X \wedge Z \subseteq X) \Rightarrow (X \setminus Z \subseteq X \setminus Y \Rightarrow Y \subseteq Z) \ . \tag{15}$$

Similarly, by (8) and I.3.2(VI) or I.3.7(2),

$$T \quad : \quad (Y \subseteq X \wedge Z \subseteq X) \Rightarrow (Y \subseteq Z \Rightarrow X \setminus Z \subseteq X \setminus Y) \ . \tag{16}$$

Finally, by I.3.2(VI), I.3.3(j), (15) and (16) , (7) follows.

Remark The reader should consider what differences would ensue, if the example had been announced thus: Prove that

Y and Z are subsets of X and $Y \subseteq Z$

is equivalent to

Y and Z are subsets of X and $X \setminus Z \subseteq X \setminus Y$.

Which formal sentence is now being proposed as a theorem? (Compare I.3.6(2).)

II.10.6 Problem Give semiformal proofs of the theorem schemas

$$T! \quad : \qquad X = (X \setminus Y) \cup (X \cap Y) \quad ,$$

$$T! \quad : \qquad (X \neq Y) \Leftrightarrow ((X \setminus Y \neq \emptyset) \vee (Y \setminus X \neq \emptyset)) \quad .$$

II.11 Power sets

II.11.1 Formal definition of power sets According to the Subset Axiom enunciated in II.12.2, if $\underline{x}$ and $\underline{y}$ denote distinct letters, the sentence

$$(\forall \underline{x})\mathrm{Coll}_{\underline{y}}(\underline{y} \subseteq \underline{x})$$

is an explicit axiom of Θ_0 , whence if follows via S5 in I.2.2, II.2.1 and II.3.4(r) that (recall the convention in II.3.12)

$$\mathrm{Coll}_{\underline{y}}(\underline{y} \subseteq X)$$

is a theorem schema. I accordingly introduce the definition schema

$$P(X) \equiv_{\mathrm{def}} \{\underline{y} : \underline{y} \subseteq X\} \quad , \tag{1}$$

wherein $\underline{y}$ denotes a letter not appearing in X , noting that $\underline{y}$ does not appear in $P(X)$. $P(X)$ is called the set of subsets of X or the power set of X .

Note the replacement rule

$$(S|\underline{u})P(X) \equiv P((S|\underline{u})X) \quad ,$$

of which a special case reads

$$(S|\underline{u})\mathcal{P}(\underline{u}) \equiv \mathcal{P}(S) \quad .$$

Note also the theorem schema

T! : $$(X = Y) \Rightarrow (\mathcal{P}(X) = \mathcal{P}(Y)) \quad .$$

(Actually, a theorem schema results on writing $\Leftrightarrow$ in place of $\Rightarrow$; cf. (6) below.)

<u>Remark</u> In the notation $\mathcal{P}(X)$, " $\mathcal{P}$ " does not denote a function in the formal sense to be described in Chapter IV, even though the notation suggests that it does. In other words, the notation is <u>not</u> a special case of that introduced in IV.1.3(1); see IV.2.3(iv). If $\underline{x}$, $\underline{y}$ denote distinct letters

$$P \equiv \{\underline{y} : \underline{y} \subset \underline{x}\}$$

denotes a set in which $\underline{x}$ appears and $\underline{y}$ does not appear; further, by II.3.4(q), the choice of $\underline{y}$ (different from $\underline{x}$) is immaterial. Further still (and in view of the freedom in the choice of $\underline{y}$), II.3.4(r) shows that what has been denoted by $\mathcal{P}(X)$ is identical with $(X|\underline{x})P$. Thus $P[\![X]\!]$ would be a preferable notation; but it is unconventional. Compare with the closing Remarks in III.2.5 and III.2.6.

II.11.2 <u>Properties of power sets</u> By II.3.3(3'),

T! : $$(\forall \underline{y})(\underline{y} \subset X \Leftrightarrow \underline{y} \in \mathcal{P}(X)) \quad . \tag{2}$$

So, by (I) , (XI) and (X) in I.3.2,

T! : $$(\forall \underline{u})(\underline{u} \subset X \Leftrightarrow \underline{u} \in \mathcal{P}(X)) \quad . \tag{2'}$$

Making use of II.5.2(11) and I.2.1, one may deduce

T! : $$\emptyset \in \mathcal{P}(X) \quad , \tag{3}$$

T! : $$X \in P(X) \quad , \tag{4}$$

T! : $$P(\emptyset) = \emptyset \quad . \tag{5}$$

Also important is the theorem schema

T! : $$(X \subseteq Y) \Leftrightarrow (P(X) \subseteq P(Y)) \quad . \tag{6}$$

Using (7) and (8) in II.4.1 and (2), one may deduce

T! : $$(\forall \underline{u})(\underline{u} \in X \Leftrightarrow \underline{u} \in P(X)) \quad ,$$

(7)

T! : $$(\forall \underline{u})(\forall \underline{v})((\underline{u} \in X \wedge \underline{v} \in X) \Leftrightarrow \{\underline{u}, \underline{v}\} \in P(X)) \quad .$$

The second of these combines with (6) and II.8(7) to imply

T! : $$(\forall \underline{u})(\forall \underline{v})((\underline{u} \in X \wedge \underline{v} \in Y) \Leftrightarrow \{\underline{u}, \underline{v}\} \in P(X \cup Y)) \quad . \tag{8}$$

The reader should supply semiformal proofs of (at least some of) (3) - (8).

II.11.3 Informal remarks Speaking informally for a moment, II.11.2(7) entails that there is a function $\underline{u} \rightsquigarrow \{\underline{u}\}$ mapping X into $P(X)$. Usually, the range of this function is only a tiny subset of $P(X)$; see IV.3.10 below.

Still speaking informally, the sets

$$\emptyset \ , \ P(\emptyset) \ , \ P(P(\emptyset)) \ , \ P(P(P(\emptyset))) \ , \ P(P(P(P(\emptyset)))) \ , \ \ldots\ldots$$

are finite sets with numbers of elements which soon become so large that it is impossible to display them individually. In due course, one could prove that the above sets have cardinal numbers

$$0 \ , \ 1 = 2^0 \ , \ 2^1 = 2 \ , \ 2^2 = 4 \ , \ 2^4 = 16 \ ,$$

while continuation of the process leads to sets with cardinal numbers

$$2^{16} = 65{,}536 \ , \ 2^{65{,}536} \ , \ \ldots .$$

These sets are of more than passing interest; see §5 of the Appendix.

II.11.4 Basic processes of set formation I have now completed the survey of six of the basic processes for forming sets, namely, those indicated by the symbols $\{ \ : \ \}$, $\{ \ \}$, $\cup$, $\cap$, $\setminus$ and $\mathcal{P}(\ )$. In later chapters more such processes will be defined, notably, the formation of cartesian products (see III.1.3), of domains and ranges (see III.2.2), of quotient sets (see III.2.8), and of functional values (see IV.1.3).

II.12 The schemas and axioms of set theory

Throughout this section I waive the conventions described in II.3.12.

II.12.1 The schemas of set theory The schemas of Θ_0 are S1 - S7 already encountered in I.2.2 and II.1.2, plus the following rather complicated one:

S8 $$(\forall \underline{x})(\exists \underline{z})(\forall \underline{y})(\underline{\underline{A}} \Rightarrow (\underline{y} \in \underline{z})) \Rightarrow (\forall \underline{w})\mathrm{Coll}_{\underline{y}}(\exists \underline{x})((\underline{x} \in \underline{w}) \wedge \underline{\underline{A}}) \ ,$$

$\underline{\underline{A}}$ denoting an arbitrary sentence, $\underline{x}$, $\underline{y}$, $\underline{z}$, $\underline{w}$ distinct letters, $\underline{z}$ and $\underline{w}$ not appearing in $\underline{\underline{A}}$.

Intuitively speaking, the hypothesis in S8 says that for every set $\underline{x}$ there is a set $\underline{z}$ (which may depend upon $\underline{x}$) such that every set $\underline{y}$ satisfying $\underline{\underline{A}}$ is an element of $\underline{z}$ (though there may be other elements of $\underline{z}$ as well); the conclusion affirms that then, if $\underline{w}$ is any set, there is a set whose elements are precisely those sets $\underline{y}$ which satisfy $\underline{\underline{A}}$ for at least one choice of $\underline{x} \in \underline{w}$.

An example (though still a very general one) may help. Let T and X denote sets, $\underline{x}$ and $\underline{y}$ distinct letters, $\underline{x}$ not appearing in X and $\underline{y}$ not appearing in X or T . The letter $\underline{x}$ may appear in T , in view of which one may choose to write $T[\![\underline{x}]\!]$ in place of T . From S8 it then follows (see (7) in §3 of the Appendix) that the sentence

$$(\exists \underline{x})(\underline{x} \in X \wedge \underline{y} = T[\![\underline{x}]\!])$$

is collectivising in $\underline{y}$. This means informally that there is a set S such that

$$(\forall \underline{y})(\underline{y} \in S \Leftrightarrow (\exists \underline{x})(\underline{x} \in X \wedge \underline{y} = T[\![\underline{x}]\!])) \qquad (1)$$

is a theorem. More informally still: one can collect into a set S all the objects of the form $T[\![\underline{x}]\!]$ corresponding to objects $\underline{x}$ belonging to X . Naturally, as one encounters more and more sets X and T of diverse sorts, the content of this example is seen to expand.

Possibilities already available are: $T[\![\underline{x}]\!] \equiv \cup \underline{x}$; $T[\![\underline{x}]\!] \equiv \cap \underline{x}$; $T[\![\underline{x}]\!] \equiv P(\underline{x})$; $T[\![\underline{x}]\!] \equiv \underline{x} \cup \{\underline{x}\}$; $T[\![\underline{x}]\!] \equiv \{\underline{z} : A[\![\underline{x}, \underline{z}]\!]\}$, where $\underline{A}$ denotes an arbitrary sentence in which the letters $\underline{x}$ and $\underline{z}$ may appear (together possibly with other letters as well); and so on. Later on, one will be able to take, for instance, $T[\![\underline{x}]\!] \equiv P(\underline{x} \times \underline{x})$ (see III.1.4) or $T[\![\underline{x}]\!] \equiv N^{\underline{x}}$ (see IV.9.3).

As instances, on taking $X \equiv N$ one could thus gain assurance of the existence of such sets as those which would be informally denoted by

$$\{N,\ N^2,\ N^3,\ \ldots\ldots.\} ,$$

$$\{N^N,\ N^{N^2},\ N^{N^3},\ \ldots\ldots.\} ,$$

$$\{P(N),\ P(N^2),\ P(N^3),\ \ldots\ldots.\} ,$$

and so on. (The reader is reminded that, formally speaking, the existence of the set N of natural numbers with the expected properties depends on the Axiom of Infinity enunciated in

II.12.2; the details appear in Chapter V below.)

On the other hand, one must not be too hasty. There would still be problems concerning the existence of the set which would be denoted informally by

$$\{N, \quad N \times N, \quad N \times (N \times N), \quad N \times (N \times (N \times N)), \quad .....\} \quad ;$$

or of the existence of a set S such that $N \in S$ and

$$(\forall \underline{x})(\underline{x} \in S \Rightarrow N \times \underline{x} \in S) \quad ,$$

where $\underline{x}$ is a letter not appearing in S . One is here faced with problems akin to that discusses in II.5.7. Once again the solution has to be deferred until Chapter V (see especially Remark (ii) in V.5.3, wherein one takes $T[\![\underline{x}]\!] \equiv N \times \underline{x}$ and then $S \equiv \text{Ran } u$).

Returning to generalities, if $\underline{x}$ does not appear in X , the set S referred to in (1) above is usually denoted by

$$\{T[\![\underline{x}]\!] : \underline{x} \in X\} \quad , \tag{2}$$

and referred to as "the set of objects of the form $T[\![\underline{x}]\!]$ where (or with, or for, or corresponding to) $\underline{x} \in X$ " . The sentence denoted by

$$(\exists \underline{x})(\underline{x} \in X \wedge \underline{y} = T[\![\underline{x}]\!])$$

is usually read as " $\underline{y}$ is of (or can be expressed in) the form $T[\![\underline{x}]\!]$ where $\underline{x} \in X$ (or: for some $\underline{x} \in X$) " ; see I.3.5(vi). I shall discuss these matters in more detail in §3 of the Appendix; see also IV.2.2 and Problem II/24.

In spite of these explanatory remarks, the schema S8 is likely to remain the most difficult of all to grasp informally. In practice, an appeal to S8 will usually be masked by an appeal to (XII) in II.3.7, possibly supplemented by the following theorem schema:

$$\text{(XIV)} \qquad (\forall \underline{x})\text{Coll}_{\underline{y}}\underline{\underline{A}} \Rightarrow (\forall \underline{w})\text{Coll}_{\underline{y}}(\exists \underline{x})((\underline{x} \in \underline{w}) \wedge \underline{\underline{A}}) \quad ,$$

$\underline{\underline{A}}$ denoting an arbitrary sentence, $\underline{x}$, $\underline{y}$, $\underline{w}$ distinct letters, $\underline{w}$ not appearing in $\underline{\underline{A}}$.

(This is an almost immediate deduction from S8 and I.3.2(III), since

$$(\forall \underline{x})\mathrm{Coll}_{\underline{y}}\underline{\underline{A}} \equiv (\forall \underline{x})(\exists \underline{z})(\forall \underline{y})(\underline{\underline{A}} \Leftrightarrow (\underline{y} \in \underline{z}))$$

$$\Rightarrow (\forall \underline{x})(\exists \underline{z})(\forall \underline{y})(\underline{\underline{A}} \Rightarrow (\underline{y} \in \underline{z}))$$

is a theorem. See also §2 of the Appendix.) Apart from these cases, there are but two occasions on which reversion to S8 or (XIV) is needed: one has appeared in II.6, and the other will appear in IV.2 (and again in §3 of the Appendix). The substance of IV.2 may also help to clarify the informal background.

Remarks (i) The hypothesis of S8 is (see Problem II/33) equivalent to that in (XIV). However, the proof of this makes use of II.3.7(XII), which in turn appeals to S8 . Thus there is no obvious justification for adopting in place of S8 the schema displayed in (XIV).

(ii) The schema S8 is the counterpart of what appears as S8 in Bourbaki (1), p.69. There is a difference in that Bourbaki does not assume that $\underline{z}$ and $\underline{w}$ are necessarily distinct. However, this difference is of no ultimate significance; that is, the implicit axioms resulting from application of our version are exactly those resulting from application of Bourbaki's version. This is because, if $\underline{x}$, $\underline{y}$, $\underline{z}$, $\underline{w}$ are as stipulated in our version, then

$$(\forall \underline{w})\mathrm{Coll}_{\underline{y}}(\exists \underline{x})((\underline{x} \in \underline{w}) \wedge \underline{\underline{A}})$$

$$\equiv (\forall \underline{z})(\underline{z}|\underline{w})\mathrm{Coll}_{\underline{y}}(\exists \underline{x})((\underline{x} \in \underline{w}) \wedge \underline{\underline{A}}) \qquad \text{(I.1.8(e))}$$

$$\equiv (\forall \underline{z})\mathrm{Coll}_{\underline{y}}(\underline{z}|\underline{w})(\exists \underline{x})((\underline{x} \in \underline{w}) \wedge \underline{\underline{A}}) \qquad \text{(II.3.4(r))}$$

$$\equiv (\forall \underline{z})\mathrm{Coll}_{\underline{y}}(\exists \underline{x})(\underline{z}|\underline{w})((\underline{x} \in \underline{w}) \wedge \underline{\underline{A}}) \qquad \text{(I.1.8(f))}$$

$$\equiv (\forall \underline{z})\mathrm{Coll}_{\underline{y}}(\exists \underline{x})((\underline{x} \in \underline{z}) \wedge \underline{\underline{A}}) \qquad \text{(I.1.4)}$$

Incidentally, the last line of Bourbaki (1), p.69 appears to contain a misprint: his Coll_X .... ahould read Coll_x .... .

II.12.2 <u>The explicit axioms of set theory</u> There are just three explicit axioms of Θ_0 , namely, the sentences denoted as follows ($\underline{x}$, $\underline{y}$, $\underline{z}$, $\underline{w}$ denoting distinct letters):

The Pairing Axiom $(\forall \underline{x})(\forall \underline{y})\mathrm{Coll}_{\underline{z}}((\underline{z} = \underline{x}) \vee (\underline{z} = \underline{y}))$,

that is,

$$(\forall \underline{x})(\forall \underline{y})(\exists \underline{w})(\forall \underline{z})((\underline{z} \in \underline{w}) \Leftrightarrow ((\underline{z} = \underline{x}) \vee (\underline{z} = \underline{y}))) \ .$$

The Subset or Power Set Axiom $(\forall \underline{x})\mathrm{Coll}_{\underline{y}}(\underline{y} \subseteq \underline{x})$,

that is,

$$(\forall \underline{x})(\exists \underline{z})(\forall \underline{y})((\underline{y} \subseteq \underline{x}) \Leftrightarrow (\underline{y} \in \underline{z})) \ .$$

The Axiom of Infinity

$$(\exists \underline{y})((\emptyset \in \underline{y}) \wedge (\forall \underline{x})(\underline{x} \in \underline{y} \Rightarrow \underline{x} \cup \{\underline{x}\} \in \underline{y})) \ .$$

The first two axioms have already been used and their content thereby indicated. The Axiom of Infinity will be discussed later in V.2 and in §5 of the Appendix; one of its functions is to resolve the type of existence problem discussed in II.5.7 and II.12.1.

Strictly speaking, the schemas and axioms should be expressed in the primitive language; cf. the corresponding remark in II.1.2. This could be done quite mechanically in principle, but the resulting strings are astronomically long. (I estimate that the sentence called the Pairing Axiom is a string of length exceeding 5×10^{20} !) This makes it clearly inconceivable that the axioms would ever have been envisaged in our primitive language: an intuitive feel had to come from informal set theory as a start. (It is only fair to point out that the astronomical length of strings which are the axioms and basic objects is in part peculiar to the particular formal system adopted in this book; other systems, less parsimonious in their primitive alphabet, would remain somewhat more manageable for somewhat longer!)

It is vital to verify and remember at all times that no letter appears in the strings named explicit axioms above: all letters are eliminated via the quantifiers $\exists$ and $\forall$. Thus, as has been said and relied upon, Θ_0 has no constants (fixed letters); all letters are variables (free letters) of Θ_0 .

There is a tedious point to be examined in the case of each of the three axioms announced above. Take for example the case of the Subset Axiom. It should be verified that the sentence denoted by

$$(\forall \underline{x})(\exists \underline{z})(\forall \underline{y})(\underline{y} \subseteq \underline{x} \Leftrightarrow \underline{y} \in \underline{z})$$

is indeed independent of the choice of the three distinct letters $\underline{x}$, $\underline{y}$, $\underline{z}$. That is, suppose one denotes by $\underline{\underline{A}}$ the sentence denoted by (1) ; suppose $\underline{x}'$, $\underline{y}'$, $\underline{z}'$ are three distinct letters and that

$$\underline{\underline{A}}' \equiv (\forall \underline{x}')(\exists \underline{z}')(\forall \underline{y}')(\underline{y}' \subseteq \underline{x}' \Leftrightarrow \underline{y}' \in \underline{z}') \quad ;$$

is it the case that $\underline{\underline{A}} \equiv \underline{\underline{A}}'$? To see that the answer is "Yes" , introduce distinct letters $\underline{x}''$, $\underline{y}''$, $\underline{z}''$ each different from all of $\underline{x}$, $\underline{y}$, $\underline{z}$, $\underline{x}'$, $\underline{y}'$, $\underline{z}'$. Let

$$\underline{\underline{A}}'' \equiv (\forall \underline{x}'')(\exists \underline{z}'')(\forall \underline{y}'')(\underline{y}'' \subseteq \underline{x}'' \Leftrightarrow \underline{y}'' \in \underline{z}'') \quad .$$

I will verify that $\underline{\underline{A}} \equiv \underline{\underline{A}}''$. A similar argument will verify that $\underline{\underline{A}}' \equiv \underline{\underline{A}}''$ and the exercise will be over.

Now

$$
\begin{aligned}
\underline{\underline{A}} &\equiv (\forall \underline{x})(\exists \underline{z})(\forall \underline{y})(\underline{y} \subseteq \underline{x} \Leftrightarrow \underline{y} \in \underline{z}) \\
&\equiv (\forall \underline{x}'')(\underline{x}''|\underline{x})(\exists \underline{z})(\forall \underline{y})(\underline{y} \subseteq \underline{x} \Leftrightarrow \underline{y} \in \underline{z}) && \text{(I.1.8(e))} \\
&\equiv (\forall \underline{x}'')(\exists \underline{z})(\forall \underline{y})(\underline{x}''|\underline{x})(\underline{y} \subseteq \underline{x} \Leftrightarrow \underline{y} \in \underline{z}) && \text{(I.1.8(f) twice)} \\
&\equiv (\forall \underline{x}'')(\exists \underline{z})(\forall \underline{y})(\underline{y} \subseteq \underline{x}'' \Leftrightarrow \underline{y} \in \underline{z}) && \text{(I.1.4)} \\
&\equiv (\forall \underline{x}'')(\exists \underline{z}'')(\underline{z}''|\underline{z})(\forall \underline{y})(\underline{y} \subseteq \underline{x}'' \Leftrightarrow \underline{y} \in \underline{z}) && \text{(I.1.8(e))} \\
&\equiv (\forall \underline{x}'')(\exists \underline{z}'')(\forall \underline{y})(\underline{z}''|\underline{z})(\underline{y} \subseteq \underline{x}'' \Leftrightarrow \underline{y} \in \underline{z}) && \text{(I.1.8(f))} \\
&\equiv (\forall \underline{x}'')(\exists \underline{z}'')(\forall \underline{y})(\underline{y} \subseteq \underline{x}'' \Leftrightarrow \underline{y} \in \underline{z}'') && \text{(I.1.4)} \\
&\equiv (\forall \underline{x}'')(\exists \underline{z}'')(\forall \underline{y}'')(\underline{y}''|\underline{y})(\underline{y} \subseteq \underline{x}'' \Leftrightarrow \underline{y} \in \underline{z}'') && \text{(I.1.8(e))} \\
&\qquad (\ \underline{y}'' \text{ does not appear in } \underline{y} \subseteq \underline{x}'' \Leftrightarrow \underline{y} \in \underline{z}''\ ) \\
&\equiv (\forall \underline{x}'')(\exists \underline{z}'')(\forall \underline{y}'')(\underline{y}'' \subseteq \underline{x}'' \Leftrightarrow \underline{y}'' \in \underline{z}'') && \text{(I.1.4)} \\
&\equiv A'' \quad .
\end{aligned}
$$

I do not suggest that any reader attempt to memorise the axioms or schemas of Θ_0 in the forms they appear above. The most one need do is to carry an idea of their intuitive content, just so that one can recognise when each is likely to be involved in any particular argument.

At this point I terminate the first stage of the review of the foundations of set theory. The subsequent steps to be covered concern relations and functions and will be dealt with separately under these chapter headings.

II.12.3 <u>General remarks</u> As has been indicated in II.3.9(v), there are several recognised versions of set theory. (A favourite example is the theory ZF described briefly in §4 of the Appendix.) These have in common the schemas S1 - S6 , which are roughly characteristic of the first order predicate calculus with equality (a purely logical system), but most lack S7 because their primitive alphabet lacks the signs τ and $\square$. Some may also lack S8 , but will have replacements for it (like the Schema of Separation, the Schema of Replacement, and the Axiom of Union mentioned in Problems II/39 and IV/25 and §4 of the Appendix). They may also differ in their explicit axioms; see again §4 of the Appendix and Suppes (1), pp.56, 238.

For example, Bourbaki (1) uses, in addition to the axioms enunciated in II.12.2, an Axiom of Extensionality (mentioned in passing in II.1.1) and an Axiom of Ordered Pairs. Our definitions of equality and ordered pairs (see II.1.1 and III.1.1) remove the need for these two axioms. (See Bourbaki (1), A1-A5, pp. 67, 69, 72, 101, 183.)

Other versions of set theory utilise an Axiom of Regularity (or Foundation), which is the sentence denoted by

$$(\forall \underline{x})(\underline{x} \neq \emptyset \Rightarrow (\exists \underline{y})(\underline{y} \in \underline{x} \wedge \underline{y} \cap \underline{x} = \emptyset)) \quad .$$

This axiom permits one to show that sentences like

$$\underline{x} \in \underline{x} \quad , \quad (\underline{x} \in \underline{y}) \wedge (\underline{y} \in \underline{x}) \quad , \quad (\underline{x} \in \underline{y}) \wedge (\underline{y} \in \underline{z}) \wedge (\underline{z} \in \underline{x}) \quad ,$$

which seem intuitively highly improbable, are indeed false. (See Problem I/33, Suppes (1), pp.53-56 and Problem II/20 below). No use of this axiom will be made in the main text of this book.

While all this variety may be a source of glee on the part of the set theory specialists, the ordinary mathematician - encouraged as he is to believe that all that he does is ultimately based on set theory - is likely to take a rather different view! However, as soon as one begins to move away from foundational matters and into everyday mathematics, the differences usually prove to be in the end not all that important. This is borne out by the fact that the vast majority of books on mathematics make no mention at all of which version of set theory lurks in the background: in most cases, what can be done in one can be done in another (though the details of the route may vary).

There is one very noteworthy exception to what is said in the last paragraph, and this relates to the so-called Axiom of Choice. This will be discussed in a little more detail in IV.5. Granted the other schemas and axioms, the Axiom of Choice can be formulated in a great variety of equivalent ways, many of which make it appear at first sight to be intuitively quite unremarkable; so unremarkable, indeed, that it remained unrecognised as an assumption for quite a while after informal set theory had emerged. One such formulation reads thus:

> If A is a disjoint set of nonvoid sets, there exists a set B such that $a \cap B$ is a singleton for all $a \in A$; in other words, it is possible to choose precisely one object from each set belonging (as an element) to A and collect these chosen objects into a set B.

There is seemingly little to object to in this! Yet, when taken in conjunction with the other schemas and axioms, themselves occasioning no immediate surprise, it has corollaries in areas pretty remote from the foundations which are intuitively almost incredible; see IV.5.3. Largely because of these corollaries, it has been the subject of more controversy and speculation and study than all the other axioms taken together.

The various versions of set theory tend to be grouped according as they do or do not include the Axiom of Choice among their schemas and axioms. Our theory Θ_0 is one in which the Axiom of Choice, although neither a schema nor an explicit axiom, is an almost trivial theorem (built in via the signs τ and $\square$

and the rules for their use). On the other hand, ZF is a theory in which the Axiom of Choice does not appear in the schemas and axioms, and (what is more) is not provable.

For further discussion of the Axiom of Choice, see IV.5, VII.3.5 and §4 of the Appendix.

I attempt no discussion of the connections between various axioms and schemas, nor of the possible superfluity in those adopted for the theory Θ_0 . The reader interested in this area should consult Fraenkel et al. (1), Chapter II.

II.13 Informal introduction of certain sets

This section is wholly informal and the conventions in II.3.12 are temporarily waived. In this section I wish merely to recall the notations used in connection with a few specific sets of the utmost importance, the purpose being merely to provide examples in the course of Chapters III and IV. (See especially the first paragraph in III.2.3.) The sets in question will be discussed more formally and at greater length in Chapters V and VI.

The sets referred to are the sets of natural numbers, of integers, of rational numbers, and of real numbers respectively, denoted throughout this book by N , Z , Q and R respectively. (To be more precise, we shall in Chapter VI preserve the proprieties by distinguishing between N and a certain "copy" $^*N$ of N imbedded in R ; see especially VI.5.2. Later we shall follow convention by playing down this distinction. At the present moment, I am overlooking this point of comparative detail.)

At the present stage I have to content myself with the "definitions"

$$N = \{0, 1, 2, ...\} ,$$

$$Z = \{....., -2, -1, 0, 1, 2, ...\} ,$$

even though the expressions on the right denote no string of the formal language.

Not much can be done at the present moment to improve the situation. What is written is merely suggestive of the facts that 0, 1, 2, ... are standard signs which name certain natural numbers, while -1, -2, ... are likewise standard signs naming certain integers.

In a somewhat similar vein, one may write

$$\mathcal{Q} = \{\underline{z} : (\exists\underline{x})(\exists\underline{y})(\underline{x} \in Z \wedge \underline{y} \in Z \wedge \underline{y} \neq 0 \wedge \underline{z} = \underline{x}\cdot\underline{y}^{-1}\} \ .$$

There is no equally simple specification of R in terms of N , Z and $\mathcal{Q}$.

As will be discussed in Chapter VI, one almost always arranges the definitions so that $N \subseteq Z \subseteq \mathcal{Q} \subseteq R$.

For illustrative purposes, I shall sometimes need to take on trust (see Chapter VI again) the existence of binary operations of addition + and multiplication $\times$ or $\cdot$ on R (these are certain functions with domains equal to $R \times R$ and ranges equal to R) and a certain subset P of R whose elements are termed positive real numbers. Following convention, $x + y$ and xy (or $x \cdot y$) denote the value (formally defined in IV.2.1) of the functions + and $\times$ at $(x, y) \in R \times R$. 0 is the unique real number such that $x + 0 = x$ for all $x \in R$. If x is a real number, $-x$ is the unique real number such that $x + (-x) = 0$; and $x - y$ is written in place of $x + (-y)$. The set P defines the natural order on R : what is conventionally written as $x > y$ (and/or $y < x$) denotes $x - y \in P$, while $x \geq y$ (and/or $y \leq x$) denotes $(x > y) \vee (x = y)$. Accordingly, $P = \{\underline{x} \in R : \underline{x} > 0\}$. (I repeat that here I am merely summarising in informal terms what will be done formally and in detail in Chapter VI.)

The reader has to be warned at this point that the signs 0 , + and $\cdot$ are used with many different meanings in mathematics, and likewise with $>$ and $<$. It is now quite hopeless to try to make notations coherent over the whole range of mathematics. In practice mathematicians are not worried about this and claim that the intended meaning is usually clear enough from the context. This matter arises again in IV.3.3.

One can arrange the definitions so that each of the sets $\mathcal{N}$, $\mathcal{Z}$, $\mathcal{Q}$, $\mathcal{R}$ is specific, that is, so that no letter appears in any of the strings so denoted; but see VI.5.

II.14 A minimising theorem

This section is devoted to a less basic metatheorem which will find essential use in V.5.2 (but nowhere else). (It could with ease be used also in V.2.1 in connection with the definition of $\mathcal{N}$, but I will there choose to work independently of it. If he prefers, the reader may ignore this section until he reaches V.2.1.)

II.14.1 Metatheorem Suppose that Θ denotes a theory, $\underline{z}$ and $\underline{w}$ distinct variables of Θ , B a set in which $\underline{z}$ does not appear, and $\underline{\underline{A}}$ a sentence. Define $\underline{\underline{A}}[\![S]\!] \equiv (S|\underline{z})\underline{\underline{A}}$ for an arbitrary string S ,

$$C \equiv \{\underline{z} \in \mathcal{P}(B) : \underline{\underline{A}}[\![\underline{z}]\!]\} ,$$

$$u \equiv \cap C .$$

Suppose that $\underline{\underline{A}}[\![B]\!]$ and

$$((\underline{w} \neq \emptyset) \wedge (\forall \underline{z})(\underline{z} \in \underline{w} \Rightarrow \underline{\underline{A}}[\![\underline{z}]\!])) \Rightarrow \underline{\underline{A}}[\![\cap \underline{w}]\!]$$

are theorems of Θ . Then $u \subseteq B$, $u \in C$, $\underline{\underline{A}}[\![u]\!]$ and

$$(\forall \underline{z})(\underline{\underline{A}}[\![\underline{z}]\!] \Rightarrow u \subseteq \underline{z})$$

are theorems of Θ .

Remark The last clause of the conclusion says that (in the theory Θ)

u has a minimising property: u is a subset of every set $\underline{z}$ such that $\underline{\underline{A}}[\![\underline{z}]\!]$. Since also $\underline{\underline{A}}[\![u]\!]$ is a theorem of Θ, one may say that (in Θ) u is the smallest set $\underline{z}$ such that $\underline{\underline{A}}[\![\underline{z}]\!]$. This is usually the vital property of u. (Cf. Problem V/52.)

II.14.2 Routine proof This might amount to little more than the following. C has the intuitively-expected properties and so $B \in C$ and therefore $C \neq \emptyset$ and $\cap C \subseteq B$, that is, $u \subseteq B$, and so $u \in P(B)$. By the second hypothesis, $\underline{\underline{A}}[\![u]\!]$. Hence $u \in C$. Assume $\underline{\underline{A}}[\![\underline{z}]\!]$. Since $\underline{\underline{A}}[\![B]\!]$, the second hypothesis yields $\underline{\underline{A}}[\![B \cap \underline{z}]\!]$. Since also $B \cap \underline{z} \in P(B)$, $B \cap \underline{z} \in C$ and so $u \equiv \cap C \subseteq B \cap \underline{z} \subseteq \underline{z}$. This completes the proof.

A reader who is somewhat familiar with set theory (even - and perhaps especially - the naive variety) would probably take this in his stride. But a novice, and especially a novice interested in the logical structure of proofs, might demand something akin to the following semiformal style.

II.14.3 Semiformal proof By II.3.7(XII), the sentence $\underline{z} \in P(B) \wedge \underline{\underline{A}}[\![\underline{z}]\!]$ is collectivising in $\underline{z}$ and so, by II.3.3(5),

$$T \quad : \qquad (\forall \underline{z})(\underline{z} \in C \Leftrightarrow (\underline{z} \in P(B)) \wedge \underline{\underline{A}}[\![\underline{z}]\!]) \quad . \tag{1}$$

Let Θ' denote the theory obtained by adjoining to Θ the axiom

$$\underline{\underline{A}}[\![B]\!] \wedge ((\forall \underline{w})(((\underline{w} \neq \emptyset) \wedge (\forall \underline{z})(\underline{z} \in \underline{w} \Rightarrow \underline{\underline{A}}[\![\underline{z}]\!]))) \Rightarrow \underline{\underline{A}}[\![\cap \underline{w}]\!]) \quad ,$$

and note that $\underline{z}$ is a variable of Θ'. By (VI) and (X) in I.3.2, it will suffice to prove that

$$T_{\Theta'} \quad : \qquad u \subseteq B \quad , \tag{2}$$

$$T_{\Theta'} \quad : \qquad u \in C \quad , \tag{3}$$

$T\Theta'$: $$\underline{\underline{A}}[\![u]\!] \ . \tag{4}$$

$T\Theta'$: $$\underline{\underline{A}}[\![\underline{z}]\!] \Rightarrow u \subseteq \underline{z} \ . \tag{5}$$

Now the hypothesis $\underline{\underline{A}}[\![B]\!]$ and (1) combine to imply that

$T\Theta'$: $$B \in C \ . \tag{6}$$

This implies that

$T\Theta'$: $$C \neq \emptyset \ . \tag{7}$$

By (6) and II.7.5(13), (2) follows.

By (1) ,

$T\Theta'$: $$\underline{z} \in C \Rightarrow \underline{\underline{A}}[\![\underline{z}]\!] \ ,$$

hence ($\underline{z}$ being a variable of Θ')

$T\Theta'$: $$(\forall \underline{z})(\underline{z} \in C \Rightarrow \underline{\underline{A}}[\![\underline{z}]\!]) \ . \tag{8}$$

By definition of Θ' and I.3.2(XI) (note that $\underline{z}$ does not appear in C)

$T\Theta'$: $$((C \neq \emptyset) \wedge (\forall \underline{z})(\underline{z} \in C \Rightarrow \underline{\underline{A}}[\![\underline{z}]\!])) \Rightarrow \underline{\underline{A}}[\![\cap C]\!] \equiv \underline{\underline{A}}[\![u]\!] \ . \tag{9}$$

Now (4) follows from (7), (8) and (9) (plus metatheorems in I.3.2 and I.3.3).

(3) now follows from (1) , (2) and (4) (recall from II.11.2(2) that $u \in \mathcal{P}(B) \Leftrightarrow u \subseteq B$ is true in Θ_0).

To prove (5) , consider the theory Θ'' obtained by adjoining to Θ' the axiom $\underline{\underline{A}}[\![\underline{z}]\!]$. Then

$$T\Theta \quad : \quad \underline{\underline{A}}[\![B]\!] \ , \tag{10}$$

$$T\Theta'' \quad : \quad \underline{\underline{A}}[\![\underline{z}]\!] \ , \tag{11}$$

Let $\underline{z}'$ denote a letter different from $\underline{z}$ and $\underline{w}$ and not appearing in $\underline{\underline{A}}$ or B . By the definition of Θ' and I.1.8(e),

$$T\Theta' \quad : \quad (\forall \underline{w})(((\underline{w} \neq \emptyset) \wedge (\forall \underline{z}')(\underline{z}' \in \underline{w} \Rightarrow \underline{\underline{A}}[\![\underline{z}']\!])) \Rightarrow \underline{\underline{A}}[\![\cap \underline{w}]\!]) \ .$$

Hence, using I.3.2(XI) to replace $\underline{w}$ by $\{B, \underline{z}\}$,

$$T\Theta' \quad : \quad (\{B, \underline{z}\} \neq \emptyset) \wedge (\forall \underline{z}')(\underline{z}' \in \{B, \underline{z}\} \Rightarrow \underline{\underline{A}}[\![\underline{z}']\!])) \Rightarrow \underline{\underline{A}}[\![B \cap \underline{z}]\!] \ .$$

Using I.3.7(1) and the fact that $\{B, \underline{z}\} \neq \emptyset$ is true in Θ_0 , it follows that

$$T\Theta' \quad : \quad ((\forall \underline{z}')(\underline{z}' \in \{B, \underline{z}\} \Rightarrow \underline{\underline{A}}[\![\underline{z}']\!])) \Rightarrow \underline{\underline{A}}[\![B \cap \underline{z}]\!] \ .$$

Hence, using II.4.3 and I.3.2(I),

$$T\Theta' \quad : \quad (\underline{\underline{A}}[\![B]\!] \wedge \underline{\underline{A}}[\![\underline{z}]\!]) \Rightarrow \underline{\underline{A}}[\![B \cap \underline{z}]\!] \ . \tag{12}$$

By (10) , (11) , I.3.3(j), I.3.2(I) and (12) ,

$$T\Theta'' \quad \underline{\underline{A}}[\![B \cap \underline{z}]\!] \ .$$

But $B \cap \underline{z} \in \mathcal{P}(B)$ is true in Θ_0 and so

$$T\Theta'' \quad : \quad (B \cap \underline{z} \in \mathcal{P}(B)) \wedge \underline{\underline{A}}[\![B \cap \underline{z}]\!] \ .$$

So, by (1) and (I) and (XI) in I.3.2,

$T\Theta''$: $$B \cap \underline{z} \in C \ .$$

This, together with II.7.5(13) and I.3.2(I), implies

$T\Theta''$: $$u \subseteq B \cap \underline{z} \ .$$

Hence (see II.2.2(3) and II.8(7))

$T\Theta''$: $$u \subseteq \underline{z} \ .$$

This combines with I.3.2(VI) to imply (5) . □

Chapter III. Relations

Introduction The term "relation" is used in everyday language, where both it and many specific examples of relations are referred to but are seldom clearly defined. Such lack of precision is probably an inevitable consequence of the compromise between precision and flexibility characteristic of informal everyday language, and one is usually content to achieve enough precision to make serious ambiguities relatively rare. Even in traditional mathematics, the term is used pretty loosely.

In current mathematics, the term "relation" is borrowed and endowed with more precision. Nevertheless, even in current mathematics, there is not complete agreement on the meaning of the term "relation" . For instance, while on most occasions most mathematicians will be happy to speak loosely of the "relation of equality" , this manner of speaking proves to be formally incorrect in several versions of formalised set theory (including that described in this book); see III.2.3(ii).

Concerning the loan of the term "relation" , see Halmos' introductory remarks ((1), p.26), where some motivating examples from everyday life are described. His stance is much the same as that to be adopted here, namely: assuming that the concept of "set" is securely founded and adequately familiar, one seeks to define the concept of "relation" in terms of sets. To quote Halmos:

> We may not know what a relation is, but we do know (or

> think we know) what a set is, and the preceding considerations (pertaining to a specific real-life relation) establish a close connection between relations and sets. The precise set theoretic treatment of relations takes advantage of that heuristic connection; ...

(The parenthesised insertions are mine.)

Several points concerning the subsequent presentation of semiformal proofs bear repetition. I shall, of course, assume all the metatheorems and theorems in Chapters I and II, but I shall not always attempt to mention explicitly every single reference to these sources. To do this would make the proofs intolerably long. Instead, I mention explicitly only those references which seem to be the most essential and helpful. It is up to the reader to check those which remain implicit. This is a tedious exercise, but it can be useful to his general understanding. Nor do I exhibit a proof of every alleged theorem: here again the reader should be prepared to contribute a personal effort, usually (but not solely) by working out the problems.

Concerning replacement rules (cf. the end of II.2.1, II.6.1, the end of II.7.2, II.10.1, II.11.1), we shall be less explicit in the future. In III.1.1, for example, the rule

$$(S|\underline{u})(X, Y) \equiv ((S|\underline{u})X, (S|\underline{u})Y)$$

is listed, but it is left to the reader to add (bearing II.3.12 in mind) the rule

$$(X|\underline{u}, Y|\underline{v})(\underline{u}, \underline{v}) \equiv (X, Y) \equiv (X|\underline{x})(Y|\underline{y})(\underline{x}, \underline{y}) \equiv (Y|\underline{y})(X|\underline{x})(\underline{x} \times \underline{y}) .$$

Similar omissions appear in III.1.3, III.1.4, III.2.4, III.2.6, III.2.7, IV.1.1, et cetera.

III.1 Ordered pairs

Informal discussion The set theoretical definition of relations is framed in terms of "ordered pairs" , and a preliminary set theoretical definition of this concept is obligatory. The formal definition, which appears in III.1.1, will be prefaced by some motivating remarks.

Ordered pairs may well have (unofficially, at least) found mention in prior school work in connection with coordinate geometry of the plane (in which, relative to a given choice of axes, each point assumes a label which is an ordered pair of real numbers). Another, somewhat more diffuse, application may have been encountered in connection with the definition of rational numbers in terms of integers: there is an almost evident connection between rational numbers a/b and ordered pairs (a, b) of integers in which $b \neq 0$. In this case, however, the connection is not so direct: all ordered pairs (ka, kb) , where (a, b) is as just specified, and k represents a nonzero integer, correspond to the same rational number a/b (reflecting the fact that $ka/kb = a/b$ if k and b are nonzero integers and a any integer).

The vital feature of ordered pairs, which has to be built in to their definition, is that with arbitrary sets X and Y (which may be identical) there is to be associated an "ordered pair" denoted by (X, Y) , again a set, in such a way that (X, Y) and (X', Y') are (better: denote) equal sets if and only if $X = X'$ and $Y = Y'$. In particular, therefore, (X, Y) and (Y, X) are to be equal as sets if and only if $X = Y$.

It is not sufficient to follow some text books and say, for example:

> An ordered pair of real numbers is a pair of real numbers written down in a definite order.

(Indeed, how can one write down a pair of real numbers, without writing them in a definite (spatial or temporal) order?) It is the sets denoted by (X, Y) and (Y, X) rather than these names or name symbols, which are to be unequal whenever

X and Y denote unequal sets. (This resurrects the point that it is sometimes vital to distinguish that which denotes from that which is denoted; see the remarks at the end of I.1.3.)

As it is used in "ordered pair" , the term "ordered" refers to an accidental feature of the way the name is written, rather than to the set denoted by that name. This is shown up by the fact that, having used the phrase "ordered pair (a, b) " , many writers (for example, Rudin (1), p.12) feel bound to add immediately a sentence of the form:

> "Ordered" means that (a, b) and (b, a) are regarded as distinct if $a \neq b$.

(Note the phrase "are regarded as" in place of "are" .) There is no doubt that the name (or name symbol) " (X, Y) " exhibits the desired asymmetry with respect to the names " X " and " Y " , but this says nothing about the objects they name.

(Another unsavoury feature transmitted by some attempted definitions is the suggestion that one has only to write down a group of symbols in order to bring into existence a mathematical object (of which the group of symbols will henceforth be a name). See also I.3.4(iv) and I.4.3(iv).)

At a formal level, one might examine the string XY as a possible bearer of the name "ordered pair associated with X and Y " ; but this will not do, since XY may fail to be a set.

Again, the unordered pair (or couple) {X, Y} (which is the set whose only members are the sets denoted by X and Y) clearly does <u>not</u> possess the vital property; see II.4.1(9). However, it is simple to check that this property <u>is</u> secured, if one defines (X, Y) to be an abbreviation for {{X}, {X, Y}} , which is the set whose only members are the sets {X} and {X, Y} , which in turn are the sets whose only members are X and X and Y respectively. (This definition was first proposed by Wiener and Kuratowski, who were among the first to show how to embrace within set theory the then-current theory of relations.)

As Halmos comments ((1), p.25), one tends to distrust this type of definition, not because there is anything obviously wrong with it, but rather because it seems to endow the new concept with features which appear "odd" . In this case, the proposed definition ensures not only the desired vital property, but also the theorem $\{X, Y\} \in (X, Y)$, which seems bizarre and unexpected, even counter-intuitive. On the other hand, the reader may feel that the adjective "ordered" is misapplied in "ordered pair" , if the latter concept is defined in the proposed fashion; he may ask what there is about the concept worthy of the term "order" . The response to this must be that, in the present development, "order" is something to be defined later (see Problem VI/11) as a special type of relation; one therefore should not expect the preconceived concept of order to appear at this stage in the theory. It would perhaps be better at this stage to speak of "asymmetric pair" in place of "ordered pair" . Later on, when natural numbers and their order have been defined, one could (see IV.9.1) rephrase the proposed definition in a way which indicates that the "order" referred to in "ordered pairs" is in a sense inherited from the order on the set of natural numbers. (As has been suggested already, however, the whole question of order in the context of ordered pairs is a red herring: the proposed definition has the merit of disclosing the fraud, and it is perhaps only traditional outlook which regards this as a disadvantage of the proposed definition. Cf. the discussion near the end of IV.7.1.)

In spite of such possible misgivings, the proposed definition will be adopted. It proves a posteriori to be entirely effective and workable; moreover, having once seen the definition and verified its efficacy, one tends to forget the seemingly irrelevant details. See also the remarks in II.1.3(ix), comparing our procedure with that of Bourbaki.

III.1.1 Formal definition of ordered pairs The ordered pair associated with X and Y (sotto voce: in that order!) is defined thus:

$$(X, Y) \equiv_{\text{def}} \{\{X\}, \{X, Y\}\} \quad . \tag{1}$$

Notice that accordingly (see especially II.4.1)

$$(S|\underline{u})(X, Y) \equiv ((S|\underline{u})X, (S|\underline{u})Y) \quad ,$$

and

$$\text{T!} \quad : \qquad (X = X' \wedge Y = Y') \Rightarrow ((X, Y) = (X', Y')) \quad .$$

(In making this (and other) definitions, one should bear in mind that all letters are sets but not conversely; so the definition (actually a definition schema) considers (X, Y) rather than $(\underline{x}, \underline{y})$; cf. II.1.3(ii).)

It can be verified (by using (*) in II.1.3(ii) and the replacement rules in II.4.1) that, if $\underline{x}$ and $\underline{y}$ denote distinct letters and $D \equiv \{\{\underline{x}\}, \{\underline{x}, \underline{y}\}\}$, then (whether or not $\underline{x}$ and/or $\underline{y}$ appears in X and/or Y) ,

$$(X, Y) \equiv (X|\underline{x}, Y|\underline{y})D \quad .$$

(On the other hand,

$$(X|\underline{x})(Y|\underline{y})D \equiv (X, (X|\underline{x})Y) \quad ,$$

which is not always equal to (X, Y) .)

III.1.2 <u>Properties of ordered pairs</u> For much of the time, almost all that one needs to remember about the definition of ordered pairs is its guarantee that

$$\text{T!} \quad : \qquad ((X, Y) = (X', Y')) \Leftrightarrow (X = X' \wedge Y = Y') \quad . \tag{1}$$

As an illustration, I will discuss a semiformal proof of (1) . The difficult part (relatively speaking) is to prove

$$\text{T} \quad : \qquad ((X, Y) = (X', Y')) \Rightarrow (X = X' \wedge Y = Y') \quad . \tag{1'}$$

To do this, I will make use of two more proof methods to be added to (I) - (XIV) in Chapters I and II, basing the justification of these new ones on free use of their predecessors. (Each of these new proof methods is a quite important "principle of logic" .) In addition, an auxiliary lemma (schema) will be required.

(XV) Let Θ denote a theory and $\underline{\underline{A}}$, $\underline{\underline{B}}$, $\underline{\underline{C}}$ arbitrary sentences. Suppose that (i) $\underline{\underline{A}} \Rightarrow \underline{\underline{C}}$, (ii) $\underline{\underline{B}} \Rightarrow \underline{\underline{C}}$ and (iii) $(\neg\underline{\underline{A}} \wedge \neg\underline{\underline{B}}) \Rightarrow \underline{\underline{C}}$ are theorems of Θ . Then $\underline{\underline{C}}$ is a theorem of Θ .

It is simple to use (VI) and (VIII) in I.3.2 to conclude that: if (i) and (ii) are true in Θ , then (iv) $\underline{\underline{A}} \vee \underline{\underline{B}} \Rightarrow \underline{\underline{C}}$ is true in Θ . On the other hand, by I.3.3(j), $(\neg\underline{\underline{A}} \wedge \neg\underline{\underline{B}}) \Leftrightarrow \neg(\underline{\underline{A}} \vee \underline{\underline{B}})$ is true in Θ_0 , hence in Θ . Thus (iii) and (III) in I.3.2 show that (v) $\neg(\underline{\underline{A}} \vee \underline{\underline{B}}) \Rightarrow \underline{\underline{C}}$ is true in Θ . One now has only to use (iv) and (v) in conjunction with (VIII) in I.3.2 to conclude that $\underline{\underline{C}}$ is true in Θ .

(XVI) Let Θ denote a theory and $\underline{\underline{A}}$, $\underline{\underline{B}}$ arbitrary sentences. Suppose that $\underline{\underline{A}} \vee \underline{\underline{B}}$ and $\neg\underline{\underline{A}}$ are theorems of Θ . Then $\underline{\underline{B}}$ is a theorem of Θ .

More loosely: If $\underline{\underline{A}} \vee \underline{\underline{B}}$ is true and $\underline{\underline{A}}$ is false, then $\underline{\underline{B}}$ is true.

Let Θ' denote the theory obtained by adjoining $\neg\underline{\underline{B}}$ to the explicit axioms of Θ . By I.3.2(VII), it will suffice to show that Θ' is contradictory. Now $\neg\underline{\underline{A}}$ and $\neg\underline{\underline{B}}$ are both theorems of Θ' , hence so too is $\neg\underline{\underline{A}} \wedge \neg\underline{\underline{B}}$, which is equivalent (in Θ_0 therefore also in Θ and Θ') to $\neg(\underline{\underline{A}} \vee \underline{\underline{B}})$. Thus $\neg(\underline{\underline{A}} \vee \underline{\underline{B}})$ is true in Θ' . Since $\underline{\underline{A}} \vee \underline{\underline{B}}$ is true in Θ , hence also in Θ' , it is manifest that Θ' is contradictory. Alternatively, use Problem I/1(2).

<u>Lemma</u> T : $(Y \neq Z) \Rightarrow (\{X\} \neq \{Y, Z\})$.

<u>Proof</u> Let Θ_1 denote the theory obtained by adjoining $Y \neq Z$ to the explicit axioms of Θ_0 . It will suffice to prove that

$T\Theta_1$: $$\{X\} \neq \{Y, Z\} .$$

If Θ_2 is the result of adjoining the explicit axiom $\neg(\{X\} = \{Y, Z\})$ - or, what is equivalent in Θ_0, $\{X\} = \{Y, Z\}$ - to Θ_1, it suffices to show that Θ_2 is contradictory. Since $Y \in \{Y, Z\}$ is true in Θ_0, it is true in Θ_2; so $Y \in \{X\}$ is true in Θ_2; hence $Y = X$ is true in Θ_2. Similarly, $Z = X$ is true in Θ_2. But then $Y = Z$ is true in Θ_2 (since $(Y = X) \wedge (Z = X) \Rightarrow (Y = Z)$ is true in Θ_0, hence true in Θ_2, and $(Y = X) \wedge (Z = X)$ is true in Θ_2). However, $Y \neq Z$, that is, $\neg(Y = Z)$, is true in Θ_1 and hence in Θ_2. Thus Θ_2 is contradictory. □

> Remark One must not at this stage yield to the temptation to say that $\{X\}$ has one element, whereas $Y \neq Z$ implies that $\{Y, Z\}$ has two elements! Cardinality is not yet defined. See V.7.2. See also the routine proof of (1) given on pp. 52-53 of Spivak (1). (Incidentally, Spivak writes (loc. cit., p.53, first line):
>
> Now $\{\{a\}, \{a, b\}\}$ contains just two members, $\{a\}$ and $\{a, b\}$; ..... .
>
> What is your reaction to this?)

Proof of (1') In the following proof, the reader should watch for numerous implicit appeals to S6 and the theorems in II.4.

Adjoin the explicit axiom

$$(X, Y) = (X', Y') \tag{2}$$

to Θ_0 to obtain a theory Θ. It will suffice to prove that $(X = X' \wedge Y = Y')$ is true in Θ. To do this, I will appeal to (XV), taking therein

$$\underline{\underline{A}} \equiv X = Y \quad , \quad \underline{\underline{B}} \equiv X' = Y' \quad , \quad \underline{\underline{C}} \equiv (X = X') \wedge (Y = Y') .$$

Stage (i) $\underline{\underline{A}} \Rightarrow \underline{\underline{C}}$ in (that is, is a theorem of) Θ. Adjoin $\underline{\underline{A}}$ to Θ

to get Θ_1 . Now (2) is true in Θ_1 and so, by definition,

$$T\Theta_1 \quad : \quad \{\{X\}, \{X, Y\}\} = \{\{X'\}, \{X', Y'\}\} \quad .$$

Since $\underline{\underline{A}}$ is true in Θ_1 , it follows that

$$T\Theta_1 \quad : \quad \{\{X\}\} = \{\{X'\}, \{X', Y'\}\} \quad .$$

Hence

$$T\Theta_1 \quad : \quad \{X'\} = \{X\} \wedge \{X', Y'\} = \{X\}$$

that is,

$$T\Theta_1 \quad : \quad X = X' \wedge X = X' \wedge X = Y' \quad .$$

Hence, since $Y = X$ is true in Θ_1 ,

$$T\Theta_1 \quad : \quad X = X' \wedge Y = Y'$$

that is,

$$T\Theta_1 \quad : \quad \underline{\underline{C}} \quad .$$

Hence

$$T\Theta \quad : \quad \underline{\underline{A}} \Rightarrow \underline{\underline{C}} \quad ,$$

as required.

<u>Stage (ii)</u> $\underline{\underline{B}} \Rightarrow \underline{\underline{C}}$ in Θ . The proof of this is exactly like that in

Stage (i); I omit the details.

<u>Stage (iii)</u> $(\neg\underline{\underline{A}} \wedge \neg\underline{\underline{B}}) \Rightarrow \underline{\underline{C}}$ in Θ . Adjoin $\neg\underline{\underline{A}} \wedge \neg\underline{\underline{B}}$ to Θ to get Θ_2 . It suffices to prove that $\underline{\underline{C}}$ is true in Θ_2 . Now (2) is true in Θ , hence true in Θ_2 , that is,

$$T\Theta_2 : \qquad \{\{X\}, \{X, Y\}\} = \{\{X'\}, \{X', Y'\}\} \ , \tag{3}$$

hence

$$T\Theta_2 : \qquad \{X\} = \{X'\} \vee \{X\} = \{X', Y'\} \ .$$

But $X' \neq Y'$ is true in Θ_2 and so, by the lemma $\{X\} \neq \{X', Y'\}$ is true in Θ_2 . Hence, by (XVI)

$$T\Theta_2 : \qquad \{X\} = \{X'\}$$

and hence

$$T\Theta_2 : \qquad X = X' \ . \tag{4}$$

Similarly, from (3) it follows that

$$T\Theta_2 : \qquad \{X, Y\} = \{X'\} \vee \{X, Y\} = \{X', Y'\} \ .$$

But $X \neq Y$ is true in Θ_2 and so, by the lemma, $\{X, Y\} \neq \{X'\}$ is true in Θ_2 . Hence, by (XVI) again,

$$T\Theta_2 : \qquad \{X, Y\} = \{X', Y'\} \ .$$

Using (4), this implies

$$T\Theta_2 \quad : \quad \{X, Y\} = \{X, Y'\}$$

and so

$$T\Theta_2 \quad : \quad Y' = X \vee Y' = Y \quad .$$

But $Y' \neq X'$ and $X' = X$ are true in Θ_2, hence $Y' \neq X$ is true in Θ_2. Another appeal to (XVI) entails

$$T\Theta_2 \quad : \quad Y' = Y \quad . \tag{5}$$

By (4) and (5),

$$T\Theta_2 \quad : \quad (X = X') \wedge (Y' = Y) \quad ,$$

that is, $\underline{\underline{C}}$ is true in Θ_2.

An appeal to (XV) now shows that $\underline{\underline{C}}$ is true in Θ, which completes the proof of (1'). □

A somewhat less "bare-handed" proof of (1') is indicated in Problem III/23.

Remark A more routine proof of III.1.2 would probably make no explicit reference to the proof methods (XV) and (XVI), which would be left for the reader to perform mentally. It would be correspondingly much less verbose.

I proceed to two simple results to be used in III.1.4 and III.2.2, namely

$$T \quad : \quad (\underline{u} \in X \wedge \underline{v} \in Y \wedge \underline{z} = (\underline{u}, \underline{v})) \Rightarrow \underline{z} \in PP(X \cup Y) \quad , \tag{6}$$

$$T \quad : \quad (\underline{x}, \underline{y}) \in \underline{w} \Rightarrow \underline{x} \in \cup\cup\underline{w} \wedge \underline{y} \in \cup\cup\underline{w} \quad . \tag{7}$$

Proof of (6) I present this in routine format. Assume $u \in X$, $v \in Y$ and $z = (u, v)$. By II.8(7) and (6), (7) and (8) in II.11.2, $\{u\} \in P(X) \subseteq P(X \cup Y)$ and $\{u, v\} \in P(X \cup Y)$. Hence, by the same token, $\{\{u\}, \{u, v\}\} \in PP(X \cup Y)$. Since $(u, v) = \{\{u\}, \{u, v\}\}$, $(u, v) \in PP(X \cup Y)$ by S6 in II.1.2. To derive (6), an appeal to (VI) in I.3.2 now suffices. □

Proof of (7) The presentation is again routine in style. Assume $(x, y) \in w$, that is, $\{\{x\}, \{x, y\}\} \in w$. Then

$$\{x\} \in (x, y) \text{ and } (x, y) \in w .$$

So, by II.6.2(6), $\{x\} \in \cup w$. Hence

$$x \in \{x\} \text{ and } \{x\} \in \cup w$$

and, by the same token, $x \in \cup\cup w$. In a similar way, one derives $y \in \cup\cup w$. Now appeal to I.3.2(VI). □

I end this subsection with the formal definition

$$z \text{ is an ordered pair} \equiv_{def} (\exists x)(\exists y)(z = (x, y)) ,$$

wherein x and y denote distinct letters each different from z (their choice is otherwise immaterial). In this connection, refer again to II.1.3(ii) and (iii); see also the remarks following III.2.1.

Some writers might introduce a special notation (perhaps OPz) in place of " z is an ordered pair" ; I, in fact, do something just like this in connection with IV.1.1 below. This is a labour-saving device, but it may also be employed in an attempt to avoid ambiguity. For instance, " z is an ordered pair" might be interpreted as asserting a (false) metastatement (namely, $z \equiv (X, Y)$ for suitable sets X and Y) rather than as a name for a formal sentence. Again, it is sometimes not totally clear whether " z is an ordered pair" is a name for a formal sentence, or whether it expresses the metastatement " $(\exists x)(\exists y)(z = (x, y))$ is a theorem" . These risks would be minimised by adopting a symbol such as

OPz . However, the phrase occurs but rarely and I rely on the context to make the meaning clear.

Notice that " z is an ordered pair" (and also later on " Z is a relation" and " f is a function") names a formal sentence. On the contrary, in this book " z is (or denotes) a set" names no such sentence; rather, it is intended always as a (true and trivial) metastatement.

In accordance with what is stated in II.1.3(iii), it is to be understood from here on that one adopts the definition schema

$$(A \text{ is an ordered pair}) \equiv_{def} (A|\underline{z})(\underline{z} \text{ is an ordered pair})$$

for arbitrary strings A ; and then

$$(A \text{ is an ordered pair}) \equiv_{def} (\exists\underline{x})(\exists\underline{y})(A = (\underline{x}, \underline{y})) \quad , \tag{9}$$

where $\underline{x}$, $\underline{y}$ denote arbitrary distinct letters not appearing in A .

III.1.3 <u>Formal definitions of first and second coordinates or projections</u> Informally, the first and second coordinates of an ordered pair $(\underline{x}, \underline{y})$ are $\underline{x}$ and $\underline{y}$ respectively. This amounts to a conditional definition of the type: if $\underline{z} = (\underline{x}, \underline{y})$, then (by definition) the first and second coordinates of $\underline{z}$ are $\underline{x}$ and $\underline{y}$ respectively. To avoid conditionality, the formal definition is framed as follows: if $\underline{x}$, $\underline{y}$, $\underline{z}$ denote distinct letters, the sets (denoted by)

$$\tau_{\underline{x}}((\exists\underline{y})(\underline{z} = (\underline{x}, \underline{y}))) \qquad \text{and} \qquad \tau_{\underline{y}}((\exists\underline{x})(\underline{z} = (\underline{x}, \underline{y})))$$

(both independent of the choice of $\underline{x}$ and $\underline{y}$) are denoted by $pr_1\underline{z}$ and $pr_2\underline{z}$ and termed the <u>first and second coordinates</u> (<u>or projections</u>) <u>of</u> $\underline{z}$. In other words,

$$pr_1\, \underline{z} \equiv_{def} \tau_{\underline{x}}((\exists\underline{y})((\underline{z} = (\underline{x}, \underline{y}))) \quad , \tag{1_1}$$

$$pr_2 \, \underline{z} \equiv_{def} \tau_{\underline{x}}((\exists \underline{x})(\underline{z} = (\underline{x}, \underline{y}))) \quad . \tag{1_2}$$

(Recall II.3.9(v): other versions of set theory employ other definitions, though agreement is reached for ordered pairs $\underline{z}$; see, for example Kelley (1), p.259 and Problem III/15 below.)

The " 1 " and " 2 " appearing in $pr_1 \, \underline{z}$ and $pr_2 \, \underline{z}$ are to be regarded as fragments of the metalanguage which serve merely to distinguish between the adopted names for certain formal objects. They are not here intended as themselves names for formal objects. This role has to be distinguished from their later use as numerals, that is, names for the formal objects otherwise denoted by $\{\emptyset\}$ and $\{\emptyset , \{\emptyset\}\}$ (see Chapter V below). In the present context they could (and might with less risk of confusion) be supplanted by other distinguishing marks; for example, one might write $p' \, \underline{z}$ and $p'' \, \underline{z}$, or even " first projection $\underline{z}$ " and " second projection $\underline{z}$ " , though this would be somewhat unconventional.

One also defines (cf. II.1.3(iii))

$$pr_i \; Z \equiv_{def} (Z|\underline{z})pr_i \; \underline{z} \quad , \tag{1_3}$$

for an arbitrary string Z and i replaced by 1 or 2 . The conventions in II.3.12 and the replacement rules I.1.4(d) and I.1.8(f) permit one to verify that then

$$pr_1 \; Z \equiv \tau_{\underline{x}}((\exists \underline{y})(Z = (\underline{x}, \underline{y}))) \tag{1_4}$$

and

$$pr_2 \; Z \equiv \tau_{\underline{y}}((\exists \underline{x})(Z = (\underline{x}, \underline{y}))) \quad , \tag{1_5}$$

$\underline{x}$ and $\underline{y}$ denoting distinct letters not appearing in Z .

One may (and should) verify the replacement rule

$$(S|\underline{u})pr_i \; Z \equiv pr_i \; (S|\underline{u})Z$$

for arbitrary strings S and Z and arbitrary letter $\underline{u}$. Moreover, one may (and should) prove the theorem schema

$$(Z = W) \Rightarrow ((\mathrm{pr}_1\, Z = \mathrm{pr}_1\, W) \wedge (\mathrm{pr}_2\, Z = \mathrm{pr}_2\, W)) \quad .$$

That the expected properties of coordinates are secured by the above definitions, is contained in the following basic theorems:

T! : $$(\mathrm{pr}_1\, (\underline{x}, \underline{y}) = \underline{x}) \wedge (\mathrm{pr}_2\, (\underline{x}, \underline{y}) = \underline{y}) \quad , \qquad (1)$$

T! : $$\underline{z} = (\underline{x}, \underline{y}) \Leftrightarrow ((\underline{z} \text{ is an ordered pair}) \wedge \underline{x} = \mathrm{pr}_1\, \underline{z} \wedge \underline{y} = \mathrm{pr}_2\, \underline{z}) \quad , \qquad (2)$$

T! : $$(\underline{z} \text{ is an ordered pair}) \Leftrightarrow (\underline{z} = (\mathrm{pr}_1\, \underline{z}, \mathrm{pr}_2\, \underline{z})) \quad . \qquad (3)$$

Semiformal proof of (1)

$$\mathrm{pr}_1\, (\underline{x}, \underline{y}) \equiv ((\underline{x}, \underline{y})|\underline{z})\mathrm{pr}_1\, \underline{z}$$

$$\equiv ((\underline{x}, \underline{y})|\underline{z})\tau_{\underline{x}'}((\exists \underline{y}')(\underline{z} = (\underline{x}', \underline{y}')) \quad ,$$

where $\underline{x}'$, $\underline{y}'$ denote distinct letters different from $\underline{x}$, $\underline{y}$, $\underline{z}$

$$\equiv \tau_{\underline{x}'}(((\underline{x}, \underline{y})|\underline{z})(\exists \underline{y}')(\underline{z} = (\underline{x}', \underline{y}')) \qquad (\text{I.1.4(d)})$$

$$\equiv \tau_{\underline{x}'}((\exists \underline{y}')((\underline{x}, \underline{y})|\underline{z})(\underline{z} = (\underline{x}', \underline{y}'))) \qquad (\text{I.1.8(f)})$$

$$\equiv \tau_{\underline{x}'}((\exists \underline{y}')((\underline{x}, \underline{y}) = (\underline{x}', \underline{y}')) \quad . \qquad ((\text{p}) \text{ in II.1.3(vi)})$$

hence

T : $$\mathrm{pr}_1\, (\underline{x}, \underline{y}) = \tau_{\underline{x}'}((\exists \underline{y}')(\underline{x} = \underline{x}' \wedge \underline{y} = \underline{y}')) \quad .$$ ((I.3.2(I), I.3.2(X), I.3.3(1), S7 in II.1.2 and III.1.2)

By I.3.3(o),

$$T \quad : \qquad (\exists \underline{y}')(\underline{x} = \underline{x}' \wedge \underline{y} = \underline{y}') \Leftrightarrow (\underline{x} = \underline{x}') \wedge (\exists \underline{y}')(\underline{y} = \underline{y}') \quad ;$$

and so, since $(\exists \underline{y}')(\underline{y} = \underline{y}')$ is true in Θ_0, I.3.7(1) entails

$$T \quad : \qquad (\exists \underline{y}')(\underline{x} = \underline{x}' \wedge \underline{y} = \underline{y}') \Leftrightarrow \underline{x} = \underline{x}' \quad .$$

So another appeal to I.3.2(I), I.3.2(X) and S7 in II.1.2 entails

$$T \quad : \qquad \mathrm{pr}_1 (\underline{x}, \underline{y}) = \tau_{\underline{x}'}(\underline{x} = \underline{x}') \quad .$$

Hence, by II.1.4(3), S7 in II.1.2, II.1.4(6) and II.1.4(4)

$$T \quad : \qquad \mathrm{pr}_1 (\underline{x}, \underline{y}) = \underline{x} \quad .$$

A similar argument proves

$$T \quad : \qquad \mathrm{pr}_2 (\underline{x}, \underline{y}) = \underline{y} \quad ,$$

and then (1) follows by appeal to I.3.3(j). □

<u>Semiformal proof of (2)</u> The proof of $\Rightarrow$ is an easy application of I.3.2(VI), S5 in I.2.2, S6 in II.1.2 and (1). To deal with $\Leftarrow$, introduce the theory Θ obtained by adjoining to Θ_0 the explicit axiom

$$(\underline{z} \text{ is an ordered pair}) \wedge \underline{x} = \mathrm{pr}_1 \, \underline{z} \wedge \underline{y} = \mathrm{pr}_2 \, \underline{z} \quad .$$

Thus ($\underline{x}'$, $\underline{y}'$ being as in the proof of (1), so that $\underline{x}'$ and $\underline{y}'$ are variables of Θ)

$$T_\Theta \ : \qquad (\exists \underline{x}')(\exists \underline{y}')(\underline{z} = (\underline{x}', \underline{y}')) \wedge \underline{x} = pr_1\ \underline{z} \wedge \underline{y} = pr_2\ \underline{z} \ .$$

By I.3.3(o), therefore,

$$T_\Theta \ : \qquad (\exists \underline{x}')(\exists \underline{y}')(\underline{z} = (\underline{x}', \underline{y}') \wedge \underline{x} = pr_1\ \underline{z} \wedge \underline{y} = pr_2\ \underline{z}) \ .$$

Hence, by (1) and I.3.3(l),

$$T_\Theta \ : \quad (\exists \underline{x}')(\exists \underline{y}')(\underline{z} = (\underline{x}', \underline{y}') \wedge \underline{x}' = pr_1\ \underline{z} \wedge \underline{x} = pr_1\ \underline{z} \wedge \underline{y}' = pr_2\ \underline{z} \wedge \underline{y} = pr_2\ \underline{z}) \ ,$$

and so, by II.1.4(4) and I.3.3(l),

$$T_\Theta \ : \qquad (\exists \underline{x}')(\exists \underline{y}')(\underline{z} = (\underline{x}', \underline{y}') \wedge \underline{x}' = \underline{x} \wedge \underline{y}' = \underline{y}) \ .$$

Hence, by III.1.2 and I.3.3(l),

$$T_\Theta \ : \qquad (\exists \underline{x}')(\exists \underline{y}')(\underline{z} = (\underline{x}', \underline{y}') \wedge (\underline{x}', \underline{y}') = (\underline{x}, \underline{y})) \ ,$$

and therefore, by II.1.4(4) and I.3.3(l),

$$T_\Theta \ : \qquad (\exists \underline{x}')(\exists \underline{y}')(\underline{z} = (\underline{x}, \underline{y})) \ ,$$

which is identical with $\underline{z} = (\underline{x}, \underline{y})$. Now use I.3.2(VI) and I.3.3(j) to derive $\Leftarrow$. □

The proof of (3) is left as an exercise for the reader.

The discussion of ordered triplets, et cetera, is more conveniently left until IV.9.

Concerning some notational vagaries, see Remark (iii) at the end of III.2.6.

III.1.4 Formal definition of Cartesian products Informally, the Cartesian product (often called simply the product) of sets X and Y (in that order) is the set of all ordered pairs, the first coordinate of which belongs to X and the second coordinate of which belongs to Y . For instance, according to this definition, one will expect that

$$\{a\} \times \{b\} = \{(a, b)\} ,$$

$$\{a\} \times \{b, c\} = \{(a, b), (a, c)\} ,$$

$$\{a, b\} \times \{c, d\} = \{(a, c), (a, d), (b, c), (b, d)\}$$

will be theorem schemas.

Regarding the phrase "in that order" , see the discussion in III.1.1 above.

Turning to formalities, I begin by considering the sentence

$$(\exists \underline{x})(\exists \underline{y})(\underline{x} \in X \wedge \underline{y} \in Y \wedge \underline{z} = (\underline{x}, \underline{y})) ,$$

which I denote temporarily by $\underline{A}$; herein $\underline{x}$, $\underline{y}$, $\underline{z}$ denote distinct letters not appearing in X or Y . Bear I.3.7(6) in mind here and in the sequel. From III.1.2(6) and two appeals to I.3.2(IX) (or I.3.3(1)), one sees that

$$\underline{A} \Rightarrow \underline{z} \in PP(X \cup Y)$$

is true. Hence, by (XII) in II.3.7, $\underline{A}$ is collectivising in $\underline{z}$. This is the signal to make the formal definition

$$X \times Y \equiv_{def} \{\underline{z} : (\exists \underline{x})(\exists \underline{y})(\underline{x} \in X \wedge \underline{y} \in Y \wedge \underline{z} = (\underline{x}, \underline{y}))\} , \qquad (1)$$

$X \times Y$ being termed the Cartesian product of X and Y (in that order).

Then (see (3) or (3') in II.3.3)

$$T! \; : \qquad \underline{z} \in X \times Y \Leftrightarrow (\exists \underline{x})(\exists \underline{y})(\underline{x} \in X \wedge \underline{y} \in Y \wedge \underline{z} = (\underline{x}, \underline{y})) \;, \tag{2}$$

Hence, by I.3.2(XI) and various replacement rules,

$$T! \qquad T \in X \times Y \Leftrightarrow (\exists \underline{x})(\exists \underline{y})(\underline{x} \in X \wedge \underline{y} \in Y \wedge T = (\underline{x}, \underline{y})) \;. \tag{2'}$$

The formal definition (1) applies for arbitrary strings X and Y and in it one may take for $\underline{x}$, $\underline{y}$, $\underline{z}$ any three distinct letters not appearing in X or Y ; but in the theorem schemas (2) and (2') it is supposed that X and Y are arbitrary sets; and, in (2) (resp. (2')), $\underline{x}$ and $\underline{y}$ are assumed not to appear in X or Y (resp. X , Y or T).

Note the replacement rule

$$(S|\underline{u})(X \times Y) \equiv (S|\underline{u})X \times (S|\underline{u})Y$$

and the theorem schema

$$T! \; : \qquad (X = X' \wedge Y = Y') \Rightarrow X \times Y = X' \times Y' \;.$$

Notice that $X \times Y$ and $Y \times X$ may be and usually are unequal sets.

The sentence schema

$$\{a, b\} \times \{c, d\} = \{(a, c), (a, d), (b, c), (b, d)\}$$

can be proved to be a theorem schema, but a "barehanded" semiformal proof is surprisingly and makes repeated use of metatheorems and theorems in I.3.3, I.3.7, II.4.1, II.6.4, Problem II/14 and III.1.4. See, however, Problem III/2, which provides an alternative route to this theorem schema.

Beware the potential confusion arising from the (entirely conventional) use of $\times$ to indicate Cartesian products and its (equally conventional) use to indicate arithmetical products of (for example) natural numbers; see the comments at the end of V.6.3 below.

III.1.5 Further notation (i) In many texts one will find III.1.4(1) abbreviated to

$$X \times Y \equiv_{def} \{(\underline{x}, \underline{y}) : \underline{x} \in X \wedge \underline{y} \in Y\} \tag{1'}$$

without further comment. This is however, not strictly in accord with the use of the set builder as explained in II.3.2. (Rather, what is involved is closely akin to what has been explained informally in connection with (1) and (2) in II.12.1.) Nevertheless, the abbreviated style is so widespread and genuinely useful that it is almost obligatory that the situation be regularised ready for use in III.2 onwards; but see Note 10.

To this end I make the formal definition schema (cf. Suppes (1), pp.35-36)

$$\{(\underline{x}, \underline{y}) : A\} \equiv_{def} \{\underline{z} : (\exists\underline{x})(\exists\underline{y})(A \wedge \underline{z} = (\underline{x}, \underline{y}))\} \tag{3}$$

wherein $\underline{x}$, $\underline{y}$, $\underline{z}$ denote distinct letters, $\underline{z}$ not appearing in A ; cf. the definition of a generalised set builder in equation (6) of §3 of the Appendix. It is a tedious but necessary exercise in the use of replacement rules to check that, if $\underline{x}'$ and $\underline{y}'$ denote distinct letters not appearing in A , then

$$\{(\underline{x}, \underline{y}) : A\} \equiv \{(\underline{x}', \underline{y}') : (\underline{x}'|\underline{x}, \underline{y}'|\underline{y})A\}$$

The virtues of (3) are mainly practical. For the purposes of semiformal proofs it is often necessary to substitute for the abbreviating symbol defined in and appearing on the left of (3) the symbol on the right, so as to be able to apply the earlier theorems and metatheorems relating to the set builder as defined in II.3.2.

In any case, one has to be careful not to fall between two stools by writing, for example,

$$X \times Y = \{\underline{z} : \underline{z} = (\underline{x}, \underline{y}) \wedge \underline{x} \in X \wedge \underline{y} \in Y\} \quad .$$

For instance, if

$$A \equiv \{\underline{z} : \underline{z} = (\underline{x}, \underline{y}) \wedge \underline{x} \in \{\emptyset\} \wedge \underline{y} \in \{\emptyset, \{\emptyset\}\}\} \quad ,$$

$$B \equiv \{\underline{z} : (\exists \underline{x})(\exists \underline{y})(\underline{z} = (\underline{x}, \underline{y}) \wedge \underline{x} \in \{\emptyset\} \wedge \underline{y} \in \{\emptyset, \{\emptyset\}\})\} \quad ,$$

Then $A = B$ is false. (Compare with a similar situation in II.3.9(iii).)

(ii) In connection with the abbreviated notation just explained in (i) immediately above, it is worth noting the theorem schema

$$\text{T! : } \quad (S = \{(\underline{x}, \underline{y}) : \underline{\underline{A}}\} \wedge \text{Coll}_{\underline{z}}(\exists \underline{x})(\exists \underline{y})(\underline{\underline{A}} \wedge \underline{z} = (\underline{x}, \underline{y}))) \Rightarrow ((a, b) \in S \Leftrightarrow (a|\underline{x}, b|\underline{y})\underline{\underline{A}}) \qquad (3_1)$$

wherein $\underline{x}$, $\underline{y}$, $\underline{z}$ denote distinct letters, $\underline{z}$ not appearing in $\underline{\underline{A}}$, and a, b denote arbitrary sets.

<u>Proof</u> Let $\underline{x}'$, $\underline{y}'$ denote distinct letters, each different from $\underline{x}$, $\underline{y}$ and $\underline{z}$ and not appearing in a, b or $\underline{\underline{A}}$. Define $\underline{\underline{A}}' \equiv (\underline{x}'|\underline{x}, \underline{y}'|\underline{y})\underline{\underline{A}}$; then $\underline{z}$ does not appear in $\underline{\underline{A}}'$. By the various replacement rules,

$$\{(\underline{x}, \underline{y}) : \underline{\underline{A}}\} \equiv \{(\underline{x}', \underline{y}') : \underline{\underline{A}}'\} \quad ,$$

$$\text{Coll}_{\underline{z}}(\exists \underline{x})(\exists \underline{y})(\underline{\underline{A}} \wedge \underline{z} = (\underline{x}, \underline{y})) \equiv \text{Coll}_{\underline{z}}(\exists \underline{x}')(\exists \underline{y}')(\underline{\underline{A}}' \wedge \underline{z} = (\underline{x}', \underline{y}')) \quad ,$$

$$(a|\underline{x}', b|\underline{y}')\underline{\underline{A}}' \equiv (a|\underline{x}, b|\underline{y})\underline{\underline{A}} \quad .$$

Thus one may suppose from the outset that $\underline{x}$ and $\underline{y}$ do not appear in a or b.

Let Θ denote the theory obtained from Θ_0 by adjoining the axiom

$$S = \{(\underline{x}, \underline{y}) : \underline{\underline{A}}\} \wedge \mathrm{Coll}_{\underline{z}}(\exists\underline{x})(\exists\underline{y})(\underline{\underline{A}} \wedge \underline{z} = (\underline{x}, \underline{y})) \quad .$$

By (3) and II.3.3(5),

$$T\Theta \quad : \qquad (\forall\underline{z})(\underline{z} \in S \Leftrightarrow (\exists\underline{x})(\exists\underline{y})(\underline{\underline{A}} \wedge \underline{z} = (\underline{x}, \underline{y}))) \quad ,$$

and so, by I.3.2(XI),

$$T\Theta \quad : \qquad ((a, b)|\underline{z})(\underline{z} \in S \Leftrightarrow (\exists\underline{x})(\exists\underline{y})(\underline{\underline{A}} \wedge \underline{z} = (\underline{x}, \underline{y}))) \quad .$$

Using the replacement rules, this is seen to be identical with

$$T\Theta \quad : \qquad (a, b) \in S \Leftrightarrow (\exists\underline{x})(\exists\underline{y})(\underline{\underline{A}} \wedge (a, b) = (\underline{x}, \underline{y})) \quad ,$$

and so, using III.1.2(1),

$$T\Theta \quad : \qquad (a, b) \in S \Leftrightarrow (\exists\underline{x})(\exists\underline{y})(\underline{\underline{A}} \wedge \underline{x} = a \wedge \underline{y} = b) \quad .$$

Two applications of Problem II/14 lead (via the replacement rules in I.1.4) to

$$T\Theta \quad : \qquad (a, b) \in S \Leftrightarrow (a|\underline{x}, b|\underline{y})\underline{\underline{A}} \quad ,$$

and it now suffices to use I.3.2(VI). □

<u>Remark</u> In (3_1), it is not generally legitimate to replace $(a|\underline{x}, b|\underline{y})\underline{\underline{A}}$ by $(a|\underline{x})(b|\underline{y})\underline{\underline{A}}$, though this is permissible if $\underline{x}$ and $\underline{y}$ do not

appear in a or b . See Problem III/21.

III.1.6 Properties of Cartesian products Only the following very simple theorem schemas will be used in the sequel.

$$T! : \quad ((S, T) \in X \times Y) \Leftrightarrow (S \in X \wedge T \in Y) , \tag{4}$$

$$T! : \quad X \subseteq X_1 \Rightarrow X \times Y \subseteq X_1 \times Y , \tag{5}$$

$$T! : \quad Y \subseteq Y_1 \Rightarrow X \times Y \subseteq X \times Y_1 , \tag{6}$$

$$T! : \quad (X \cup X') \times Y = (X \times Y) \cup (X' \times Y) . \tag{7}$$

The proofs of all these are based upon III.1.4(2). I will give some details in the case of (4) , leaving the remainder to the reader; see Problem III/1, which enunciates a few other similar theorem schemas to be proved.

Semiformal proof of (4) By I.3.3(j), it suffices to prove that

$$T : \quad ((S, T) \in X \times Y) \Rightarrow (S \in X \wedge T \in Y) , \tag{8}$$

and

$$T : \quad (S \in X \wedge T \in Y) \Rightarrow ((S, T) \in X \times Y) . \tag{9}$$

To deal with (8) , adjoin to Θ_0 the axiom $(S, T) \in X \times Y$ to obtain a theory Θ . By III.1.4(2) and I.3.2(II) (and various replacement rules)

$$T\Theta : \quad (\exists \underline{x})(\exists \underline{y})(\underline{x} \in X \wedge \underline{y} \in Y \wedge (S, T) = (\underline{x}, \underline{y})) ,$$

where $\underline{x}$ and $\underline{y}$ denote distinct letters not appearing in S , T , X or Y . Then, using III.1.2(1),

$$T\Theta \quad : \qquad (\exists \underline{x})(\exists \underline{y})(\underline{x} \in X \wedge \underline{y} \in Y \wedge S = \underline{x} \wedge T = \underline{y}) \quad . \qquad (10)$$

On the other hand, by S6 in II.1.2,

$$T\Theta \quad : \qquad \underline{x} \in X \wedge \underline{y} \in Y \wedge S = \underline{x} \wedge T = \underline{y} \Rightarrow S \in X \wedge T \in Y \quad .$$

This, (10) and I.3.3(l) entail

$$T\Theta \quad : \qquad (\exists \underline{x})(\exists \underline{y})(S \in X \wedge T \in Y) \quad . \qquad (11)$$

Since $\underline{x}$ and $\underline{y}$ do not appear in $S \in X \wedge T \in Y$, (11) is identical with $S \in X \wedge T \in Y$. An appeal to I.3.2(VI) now entails (8) .

Turning to (9) , adjoin to Θ_0 the axiom $S \in X \wedge T \in Y$ to obtain a theory Θ' . From II.1.4(2) and I.3.3(j),

$$T\Theta' \ : \qquad S = S \wedge T = T \wedge S \in X \wedge T \in Y \quad ,$$

that is,

$$T\Theta' \ : \qquad (S|\underline{x})(T|\underline{y})(S = \underline{x} \wedge T = \underline{y} \wedge \underline{x} \in X \wedge \underline{y} \in Y) \quad .$$

So, by III.1.2(1), I.3.3(j), S6 and I.3.2(I),

$$T\Theta' \ : \qquad (S|\underline{x})(T|\underline{y})(\underline{x} \in X \wedge \underline{y} \in Y \wedge (S, T) = (\underline{x}, \underline{y})) \quad .$$

A double appeal to S5 in I.2.2 and I.3.3(n) now implies

$$T\Theta' \ : \qquad (\exists \underline{x})(\exists \underline{y})(\underline{x} \in X \wedge \underline{y} \in Y \wedge (S, T) = (\underline{x}, \underline{y}))$$

and so, by III.1.4(2'),

$T\Theta'$: $$(S, T) \in X \times Y \ .$$

Use of I.3.2(VI) now entails (9) and completes the proof.

Alternatively, and in more condensed style : choose distinct letters $\underline{x}$ and $\underline{y}$ not appearing in S , T , X or Y . By III.1.4(2'),

$$(S, T) \in X \times Y \Leftrightarrow (\exists \underline{x})(\exists \underline{y})(\underline{x} \in X \wedge \underline{y} \in Y \wedge (S, T) = (\underline{x}, \underline{y})) \ .$$

Hence, using III.1.2(1), I.3.2(1) and I.3.3(1),

$$(S, T) \in X \times Y \Leftrightarrow (\exists \underline{x})(\exists \underline{y})(\underline{x} \in X \wedge \underline{y} \in Y \wedge S = \underline{x} \wedge T = \underline{y}) \ .$$

Then, using I.3.2(1) and Problem II/14,

$$(S, T) \in X \times Y \Leftrightarrow S \in X \wedge T \in Y \ .$$

III.1.7 Problem Verify that the sentence schema

$$X \cup (Y \times Z) = (X \times Y) \cup (X \times Z) \tag{12}$$

is not true.

III.1.8 Remarks The preceding problem might, in spite of the remarks in I.3.8(vi), be interpreted in various ways, of which I mention but three

(i) It might be taken to mean: Exhibit sets X , Y , Z such that the sentence (denoted by)

$$X \cup (Y \times Z) \neq (X \times Y) \cup (X \times Z) \tag{13}$$

is true.

(ii) It might be taken to mean: Prove

$$(\exists \underline{x})(\exists \underline{y})(\exists \underline{z})(\underline{x} \cup (\underline{y} \times \underline{z}) \neq (\underline{x} \times \underline{y}) \cup (\underline{x} \times \underline{z})) \ , \tag{14}$$

$\underline{x}$, $\underline{y}$, $\underline{z}$ denoting distinct letters.

(iii) It might be taken to mean: Exhibit sets X , Y , Z such that the sentence (denoted by) (12) is not true.

A solution to version (i) leads to a solution to version (ii) . (For it may be assumed that $\underline{x}$, $\underline{y}$, $\underline{z}$ do not appear in X , Y or Z and then, defining

$$\underline{\underline{A}} \equiv \underline{x} \cup (\underline{y} \times \underline{z}) \neq (\underline{x} \times \underline{y}) \cup (\underline{x} \times \underline{z}) \quad ,$$

(13) is identical with

$$(X|\underline{x})(Y|\underline{y})(Z|\underline{z})\underline{\underline{A}} \quad .$$

On the other hand (cf. Problem I/17(ii))

$$(X|\underline{x})(Y|\underline{y})(Z|\underline{z})\underline{\underline{A}} \Rightarrow (\exists\underline{x})(\exists\underline{y})(\exists\underline{z})\underline{\underline{A}}$$

is a theorem, and I.3.2(I) leads to (14) .) On the other hand, a solution to version (ii) leads to a solution of version (i) (cf. Problem I/17(iv)), though only at the expense of explicit use of the selector τ ; on this point, see I.3.4(viii). Versions (i) and (ii) are indeed the "constructional" and "existential" interpretations of the problem; see again I.3.4(viii).

Version (iii) is the "odd man out" inasmuch as its solution hinges on the consistency of set theory. Granted consistency, a solution to version (i) provides a solution to version (iii). Also, if the sentence (12) is known to be decidable, a solution to version (iii) provides a solution to version (i) .

III.2 Relations

In terms of ordered pairs, one defines a <u>(binary) relation</u> simply as a set of ordered pairs. The formal definition is as follows.

III.2.1 Formal definition of relations

$$\underline{w} \text{ is a relation } \equiv_{\text{def}} (\forall\underline{z})(\underline{z} \in \underline{w} \Rightarrow (\underline{z} \text{ is an ordered pair})) \qquad (1)$$

$$\equiv (\forall\underline{z})(\underline{z} \in \underline{w} \Rightarrow (\exists\underline{x})(\exists\underline{y})(\underline{z} = (\underline{x}, \underline{y}))) \quad ,$$

wherein $\underline{w}$, $\underline{x}$, $\underline{y}$, $\underline{z}$ denote distinct letters. Then, when $\underline{w}$ is chosen, the right hand side of this definition denotes precisely one sentence in which $\underline{w}$

is the only letter appearing and which is independent of the choice of the distinct letters $\underline{x}$, $\underline{y}$, $\underline{z}$, each different from $\underline{w}$.

In accordance with II.1.3(iii), if Z denotes an arbitrary string, one further defines

$$Z \text{ is a relation} \equiv_{def} (Z|\underline{w})(\underline{w} \text{ is a relation}) ,$$

which is independent of the choice of $\underline{w}$. The replacement rules show that, if $\underline{x}$, $\underline{y}$, $\underline{z}$ are distinct letters, and if $\underline{z}$ does not appear in Z , then

$$Z \text{ is a relation} \equiv (\forall \underline{z})(\underline{z} \in Z \Rightarrow (\exists \underline{x})(\exists \underline{y})(\underline{z} = (\underline{x}, \underline{y}))) .$$

It can be verified that

$$(S|\underline{u})(Z \text{ is a relation}) \equiv (S|\underline{u})Z \text{ is a relation},$$

for arbitrary strings S and Z and an arbitrary letter $\underline{u}$.

The word "graph" would be a very happy replacement for the name "relation" , and it (or rather its French equivalent "graphe") is so used in French. But the term is not frequently so used by English writers.

The above use of the term "relation" is totally different from Bourbaki's (his "relation" is synonymous with our "sentence"").

One often writes $\underline{x}Z\underline{y}$ in place of $(\underline{x}, \underline{y}) \in Z$, though the notation is actually in conflict with that used in connection with the juxtaposition of strings.

Examples of relations (for some of which see III.2.3) stem from the theorem schema

$$T! : \quad (Z \subseteq (X \times Y)) \Rightarrow (Z \text{ is a relation}) . \quad (3)$$

When the hypothesis of (3) is true, Z is usually said to be a relation <u>from</u> X <u>into</u> Y or to be a relation <u>in</u> X × Y . As will be seen in III.2.2(6), there is a converse to (3) ; see also Problem III/24.

III.2.2 Formal definitions of domain and range In informal set theory, the domain of a relation Z is defined to be the set of all $\underline{x}$ such that $(\underline{x}, \underline{y}) \in Z$ is true for some (that is, for at least one) $\underline{y}$; and the range of Z to be the set of all $\underline{y}$ such that $(\underline{x}, \underline{y}) \in Z$ is true for some $\underline{x}$. The formal definitions are modelled on these, but to ensure that the definitions will be useful, one first needs "collectivising" theorems.

From III.1.2(7) and an appeal to I.3.2(IX) (or I.3.3(1)) it follows that

$$T \quad : \qquad (\exists\underline{y})((\underline{x}, \underline{y}) \in \underline{w}) \Rightarrow \underline{x} \in \cup\cup\underline{w}$$

and

$$T \quad : \qquad (\exists\underline{x})((\underline{x}, \underline{y}) \in \underline{w}) \Rightarrow \underline{y} \in \cup\cup\underline{w} \ .$$

Hence, in view of (XII) in II.3.7, both

$$\text{Coll}_{\underline{x}}(\exists\underline{y})((\underline{x}, \underline{y}) \in \underline{w}) \tag{1}$$

and

$$\text{Coll}_{\underline{y}}(\exists\underline{x})((\underline{x}, \underline{y}) \in \underline{w}) \tag{1'}$$

are true. In (1) and (1') , one has the appropriate "collectivising" theorems.

At this point one makes the formal definitions

$$\text{Dom}\ \underline{w} \equiv_{\text{def}} \{\underline{x} : (\exists\underline{y})((\underline{x}, \underline{y}) \in \underline{w})\} \ , \tag{2}$$

$$\text{Ran}\ \underline{w} \equiv_{\text{def}} \{\underline{y} : (\exists\underline{x})((\underline{x}, \underline{y}) \in \underline{w})\} \ , \tag{3}$$

the sets thus defined being called the <u>domain</u> and <u>range</u> of $\underline{w}$, respectively. (The choice of the distinct letters $\underline{x}$ and $\underline{y}$, each different from $\underline{w}$, is immaterial.)

Reverting to (1) , (1') , (2) and (3) , it may be deduced from II.3.3(3) that

$$T_! \quad : \qquad (\underline{x} \in \text{Dom } \underline{w}) \Leftrightarrow (\exists \underline{y})((\underline{x}, \underline{y}) \in \underline{w}) \tag{4}$$

and

$$T_! \quad : \qquad (\underline{y} \in \text{Ran } \underline{w}) \Leftrightarrow (\exists \underline{x})((\underline{x}, \underline{y}) \in \underline{w}) \ . \tag{5}$$

It follows that

$$T_! \quad : \qquad (\underline{w} \text{ is a relation}) \Leftrightarrow \underline{w} \subseteq (\text{Dom } \underline{w}) \times (\text{Ran } \underline{w}) \ . \tag{6}$$

Proof of (6) The implication $\Leftarrow$ is a special case of III.2.1(3). It remains to prove $\Rightarrow$. To do this, introduce the theory Θ obtained from Θ_0 by adjoining the axiom " $\underline{w}$ is a relation" ; by (VI) in I.3.2, it will suffice to prove that

$$T_! \quad : \qquad \underline{w} \subseteq (\text{Dom } \underline{w}) \times (\text{Ran } \underline{w}) \ .$$

Since $\underline{z}$ is not a constant of Θ , I.3.2(X) shows it is sufficient for this to prove that

$$T_\Theta \quad : \qquad \underline{z} \in \underline{w} \Rightarrow \underline{z} \in (\text{Dom } \underline{w}) \times (\text{Ran } \underline{w}) \ .$$

If Θ' denotes the theory obtained from Θ by adjoining the axiom $\underline{z} \in \underline{w}$, another appeal to (VI) shows that it is enough to prove that

$$T_{\Theta'} \quad : \qquad \underline{z} \in (\text{Dom } \underline{w}) \times (\text{Ran } \underline{w}) \ . \tag{7}$$

Now " $\underline{w}$ is a relation" is true in Θ , hence true in Θ' , and so

$T\Theta'$: $\underline{z} \in \underline{w} \Rightarrow (\exists\underline{x})(\exists\underline{y})(\underline{z} = (\underline{x}, \underline{y}))$. (8)

Also, by definition of Θ' ,

$T\Theta'$: $\underline{z} \in \underline{w}$. (9)

By (8) and (9) ,

$T\Theta'$: $(\exists\underline{x})(\exists\underline{y})(\underline{z} = (\underline{x}, \underline{y}))$. (10)

By (9) , (10) and (o) in I.3.3,

$T\Theta'$: $(\exists\underline{x})(\exists\underline{y})(\underline{z} = (\underline{x}, \underline{y}) \wedge \underline{z} \in \underline{w})$. (11)

Now $\underline{z} = (\underline{x}, \underline{y}) \wedge \underline{z} \in \underline{w} \Rightarrow (\underline{x}, \underline{y}) \in \underline{w}$ is true in Θ_0 , hence true in Θ' . Using (4) and (5) it follows that

$T\Theta$: $\underline{z} = (\underline{x}, \underline{y}) \wedge \underline{z} \in \underline{w} \Rightarrow \underline{x} \in \text{Dom } \underline{w} \wedge \underline{y} \in \text{Ran } \underline{w}$.

This, with (11) and (1) in I.3.3, entails

$T\Theta'$ $(\exists\underline{x})(\exists\underline{y})(\underline{z} = (\underline{x}, \underline{y}) \wedge \underline{x} \in \text{Dom } \underline{w} \wedge \underline{y} \in \text{Ran } \underline{w})$.

By III.1.4(2), therefore,

$T\Theta'$: $\underline{z} \in (\text{Dom } \underline{w}) \times (\text{Ran } \underline{w})$,

which is what had to be proved. □

For an arbitrary string W , Dom W and Ran W are defined to be $(W|\underline{w})\text{Dom } \underline{w}$ and $(W|\underline{w})\text{Ran } \underline{w}$ respectively, the choice of the letter $\underline{w}$ being immaterial; see II.1.3(iii). It then follows (by use of the appropriate replacement rules) that

$$\text{Dom } W \equiv \{\underline{x} : (\exists \underline{y})((\underline{x}, \underline{y}) \in W)\} \quad ,$$

$$\text{Ran } W \equiv \{\underline{y} : (\exists \underline{x})((\underline{x}, \underline{y}) \in W)\} \quad ,$$

where $\underline{x}$ and $\underline{y}$ denote distinct letters not appearing in W (see also Problem III/19); and that

$$(S|\underline{u})\text{Dom } W \equiv \text{Dom } (S|\underline{u})W \quad ,$$

$$(S|\underline{u})\text{Ran } W \equiv \text{Ran } (S|\underline{u})W \quad .$$

It follows also that any relation Z is a relation from $\text{Dom } Z$ into $\text{Ran } Z$, that is, a relation in $\text{Dom } Z \times \text{Ran } Z$.

It should be proved that

T! : $$(X = X') \Rightarrow ((\text{Dom } X = \text{Dom } X') \wedge (\text{Ran } X = \text{Ran } X')) \quad ,$$

and, more generally,

T! : $$(X \subseteq X') \Rightarrow ((\text{Dom } X \subseteq \text{Dom } X') \wedge (\text{Ran } X \subseteq \text{Ran } X')) \quad . \tag{12}$$

Furthermore,

T! : $$(Y \neq \emptyset) \Rightarrow (\text{Dom } (X \times Y) = X) \tag{13}$$

and

T! : $$(X \neq \emptyset) \Rightarrow (\text{Ran } (X \times Y) = Y) \quad ; \tag{14}$$

see Problem III/3.

Remarks (i) The reader has to be on his guard in connection with the details of terminology and definitions. The definitions of relations, domains and ranges adopted above are not universally adopted. For instance, Hersee (1), Item 6, records a Teaching Committee recommendation that for school use a relation be defined as an ordered triplet (S, X, Y) (understood to denote ((S, X), Y), for example) in which S, X and Y are sets and $S \subseteq X \times Y$, the sets X and Y being termed respectively the domain and the codomain of the said relation. There are clear and significant differences between these definitions and ours. (We make no use of the concept of codomain.)

What we term a relation, Bourbaki (1), p.75 terms a graph. A triplet (S, X, Y) of the type just spoken of is termed by Bourbaki (loc. cit., p.76) a correspondence; X is termed the source and Y the target of the correspondence (S, X, Y).

In category theory, it is usual to define "range" in a way different from ours, and to adopt the term "image of f" or "Im f" in place of our Ran f; such usage also appears sometimes in other areas of mathematics. As always, a reader has to seek out the author's meaning and accept it (gracefully or otherwise) while following that author.

(ii) It is important to stress that (in our formalism) a relation determines its domain and range completely. Consequently, although two relations may be alike to the extent that (intuitively speaking) the elements of their constituent pairs are linked by the same "formula" or "rule", they are not the same relation unless their respective domains and ranges agree. A thorough appreciation of this point is vital, the more so because many relations figuring in high school work are likely to be generated from algebraic equations and inequalities; see III.2.3(vii). An equation such as $x^2 + 2x - 3y^2 = 0$ serves to generate infinitely many different relations corresponding to the liberty in specifying domains and ranges. (It should be remembered that the formula per se may contain no indication as to the interpretation of the symbols x and y appearing in it: given the right context, the above example is as "meaningful" for (say) square matrices of given order as it is for real numbers.) We return to these ideas in III.2.4.

III.2.3 Discussion of some examples Having indicated how the language of set theory is used to define the general concept of relation, it is desirable to provide some specific instances of relations. Here, however, a problem arises (and will arise again in connection with functions at the outset of IV.3) namely: almost every example to be mentioned presupposes the definitions and properties of certain basic mathematical objects, such as natural numbers, integers, real numbers, order relations, and so forth. A logical development would demand that all these things be dealt with in their proper order. Plainly, this state of affairs is not realised in the traditional order of teaching mathematics at any level, and rarely is any attempt made to realise this state of affairs; quite possibly it is best that no attempt _should_ be made. Thus, at this point it has to be assumed that the informal versions of the concepts just mentioned (and many others like them) are good enough for use in illustrative examples. This explains the presence of II.13 above.

With this point made and accepted, progress is possible.

In the following examples, $\underline{x}$, $\underline{y}$, $\underline{z}$, $\underline{w}$, $\underline{u}$ denote distinct letters not appearing in X , Y , S , N , Z , R , nor the sets denoted by $+$, $\cdot$, $-$, $<$; see VI.5 below.

(i) $\emptyset$ is a relation (because every member of $\emptyset$, of which there is not a single one, is an ordered pair) and $\mathrm{Dom}\,\emptyset = \mathrm{Ran}\,\emptyset = \emptyset$. Not very interesting. All the same, a meticulous reader will construct a semiformal proof of the sentence

$$(\forall \underline{z})((\underline{z} \in \emptyset) \Rightarrow (\exists \underline{x})(\exists \underline{y})(\underline{z} = (\underline{x}, \underline{y}))) ,$$

$\underline{x}$, $\underline{y}$, $\underline{z}$ denoting distinct letters.

(ii) If X denotes an arbitrary set, the relation

$$I_X \equiv_{\mathrm{def}} \{\underline{z} : (\exists \underline{x})(\underline{x} \in X \wedge \underline{z} = (\underline{x}, \underline{x}))\}$$

is called the _identity relation in_ X , or the _equality relation in_ X . In abbreviated notation (see III.1.5(i)), one would write

$$I_X \equiv \{(\underline{x}, \underline{x}) : \underline{x} \in X\} \quad .$$

(Often, a relation from X into X is spoken of as a relation <u>in</u> X, even though this may conflict with the terminology described at the end of III.2.1.)

Evidently,

$$\text{Dom } I_X = \text{Ran } I_X = X \quad .$$

(Nevertheless, provide a proof!)

It should perhaps be noted that what is frequently spoken of as "the relation of equality", with no reference set X specified, may not correspond to a relation in the technical sense, that is, there may exist no relation S in the technical sense such that

$$(\underline{x}, \underline{y}) \in S \Leftrightarrow (\underline{x} = \underline{y})$$

is true ($\underline{x}$ and $\underline{y}$ denoting distinct letters not appearing in S): the verdict would depend on the version of set theory adopted. In the version described in this book, there is no set S such that

$$(\underline{x} = \underline{y}) \Rightarrow (\underline{x}, \underline{y}) \in S$$

is a theorem. Put more forcibly,

$$T \quad : \qquad \neg \text{Coll}_{\underline{z}}((\exists \underline{x})(\underline{z} = (\underline{x}, \underline{x}))) \quad ; \qquad (1)$$

or, equivalently,

$$T \quad : \qquad \neg(\exists \underline{y})(\forall \underline{x})((\underline{x}, \underline{x}) \in \underline{y}) \quad . \qquad (2)$$

The reader may (and should) wish to formulate a semiformal proof of this; see

Problem III/4.

Similar remarks apply with $\in$ or $\subseteq$ in place of $=$ throughout.

(iii) If R denotes the set of real numbers (see II.13 and Chapter VI below), the set

$$\{(\underline{x}, \underline{y}) : \underline{x} \in R \wedge \underline{y} \in R \wedge \underline{x} < \underline{y}\} \equiv$$

$$\{\underline{z} : (\exists\underline{x})(\exists\underline{y})(\underline{x} \in R \wedge \underline{y} \in R \wedge \underline{x} < \underline{y} \wedge \underline{z} = (\underline{x}, \underline{y}))\}$$

is a relation in R (see (ii) above); its domain and range are each equal to R. One might describe it as "the relation of greater than (for real numbers)", but this name might, with equal justification, be applied to the inverse relation (see III.2.5 below). It would perhaps be safer to speak of it as "the standard relation of order for real numbers", but the ambiguity remains. (Here is an instance in which everyday language is flexible rather than precise.)

(iv) Denote by $\mathring{N} \equiv N \setminus \{0\}$ the set of nonzero natural numbers (cf. II.13 and Chapters V and VI below). The set

$$E \equiv \{\underline{w} : (\exists\underline{x})(\exists\underline{y})(\exists\underline{z})(\underline{x} \in \mathring{N} \wedge \underline{y} \in \mathring{N} \wedge \underline{z} \in \mathring{N} \wedge (\underline{x} \text{ is prime})$$

$$\wedge (\underline{y} \text{ is prime}) \wedge (\underline{z} = \underline{x} + \underline{y}) \wedge (\underline{w} = ((\underline{x}, \underline{y}), \underline{z})))\}$$

is a relation from $\mathring{N} \times \mathring{N}$ into $\mathring{N}$. (The "meaning" of the phrase " $\underline{x}$ is prime" is here taken to be intuitively clear; its formal definition will be discussed more fully in V.9.2(ii).) It is easy to prove that

$$\text{Dom } E = P \times P \quad ,$$

where P is the set of all primes; and that Ran E is the set of all natural numbers expressible as the sum of two primes (that is, natural numbers of the form $\underline{x} + \underline{y}$, where $\underline{x} \in P$ and $\underline{y} \in P$; see I.3.5(vi) and II.12.1). But it is a very difficult problem to describe Ran E in a way which effectively exhibits its elements, or makes it easy to decide whether or not a "randomly chosen" large natural number is an element of Ran E .

In 1742, Goldbach (in a letter to Euler) conjectured that every even natural number greater than 2 is an element of Ran E (though he did not express himself in precisely these words). This has come to be known as Goldbach's conjecture. So far, the corresponding sentence:

$$\text{Ran E} = \{\underline{n} \in N : (\underline{n} \text{ is even}) \wedge (2 < \underline{n})\} ,$$

wherein " $\underline{n}$ is even" is a name for the sentence

$$(\exists \underline{m})(\underline{m} \in N \wedge \underline{n} = 2 \cdot \underline{m}) ,$$

$\underline{m}$ denoting a letter different from $\underline{n}$ (see II.3.9(iii) and V.6.3), has neither been proved nor refuted. (It may even be undecidable, I suppose.) In 1930, Schnirelmann proved that there exists a nonzero natural number α such that every natural number greater than α is expressible as the sum of at most α primes; and, in 1937, Vinogradov proved that there exists a natural number β such that every odd natural number greater than β is expressible as the sum of three primes. Also, in 1953, Buhštab proved that every sufficiently large natural number is expressible as the sum a + b of two natural numbers a and b , each of which has at most four prime divisors. By the standards of this book, the proofs of all these partial results are extremely difficult.

(v) $$D \equiv \{\underline{z} : (\exists \underline{x})(\exists \underline{y})(\exists \underline{u})(\underline{x} \in N \wedge \underline{y} \in N \wedge \underline{u} \in N \wedge (\underline{y} = \underline{x} \cdot \underline{u}) \wedge (\underline{z} = (\underline{x}, \underline{y}))\}$$

is a relation in N ; $\underline{x}D\underline{y}$ signifies that $\underline{x} \in N$, $\underline{y} \in N$ and that $\underline{x}$ is a divisor of $\underline{y}$ in N ; $\mathrm{Dom}\ D = \mathrm{Ran}\ D = N$. D might be termed "the relation of divisibility in N " , although here again (as in (iii)) this name might be attached to the inverse relation. (Divisibility in N is discussed more formally in V.9.1, where D is denoted by $|_N$.)

(vi) $$C \equiv \{(\underline{x}, \underline{y}) : \underline{x} \in R \wedge \underline{y} \in R \wedge (\underline{x} - \underline{y} \in Z)\}$$

$$\equiv \{\underline{z} : (\exists \underline{x})(\exists \underline{y})(\underline{x} \in R \wedge \underline{y} \in R \wedge (\underline{x} - \underline{y} \in Z) \wedge (\underline{z} = (\underline{x}, \underline{y}))\}$$

is a relation in R , termed "the relation of congruence modulo one in R " .

(vii) The foregoing examples are all of the following general type. Take a sentence $\underline{\underline{A}} \equiv \underline{\underline{A}}[\![\underline{x}, \underline{y}]\!]$ (the notation is as in II.1.3(iv)). Choose "reference sets" X and Y , in which the letters $\underline{x}$, $\underline{y}$, $\underline{z}$ do not appear. By III.1.6(4),

$$\underline{x} \in X \wedge \underline{y} \in Y \Rightarrow (\underline{x}, \underline{y}) \in X \times Y$$

is true; hence

$$\underline{x} \in X \wedge \underline{y} \in Y \quad \underline{\underline{A}}[\![\underline{x}, \underline{y}]\!] \wedge (\underline{z} = (\underline{x}, \underline{y})) \Rightarrow \underline{z} \in X \times Y$$

is true, and so (by (IX) in I.3.2)

$$(\exists \underline{x})(\exists \underline{y})(\underline{x} \in X \wedge \underline{y} \in Y \wedge \underline{\underline{A}}[\![\underline{x}, \underline{y}]\!] \wedge (\underline{z} = (\underline{x}, \underline{y}))) \Rightarrow \underline{z} \in X \times Y$$

is true. From (XII) in II.3.7 it follows that the sentence

$$(\exists \underline{x})(\exists \underline{y})(\underline{x} \in X \wedge \underline{y} \in Y \wedge \underline{\underline{A}}[\![\underline{x}, \underline{y}]\!] \wedge (\underline{z} = (\underline{x}, \underline{y})))$$

is collectivising in $\underline{z}$; and from the theorems in II.3 it follows that

$$R \equiv \{(\underline{x}, \underline{y}) : \underline{x} \in X \wedge \underline{y} \in Y \wedge \underline{\underline{A}}[\![\underline{x}, \underline{y}]\!]\} \equiv \{\underline{z} : (\exists \underline{x})(\exists \underline{y})$$

$$(\underline{x} \in X \wedge \underline{y} \in Y \wedge \underline{\underline{A}}[\![\underline{x}, \underline{y}]\!] \wedge (\underline{z} = (\underline{x}, \underline{y})))\} \subseteq X \times Y \quad .$$

Thus the set on the left is a relation with domain equal to a subset of X and range equal to a subset of Y . Furthermore, the sentence

$$(\forall \underline{z})(\underline{z} \in R \Leftrightarrow (\exists \underline{x})(\exists \underline{y})(\underline{x} \in X \wedge \underline{y} \in Y \wedge \underline{\underline{A}}[\![\underline{x}, \underline{y}]\!] \wedge (\underline{z} = (\underline{x}, \underline{y})))$$

is true.

Virtually all the relations encountered in day-to-day mathematics are obtainable by following the above recipe.

Another instance is the relation

$$\in_X \equiv \{\underline{z} : (\exists \underline{x})(\exists \underline{y})(\underline{x} \in X \wedge \underline{y} \in X \wedge \underline{x} \in \underline{y} \wedge \underline{z} = (\underline{x}, \underline{y}))\} \quad ,$$

X denoting an arbitrary set and $\underline{x}$, $\underline{y}$, $\underline{z}$ distinct letters not appearing in X . This is the relation of membership (or elementhood) in X . In spite of this terminology, and in spite of the fact that $\in$ is almost always spoken of as a "relational sign" , the situation is much as in (ii) above. In particular,

$$T \quad : \qquad \neg \mathrm{Coll}_{\underline{z}}((\exists \underline{x})(\exists \underline{y})(\underline{x} \in \underline{y} \wedge \underline{z} = (\underline{x}, \underline{y}))) \quad ; \qquad (3)$$

see Problem III/4.

(viii) <u>Diophantine sets and relations</u> As a special case of (vii), take X and Y to be $\dot{N}$, the set of positive natural numbers, and $\underline{\underline{A}}[\![\underline{x}, \underline{y}]\!]$ to be of the form

$$(\exists \underline{z}_1) \ldots\ldots (\exists \underline{z}_k)(\underline{z}_1 \in \dot{N} \wedge \ldots \wedge \underline{z}_k \wedge \dot{N} \wedge P(\underline{x}, \underline{y}, \underline{z}_1, \ldots, \underline{z}_k) = 0) \quad ,$$

where k is a positive natural number and P is a "special polynomial of rank $k + 2$ " - a concept defined in a moment. The resulting relation (a subset of $\dot{N} \times \dot{N}$) is then an instance of a Diophantine relation in $\dot{N}$ (or Diophantine subset of $\dot{N} \times \dot{N}$) . The name is adopted because Diophantus initiated the study of particular equations of the sort $P = 0$, the unknowns being sought among positive natural numbers; a section of classical number theory is concerned with such "Diophantine equations" . (Fermat's famous last conjecture, mentioned in II.3.1(i), is concerned with perhaps the most celebrated particular Diophantine equation; he conjectured that the equation

$$x^{\underline{a}+2} + y^{\underline{a}+2} - z^{a+2} = 0$$

is insoluble for $\underline{x}$, $\underline{y}$, $\underline{z}$, $\underline{a} \in \dot{N}$. This conjecture has so far been neither proved nor refuted.) As is mentioned in V.13 below, Hilbert raised a new type of "decision problem" concerning the solubility of Diophantine equations in general, a solution to which had to wait for seventy years. Diophantine sets play a capital role in this solution.

Coming to the missing definition: if r is a positive natural number, a special polynomial of rank r is defined to be a polynomial function with domain $\dot{N} \times \ldots \times \dot{N}$ (r factors) and integer coefficients. (This definition is informal, of course; cf. IV.3.4 and IV.3.7.)

To this may now be added the definition: a Diophantine subset of $\dot{N}$ is a set S of the form

$$S = \{\underline{x} \in \dot{N} : (\exists \underline{z}_1) \ldots (\exists \underline{z}_k)(\underline{z}_1 \in \dot{N} \wedge \ldots \wedge \underline{z}_k \in \dot{N}$$

$$\wedge\ P(\underline{x}, \underline{z}_1, \ldots, \underline{z}_k) = 0)\} \quad ,$$

where P is a special polynomial of rank $k + 1$ (P possibly depending on the set in question, of course; the definition is again informal).

If a Diophantine relation in $\mathring{N}$ happens to be a function (see Chapter IV) with domain $\mathring{N}$ (hence a function $\mathring{N} \to \mathring{N}$), it is said to be a <u>Diophantine function on</u> $\mathring{N}$. Thus, a function $f : \mathring{N} \to \mathring{N}$ is Diophantine, if and only if, for some positive natural number k, there is a special polynomial P of rank $k + 2$ such that

$$(\underline{x} \in \mathring{N} \wedge \underline{y} = f(\underline{x})) \Leftrightarrow \underline{x} \in \mathring{N} \wedge \underline{y} \in \mathring{N} \wedge (\exists \underline{z}_1) \ldots (\exists \underline{z}_k)(\underline{z}_1 \in \mathring{N} \wedge \ldots \wedge \underline{z}_k \in \mathring{N}$$

$$\wedge\, P(\underline{x}, \underline{y}, \underline{z}_1, \ldots, \underline{z}_k) = 0) \quad .$$

(In more formal terms, one would stipulate here and above that $\underline{x}$, $\underline{y}$, $\underline{z}_1$, ..., $\underline{z}_k$ denote distinct letters not appearing in f, P or S.)

Diophantine subsets of, and Diophantine functions with domain, $\mathring{N} \times \ldots \times \mathring{N}$ (any finite number of factors) are defined likewise.

A few simple instances follow (see also Problems III/11 - III/13).

(a) The set of numbers $\underline{x} \in \mathring{N}$ which are <u>not</u> powers of 2 is Diophantine, being equal to

$$\{\underline{x} \in \mathring{N} : (\exists \underline{z}_1)(\exists \underline{z}_2)(\underline{z}_1 \in \mathring{N} \wedge \underline{z}_2 \in \mathring{N} \wedge (\underline{x} - \underline{z}_1(2\underline{z}_2 + 1) = 0))\} \quad .$$

(b) The set of $\underline{x} \in \mathring{N}$ which are composite is Diophantine, since it equals

$$\{\underline{x} \in \mathring{N} : (\exists \underline{z}_1)(\exists \underline{z}_2)(\underline{z}_1 \in \mathring{N} \wedge \underline{z}_2 \in \mathring{N} \wedge (\underline{x} - (\underline{z}_1 + 1)(\underline{z}_2 + 1) = 0))\} \;.$$

(c) The usual order relation $<$ in $\mathring{N}$ is Diophantine, since

$$(\underline{x} \in \mathring{N} \wedge \underline{y} \in \mathring{N} \wedge (\underline{x} < \underline{y})) \Leftrightarrow \underline{x} \in \mathring{N} \wedge \underline{y} \in \mathring{N} \wedge (\exists \underline{z})(\underline{z} \in \mathring{N} \wedge (\underline{x} - \underline{y} + \underline{z} = 0)) \quad ;$$

so too is the relation of divisibility in $\mathring{N}$, since

$$(x \in \mathring{N} \wedge y \in \mathring{N} \wedge (x \text{ divides } y)) \Leftrightarrow x \in \mathring{N} \wedge y \in \mathring{N} \wedge (\exists z)(z \in \mathring{N} \wedge (x \cdot z - y = 0)) \quad .$$

(d) If Q is a special polynomial of rank 1 and $\operatorname{Ran} Q \subseteq \mathring{N}$, then Q is a Diophantine function on $\mathring{N}$, since

$$(x \in \mathring{N} \wedge y = Q(x)) \Leftrightarrow x \in \mathring{N} \wedge y \in \mathring{N} \wedge (\exists z)(z \in \mathring{N} \wedge (y - Q(x) + (z - 1)^2 = 0)) \quad .$$

It is easy to gain the impression that Diophantine sets and functions are exceedingly special. In some senses they _are_ so; but in other senses they are not, though to verify this demands a lot of ingenuity. For example, the functions $n \rightsquigarrow n!$ with domain $\mathring{N}$ and $(n, m) \rightsquigarrow n^m$ with domain $\mathring{N} \times \mathring{N}$ can with some difficulty be proved to be Diophantine; and the set of all $x \in \mathring{N}$ which are powers of 2 proves to be Diophantine. (The notation $\rightsquigarrow$ is as in IV.2.2.)

For details and devices, see Davis (1) and the article by Davis, Matiyesavič and Robinson in Browder (1), pp.323-378. See also the problems at the end of this chapter. The piquancy of the situation is that, although the concept of Diophantine sets and functions may at first seem rather superficial, it is in fact tied up with profound problems; see V.13 below.

III.2.4 _Extension and restriction of relations_ It is simple to formalise some of the ideas which cropped up in Remark (ii) following III.2.2.

If T and Z denote relations, T is said to be a _restriction_ or _contraction_ of Z , and Z to be an _extension_ of T , if and only if $T \subseteq Z$. Thus, nothing really new is involved here: all that is being done is to introduce alternative names for subsets, with the informal intention that they be reserved for special occasions.

For example, if Z is a relation and X a set, the relation

$$Z \S X \equiv_{\text{def}} \{(x, y) \in Z : x \in X\} \equiv \{z \in Z : (\exists x)(\exists y)(x \in X \wedge z = (x, y))\} \tag{1}$$

is a restriction of Z , usually termed the <u>restriction of</u> Z <u>to</u> X . In particular, $Z \S \emptyset = \emptyset$ and $Z \S \text{Dom}\, Z = Z$.

In spite of informal intentions, (1) is to be regarded as an unconditional definition of $Z \S X$ applying for arbitrary strings X and Z ; when X and Z denote sets, $Z \S X$ denotes a set which is a relation (even though Z may not be a relation). The notations

$$Z \mid X \quad \text{and} \quad Z_{|X}$$

are frequently used by other authors in place of $Z \S X$.

It can be verified that

$$(S|\underline{u})(Z\S X) \equiv ((S|\underline{u})Z)\S((S|\underline{u})X)$$

for arbitrary strings S , X , Z and an arbitrary letter $\underline{u}$; and one may prove the theorem schema

$$(Z = Z' \wedge X = X') \Rightarrow (Z \S X = Z' \S X') \quad .$$

I leave the reader to prove that

$$\text{T!} \;:\quad Z \S X = Z \cap (X \times \text{Ran}\, Z) \;; \tag{2}$$

that

$$\text{T!} \;:\quad \text{Dom}\,(Z \S X) = X \cap \text{Dom}\, Z \;; \tag{3}$$

and that

$$\text{T!} \;:\quad \begin{aligned}(\underline{z} \in Z \S X) \Leftrightarrow ((\underline{z} &\text{ is an ordered pair)} \\ &\wedge (\underline{z} \in Z) \wedge (\text{pr}_1\, \underline{z} \in X)) \quad .\end{aligned} \tag{4}$$

Similarly, if Z is a relation and Y a set, then

$$\{(\underline{x}, \underline{y}) \in Z : \underline{y} \in Y\} \equiv \{\underline{z} \in Z : (\exists \underline{x})(\exists \underline{y})(\underline{y} \in Y \wedge \underline{z} = (\underline{x}, \underline{y}))\}$$

is a restriction of Z . (There appears to be no commonly accepted notation or name for this particular type of restriction.)

In practice, the concept of restriction comes to the fore in connection with functions (see Chapter IV), inasmuch as a relation which is not itself a function, may possess many restrictions which are functions (see IV.5.2 below); a fair number of important functions are obtained in this fashion (see IV.3.8).

III.2.5 Inverse relations If Z is a relation, the set denoted (see III.1.5) by

$$\{(\underline{y}, \underline{x}) : (\underline{x}, \underline{y}) \in Z\}$$

is a relation, termed the inverse of Z and denoted by Inv Z or Inv(Z) . Expressed formally,

$$\begin{aligned} \mathrm{Inv}\, Z &\equiv_{\mathrm{def}} \{(\underline{x}, \underline{y}) : (\underline{y}, \underline{x}) \in Z\} \\ &\equiv \{\underline{z} : (\exists\underline{x})(\exists\underline{y})((\underline{y}, \underline{x}) \in Z \wedge \underline{z} = (\underline{x}, \underline{y}))\} \end{aligned} \qquad (1)$$

which is understood to apply for an arbitrary string Z and which is independent of the choice of the distinct letters $\underline{x}$, $\underline{y}$, $\underline{z}$ not appearing in Z . Since

$$(\exists\underline{x})(\exists\underline{y})((\underline{y}, \underline{x}) \in Z \wedge \underline{z} = (\underline{x}, \underline{y}))$$

may be proved via III.2.2 to imply that $\underline{z} \in \mathrm{Ran}\, Z \times \mathrm{Dom}\, Z$, the sentence appearing after the second colon in (1) , is collectivising in $\underline{z}$ (see (XII) in II.3.7, remembering that $\underline{z}$ does not appear in Z) and so (see II.3.3(5))

$$\mathrm{T!} \quad : \qquad (\forall\underline{z})(\underline{z} \in \mathrm{Inv}\, Z \Leftrightarrow (\exists\underline{x})(\exists\underline{y})((\underline{y}, \underline{x}) \in Z \wedge \underline{z} = (\underline{x}, \underline{y})) \quad . \qquad (2)$$

As one would expect, $(S|\underline{u})\mathrm{Inv}\, Z \equiv \mathrm{Inv}\, (S|\underline{u})Z$; and

$$Z = Z' \Rightarrow \text{Inv } Z = \text{Inv } Z'$$

is a theorem schema.

Thus, if $\underline{x}Z\underline{y}$ signifies $\underline{x} > \underline{y}$ (cf. III.2.3(iii)), $\underline{x}$ Inv Z $\underline{y}$ signifies $\underline{x} < \underline{y}$; if $\underline{x}D\underline{y}$ signifies $\underline{x}$ is a divisor of $\underline{y}$ (see III.2.3(v)), $\underline{x}$(Inv D)$\underline{y}$ signifies $\underline{x}$ is a multiple of $\underline{y}$; if $\underline{x}C\underline{y}$ signifies $\underline{x} - \underline{y}$ is an integer (see III.2.3(vi)), $\underline{x}$(Inv C)$\underline{y}$ signifies $-\underline{x} + \underline{y}$ is an integer. Of these, the first two examples show that (even if Z is a relation) Z and Inv Z may be quite different relations (in spite of any similarity in their everyday names; see III.2.3 again); and the third shows that it can happen that Z = Inv Z in which case the relation Z is said to be <u>symmetric</u>. (See III.2.8 below.)

The theorem schema (2) implies

$$\text{Inv } Z \text{ is a relation} . \tag{2'}$$

(Note the absence of any hypothesis to the effect that Z is a relation.) Furthermore

T! : $$\text{Dom Inv } Z = \text{Ran } Z , \tag{3}$$

T! : $$\text{Ran Inv } Z = \text{Dom } Z . \tag{4}$$

If Z is a relation from X into Y , then Inv Z is a relation from Y into X .

As a preliminary to the proofs of (3) and (4) , note that a corollary of III.1.5(3_1) is

T! : $$(\underline{x}, \underline{y}) \in \text{Inv } Z \Leftrightarrow (\underline{y}, \underline{x}) \in Z . \tag{5}$$

<u>Proof of (3)</u> Suppose that $\underline{x} \in$ Dom Inv Z . Then also

$$(\exists \underline{y})((\underline{x}, \underline{y}) \in \text{Inv } Z) \quad .$$

Using (5) and (j) in I.3.3, it follows that

$$(\exists \underline{y})((\underline{y}, \underline{x}) \in Z) \quad .$$

But this signifies precisely that $\underline{x} \in \text{Ran } Z$. Using the criterion of deduction I.3.2(VI),

$$T \quad : \qquad \underline{x} \in \text{Dom Inv } Z \Rightarrow \underline{x} \in \text{Ran } Z \quad .$$

In a similar way, one proves that

$$T \quad : \qquad \underline{x} \in \text{Ran } Z \Rightarrow \underline{x} \in \text{Dom Inv } Z \quad .$$

Hence, (by (j) in I.3.3)

$$T \quad : \qquad \underline{x} \in \text{Dom Inv } Z \Leftrightarrow \underline{x} \in \text{Ran } Z \quad ,$$

and so (by (X) in I.3.2)

$$T \quad : \qquad (\forall \underline{x})(\underline{x} \in \text{Dom Inv } Z \Leftrightarrow \underline{x} \in \text{Ran } Z) \quad .$$

Thus (see II.1.1(1))

$$T \quad : \qquad \text{Dom Inv } Z = \text{Ran } Z$$

and (3) is proven. $\square$

The proof of (4) is similar and is left to the reader.

Remarks (i) Familiar alternatives to the notation Inv Z are Z^{-1} and $\check{Z}$. If the former is adopted, care must be taken to avoid confusion with the same notation as used in connection with reciprocals or numerical inverses of real-valued functions, as in IV.3.3 below. (This is one reason for our choice of the more cumbersome notation Inv Z .) See also VI.8.

(ii) In writing Inv(Z) , the parentheses are a device to reduce possible ambiguities in cases where Z is replaced by a complicated group of symbols. As in the Remark at the end of II.11.1, Inv(Z) is not intended as an instance of the functional value notation formally defined in IV.1.3 below; and Inv⟦Z⟧ would be a more appropriate, but totally unconventional, notation.

III.2.6 Image of a set by a relation If S and A denote arbitrary strings, I formally define

$$S\langle A\rangle \equiv_{\text{def}} \{\underline{y} : (\exists\underline{x})(\underline{x} \in A \wedge (\underline{x}, \underline{y}) \in S)\} \quad . \tag{1}$$

In connection with (1) , recall the conventions in II.3.12: it is a routine matter (depending on use of the replacement rules in I.1.4, I.1.8 and II.3.4) to check that (1) is independent of the choice of the distinct letters $\underline{x}$ and $\underline{y}$, neither appearing in S or A . Moreover, one has the further replacement rule

$$(T|\underline{u})S\langle A\rangle \equiv ((T|\underline{u})S)\langle (T|\underline{u})A)\rangle$$

and the theorem schema

$$(S = S' \wedge A = A') \Rightarrow (S\langle A\rangle = S'\langle A'\rangle) \quad .$$

Informally, if S and A are sets, $S\langle A\rangle$ is the set of second coordinates of ordered pairs belonging to S whose first coordinates are elements of A ; see Problem III/20. For instance, if $S \equiv \{1, 2, (2, 3), (2, 4), (5, 6)\}$, then

$$S\langle\{1\}\rangle = \emptyset \quad , \quad S\langle\{2\}\rangle = \{3, 4\} = S\langle\{1, 2\}\rangle \quad ,$$

$$S\langle \{2, 5\} \rangle = \{3, 4, 6\}$$

will prove to be theorems.

Returning to formalities, since $(\underline{x}, \underline{y}) \in S$ implies $\underline{y} \in \cup\cup S$ (see III.1.2(7)), it follows that if A and S are sets, then the sentence after the colon in (1) is collectivising in $\underline{y}$ (see (IX) in I.3.2, (XII) in II.3.7 and recall II.3.12). So, by II.3.3(5), one has the theorem schema

T! : $$(\forall \underline{y})(\underline{y} \in S\langle A \rangle \Leftrightarrow (\exists \underline{x})(\underline{x} \in A \wedge (\underline{x}, \underline{y}) \in S)) \quad . \qquad (2)$$

$S\langle A \rangle$ is termed the <u>image of</u> A <u>by</u> (or <u>under</u>) S . One is usually interested in such images when S is a relation or (more especially) a function and A is a subset of Dom S . However, there is no compulsion for this to be the case, and there is no necessary connection between A and Dom S .

The following theorem schemas can be proved :

T! : $$S\langle A \rangle = S\langle A \cap \text{Dom } S \rangle \subseteq S\langle \text{Dom } S \rangle = \text{Ran } S \quad , \qquad (3)$$

T! : $$(\text{Inv } S)\langle A \rangle \subseteq \text{Ran Inv } S = \text{Dom } S \quad , \qquad (4)$$

T! : $$S\langle \emptyset \rangle = \emptyset \quad , \qquad (5)$$

T! : $$(S\S B)\langle A \rangle = S\langle A \cap B \rangle \quad , \qquad (6)$$

T! : $$\text{Ran } (S\S B) = S\langle B \rangle \quad , \qquad (7)$$

T! : $$S\langle A \cup B \rangle = S\langle A \rangle \cup S\langle B \rangle \quad , \qquad (8)$$

T! : $$S\langle A \cap B \rangle \subseteq S\langle A \rangle \cap S\langle B \rangle \quad , \qquad (9)$$

T! : $$A \subseteq B \Rightarrow S\langle A \rangle \subseteq S\langle B \rangle \quad , \qquad (10)$$

T! : $$S\langle A\rangle \setminus S\langle B\rangle \subseteq S\langle A \setminus B\rangle \quad , \tag{11}$$

wherein S , A , B denote arbitrary sets.

A semiformal proof of (3) , for example, would rely on the fact (that is, theorem or theorem schema) that $(\underline{x}, \underline{y}) \in S \Rightarrow \underline{x} \in \mathrm{Dom}\, S$, and so that

$$\underline{x} \in A \wedge (\underline{x}, \underline{y}) \in S \Leftrightarrow \underline{x} \in A \cap \mathrm{Dom}\, S \wedge (\underline{x}, \underline{y}) \in S \quad ;$$

see I.3.7(3). One then makes use of II.3.6(18). The reader should write out the details.

See also Problems III/5, III/20, IV/6 and IV/22.

Remarks (i) In place of $S\langle A\rangle$, one often sees written S(A) or S[A] ; cf., for example, Gleason (1), p.46. I have avoided the former alternative because it invites confusion with IV.1.3; see the discussion in IV.1.6.

(ii) Care should be taken not to confuse $(\mathrm{Inv}\ S)\langle A\rangle$ and $\mathrm{Inv}(S\langle A\rangle)$. For example, define (as in Chapter V),

$$0 \equiv \emptyset \quad , \quad 1 \equiv 0^+ \equiv \{\emptyset\} \quad , \quad 2 \equiv 1^+ \equiv 1 \quad \{1\} = \{\emptyset, \{\emptyset\}\} \quad ,$$

$$S \equiv \{(0, 0), (1, 0), (2, 2)\} \quad ,$$

$$A \equiv \{0, 1\} \quad ;$$

then

$$(\mathrm{Inv}\ S)\langle A\rangle = \{0, 1\}$$

and

$$\mathrm{Inv}(S\langle A\rangle) = \emptyset$$

are unequal sets.

As has been remarded on several analogous occasions already, the use of parentheses in $(\mathrm{Inv}\ S)\langle A\rangle$ and $\mathrm{Inv}(S\langle A\rangle)$ is precisely so as to avoid the ambiguity residing in the notation $\mathrm{Inv}\ S\langle A\rangle$, not to indicate functional evaluation in the sense of IV.1.3; see again IV.1.6.

(iii) The notation $S\langle A\rangle$, used above for images of sets by relations, is sometimes mimicked in other situations. For example, $\mathrm{pr}_1\langle S\rangle$ and $\mathrm{pr}_2\langle S\rangle$ are often used to denote the sets

$$\{\mathrm{pr}_1\, \underline{z} : \underline{z} \in S\}$$

and

$$\{\mathrm{pr}_2\, \underline{z} : \underline{z} \in S\} \quad ,$$

defined as in II.12.1, $\underline{z}$ denoting a letter not appearing in

S . This is a very reasonable adaptation. However, one frequently finds $pr_1 S$ and $pr_2 S$ written in place of

$pr_1 \langle S \rangle$ and $pr_2 \langle S \rangle$, which is not so reasonable and which

can hardly fail to be very confusing when placed alongside the notation appearing in III.1.3.

See also Problems III/19 and IV/1.

III.2.7 Composition of relations I shall say just enough on this topic to facilitate the discussion of composite functions in IV.6, this being the case which is of primary importance in high school and much tertiary level work.

Informally, if S and T are sets (in practice almost invariably relations in this context) the composite $S \circ T$ is the relation which is the set of ordered pairs $(\underline{x}, \underline{z})$ such that, for some $\underline{y}$ (which may depend on $\underline{x}$ and $\underline{z}$), $(\underline{x}, \underline{y}) \in T$ and $(\underline{y}, \underline{z}) \in S$. Thus $\underline{x}$ and $\underline{z}$ are related via an intermediary $\underline{y}$. Unclehood is the composite (in a suitable order) of fatherhood and brotherhood. (Incidentally, the expression of various human relationships in set theoretical terms can be an amusing exercise.) Again, if

$$S \equiv \{(1, a), (2, b), (3, c)\} \quad , \quad T \equiv \{(d, 2), (e, 3), (f, 4)\} \quad ,$$

then

$$S \circ T = \{(d, b), (e, c)\} \quad .$$

Getting down to formalities, for arbitrary strings S and T (not necessarily different and not necessarily relations) I formally define

$$S \circ T \equiv_{def} \{\underline{w} : (\exists \underline{x})(\exists \underline{y})(\exists \underline{z})((\underline{x}, \underline{y}) \in T \wedge (\underline{y}, \underline{z}) \in S \wedge \underline{w} = (\underline{x}, \underline{z}))\} \quad ; \qquad (1)$$

recall the conventions in II.3.12. In the shorthand notation discussed in III.1.5, (1) would appear as

$$S \circ T \equiv_{def} \{(\underline{x}, \underline{z}) : (\exists \underline{y})((\underline{x}, \underline{y}) \in T \wedge (\underline{y}, \underline{z}) \in S)\} \quad , \qquad (1')$$

which is certainly easier to remember.

In the definition (1), the choice of the distinct letters $\underline{x}$, $\underline{y}$, $\underline{z}$, $\underline{w}$ not appearing in S or T is immaterial. Moreover, one can verify the replacement rule

$$(U|\underline{u})(S \circ T) \equiv (U|\underline{u})S \circ (U|\underline{u})T$$

for arbitrary strings S, T, U and an arbitrary letter $\underline{u}$, and prove the theorem schema

$$(S = S' \wedge T = T') \Rightarrow (S \circ T = S' \circ T') \quad ,$$

wherein S, S', T, T' denote arbitrary sets.

If S and T are sets, $S \circ T$ is a set. Moreover, in this case the sentence

$$(\underline{x}, \underline{y}) \in T \wedge (\underline{y}, \underline{z}) \in S \wedge \underline{w} = (\underline{x}, \underline{z})$$

implies (see III.1.2(7)) that $\underline{x} \in \cup\cup T$ and $\underline{z} \in \cup\cup S$ and hence that $\underline{w} \in (\cup\cup T) \times (\cup\cup S)$. So, by I.3.2(IX), II.3.7(XII) and III.3.12, the sentence appearing after the colon in (1) is collectivising in $\underline{w}$; hence, by II.3.3(5), one has the theorem schema

$$\text{T!} \;:\; \begin{aligned}&(\forall\underline{w})(\underline{w} \in S \circ T \Leftrightarrow (\exists\underline{x})(\exists\underline{y})(\exists\underline{z})((\underline{x}, \underline{y}) \in T \\ &\qquad \wedge (\underline{y}, \underline{z}) \in S \wedge \underline{w} = (\underline{x}, \underline{z})))\end{aligned} \quad . \tag{2}$$

From this, combined with the metatheorems in I.3.2 and I.3.3, III.1.2(1) and Problem II/14, plus various replacement rules, one may deduce the theorem schema

$$\text{T!} \;:\; (a, b) \in S \circ T \Leftrightarrow (\exists\underline{y})((a, \underline{y}) \in T \wedge (\underline{y}, b) \in S) \tag{2'}$$

where a, b, S, T denote arbitrary sets and $\underline{y}$ an arbitrary letter not appearing in a , b , S or T .

From (2) it is clear that $S \circ T$ is always a relation (if S and T are sets). Further (see (3) and (4) below), if S is a relation from Y into Z and T a relation from X into Y , then $S \circ T$ is a relation from X into Z .

Although it can be proved that $\circ$ is associative (see Suppes (1), pp. 63-64; he writes T/S in place of $S \circ T$), it is not commutative: it is often the case that $S \circ T$ and $T \circ S$ are different (unequal) sets.

Cancellation has to be treated with caution. Indeed,

$$(S \cup \{\emptyset\}) \circ T = S \circ T = S \circ (T \cup \{\emptyset\})$$

is a theorem schema. Inasmuch as $\{\emptyset\}$ is not a relation (prove this!), this is perhaps not totally discouraging: the reader should do a little research at this point.

The following theorem schemas should be noted:

T! : $$\mathrm{Dom}(S \circ T) = (\mathrm{Inv}\ T)\langle \mathrm{Dom}\ S \rangle \quad , \tag{3}$$

T! : $$\mathrm{Ran}(S \circ T) = S\langle \mathrm{Ran}\ T \rangle \quad , \tag{4}$$

T! : $$\mathrm{Inv}(S \circ T) = \mathrm{Inv}\ T \circ \mathrm{Inv}\ S \quad ; \tag{5}$$

in (5) the interchange of the order of the "factors" is noteworthy. Further, if T is a relation in X (that is, from X into X), then $T \circ \mathrm{Inv}\ T$, $\mathrm{Inv}\ T \circ T$ and I_X may be all different.

<u>Proof of (3)</u> Repeated use will be made of the metatheorems in I.3.2 and I.3.3, especially I.3.3(1), and also I.3.7(1). One has

T :	$\underline{x} \in \mathrm{Dom}(S \circ T)$	$\Leftrightarrow (\exists \underline{z})((\underline{x}, \underline{z}) \in S \circ T)$	(III.2.2(4))
		$\Leftrightarrow (\exists \underline{z})(\exists \underline{y})((\underline{x}, \underline{y}) \in T \wedge (\underline{y}, \underline{z}) \in S)$	((2'))
		$\Leftrightarrow (\exists \underline{y})(\exists \underline{z})((\underline{x}, \underline{y}) \in T \wedge (\underline{y}, \underline{z}) \in S)$	(I.3.3(n))
		$\Leftrightarrow (\exists \underline{y})((\underline{x}, \underline{y}) \in T \wedge (\exists \underline{z})((\underline{y}, \underline{z}) \in S))$	(I.3.3(o))
		$\Leftrightarrow (\exists \underline{y})((\underline{x}, \underline{y}) \in T \wedge \underline{y} \in \mathrm{Dom}\ S)$	(II.2.2(4))
		$\Leftrightarrow (\exists \underline{y})((\underline{y}, \underline{x}) \in \mathrm{Inv}\ T \wedge \underline{y} \in \mathrm{Dom}\ S)$	(III.2.5(5))
		$\Leftrightarrow \underline{x} \in (\mathrm{Inv}\ T)\langle \mathrm{Dom}\ S \rangle$.	(III.2.6(2))

Now use I.3.2(X) and II.1.1(1). □

<u>Proof of (4)</u> This proceeds in similar fashion. One has

T :	$\underline{z} \in \mathrm{Ran}(S \circ T)$	$\Leftrightarrow (\exists \underline{x})((\underline{x}, \underline{z}) \in S \circ T)$	(III.2.2(5))
		$\Leftrightarrow (\exists \underline{x})(\exists \underline{y})((\underline{x}, \underline{y}) \in T \wedge (\underline{y}, \underline{z}) \in S)$	((2'))
		$\Leftrightarrow (\exists \underline{y})(\exists \underline{x})((\underline{x}, \underline{y}) \in T \wedge (\underline{y}, \underline{z}) \in S)$	(I.3.3(n))
		$\Leftrightarrow (\exists \underline{y})(\exists \underline{x})((\underline{x}, \underline{y}) \in T) \wedge (\underline{y}, \underline{z}) \in S)$	(I.3.3(o))
		$\Leftrightarrow (\exists \underline{y})(\underline{y} \in \mathrm{Ran}\ T \wedge (\underline{y}, \underline{z}) \in S)$	(III.2.2(5))
		$\Leftrightarrow \underline{z} \in S\langle \mathrm{Ran}\ T \rangle$,	(III.2.6(2))

and once again an appeal to I.3.2(X) and II.1.1(1) ends the proof. □

Remarks (i) The above proofs have been shortened by use of the metatheorems appearing in Problem I/18; they are, in fact, becoming close in style to routine proofs. When expressed more fully, the first two lines of the proof of (3) might appear thus:

$$T \quad : \qquad \underline{x} \in \mathrm{Dom}(S \circ T) \Leftrightarrow (\exists \underline{z})((\underline{x}, \underline{z}) \in S \circ T)$$

by III.2.2(4) and the appropriate replacement rules (making use of the convention in II.3.12); also,

$$T \quad : \quad (\exists \underline{z})((\underline{x}, \underline{z}) \in S \circ T) \Leftrightarrow (\exists \underline{z})(\exists \underline{y})((\underline{x}, \underline{y}) \in T \wedge (\underline{y}, \underline{z}) \in S)$$

by (2') and the appropriate replacement rules; hence also

$$T \quad : \quad \underline{x} \in \mathrm{Dom}(S \circ T) \Leftrightarrow (\exists \underline{z})(\exists \underline{y})((\underline{x}, \underline{y}) \in T \wedge (\underline{y}, \underline{z}) \in S)$$

by I.3.7(1). Each pair of adjacent lines on the proof of (3) has to be treated in the same way, the effect being that of telescoping the intermediate steps and arriving finally at

$$T \quad : \qquad \underline{x} \in \mathrm{Dom}(S \circ T) \Leftrightarrow \underline{x} \in (\mathrm{Inv}\ T)\langle \mathrm{Dom}\ S \rangle \quad .$$

Similar remarks apply whenever one makes use of chains of equivalences.

(ii) As always, the reader has to be ready to adjust to alternative notations. As an instance, Hersee (1), Item 8 records a Teaching Committee recommendation that (at least in the case of functions f and g) the notation fg be adopted in place of $f \circ g$, at the same time as an adoption of $f \times g$ in place of fg or $f.g$ to denote the pointwise product of real or complex valued functions f and g (see IV.3.3 and IV.6.6 below). Cf. IV.1.4(viii).

III.2.8 Equivalence relations and quotient sets I start by considering informally two relations in the set Z of integers, namely,

$$S_1 \equiv \{\underline{z} : (\exists \underline{x})(\exists \underline{y})(\underline{x} \in Z \wedge \underline{y} \in Z \wedge (\underline{x} - \underline{y} \text{ is divisible by } 3) \wedge \underline{z} = (\underline{x}, \underline{y}))\}$$

and

$$S_2 \equiv \{\underline{z} : (\exists \underline{x})(\exists \underline{y})(\underline{x} \in Z \wedge \underline{y} \in Z \wedge \underline{x}^2 = \underline{y}^2 \wedge \underline{z} = (\underline{x}, \underline{y}))\} \quad ;$$

cf. III.2.3(viii). If S denotes either S_1 or S_2 , the following sentences are then true:

(i) S is a relation in Z ;

(ii) S is reflexive in Z ($\equiv I_Z \subseteq S$) ;

(iii) S is symmetric ($\equiv$ Inv S = S) ;

(iv) S is transitive ($\equiv S \circ S \subseteq S$) .

The truth of (i) , (ii) , (iii) and (iv) is usually expresses by saying that S is an <u>equivalence relation in</u> Z .

Equivalence relations are of great importance in mathematics and merit individual attention. In everyday life, they arise whenever one classifies the elements of a given set according to a certain "relation of similarity" . A rather dull example is I_X : this is always an equivalence relation in X ; the "relation of similarity" is here equality of elements of X .

At the formal level I define

$$\begin{aligned}&(S \text{ is an equivalence relation in } X)\\ &\equiv_{def} (S \text{ is a relation in } X) \wedge (I_X \subseteq S) \wedge (\text{Inv } S = S) \wedge (S \circ S \subseteq S) .\end{aligned} \tag{1}$$

To facilitate further discussion I also define, for arbitrary strings S and A

$$S((A)) \equiv_{def} S\langle \{A\} \rangle , \tag{2}$$

(which is not to be confused with S(A) defined in accordance with IV.1.3 below); and I will abbreviate " S is an equivalence relation in X " to &(S, X) (though &⟦S, X⟧ would be a better notation).

I leave to the reader the task of proving the following simple theorem schema ($\underline{x}$ denoting a letter not appearing in X or S):

$$\text{T} \quad : \quad \begin{aligned}&\&(S, X) \Rightarrow (\text{Dom } S = X) \wedge (\text{Ran } S = X)\\ &\qquad \wedge (\forall \underline{x})(\underline{x} \in X \Rightarrow (\underline{x} \in S((\underline{x})) \wedge S((\underline{x})) \subseteq X)) .\end{aligned} \tag{3}$$

The set $S((\underline{x}))$ is known as the <u>equivalence class</u> (or <u>coset</u>) of $\underline{x}$ <u>modulo</u> S , or <u>coset modulo</u> S <u>of</u> (or <u>containing</u> or <u>determined by</u>) $\underline{x}$. The principal theorem (schema) concerns these equivalence classes. Define

$$Q \equiv \{\underline{q} : (\exists \underline{x})(\underline{x} \in X \wedge \underline{q} = S((\underline{x})))\} \quad , \tag{4}$$

where $\underline{q}$ denotes a letter different from $\underline{x}$ and not appearing in X or S (cf. the conventions in II.3.12). Q is called the <u>quotient set of</u> X <u>modulo</u> S and is often denoted by X/S . In the notation explained in II.12.1 and §3 of the Appendix,

$$Q \equiv \{S((\underline{x})) : \underline{x} \in X\} \quad .$$

The main theorem (schema) reads

$$T \quad : \quad \begin{aligned} &\&(S, X) \Rightarrow ((\cup Q = X) \wedge (\forall \underline{q})(\underline{q} \in Q \Rightarrow \underline{q} \neq \emptyset \wedge \underline{q} \subseteq X) \\ &\qquad \wedge (\forall \underline{p})(\forall \underline{q})((\underline{p} \in Q \wedge \underline{q} \in Q \wedge \underline{p} \neq \underline{q}) \Rightarrow \underline{p} \cap \underline{q} = \emptyset)) \end{aligned} \tag{5}$$

where $\underline{p}$ and $\underline{q}$ denote distinct letters not appearing in S or X .

Note first of all that the informal content of the conclusion of (5) is that each element of Q (that is, each equivalence class modulo S of some element of X) is a nonvoid subset of X , and that two such equivalence classes are either equal or disjoint. In other words, the set Q is disjoint (see II.9) and its union is equal to X . This situation is expressed by saying that the set Q of these equivalence classes is a <u>partition of</u> X .

A corollary is (when informally expressed) that there is just one coset modulo S containing (as an element) any given element x of X , this coset being equal to S((x)) .

In the case of S_1 , the quotient set Q is equal to

$$\{\{\ldots, -6, -3, 0, 3, 6, \ldots\} \ , \ \{\ldots, -8, -5, -2, 1, 4, 7, \ldots\} \ , \\ \{\ldots, -7, -4, -1, 2, 5, 8, \ldots\}\}$$

having just three elements. In the case of S_2, the quotient set Q is equal to

$$\{\{0\}, \{-1, 1\}, \{-2, 2\}, \{-3, 3\}, ...\} ,$$

the set of all sets of the form $\{-\underline{x}, \underline{x}\}$, where $\underline{x}$ is a natural number.

As a preliminary to the proof of (5), note that from (3) it follows that

$$T \quad : \quad \&(S, X) \Rightarrow ((\forall \underline{x})(\underline{x} \in X \wedge \underline{q} = S((\underline{x}))) \Rightarrow \underline{q} \in P(X)) \ . \tag{6}$$

Hence, if one assumes $\&(S, X)$, then (by (XII) in II.3.7) the sentence after the colon in (4) is collectivising in $\underline{q}$ and so Q has the intuitively expected property. Thus

$$T \quad : \quad \&(S, X) \Rightarrow ((\forall \underline{q})(\underline{q} \in Q \Leftrightarrow (\exists \underline{x})(\underline{x} \in X \wedge \underline{q} = S((\underline{x})))) \wedge Q \subseteq P(X)) \ . \tag{7}$$

Proof of (5) Let Θ denote the theory obtained by adjoining to Θ_0 the axiom $\&(S, X)$. In what follows, $\underline{p}$, $\underline{q}$, $\underline{r}$, $\underline{x}$, $\underline{y}$ denote distinct letters, no one of which appears in S or X; they are therefore all variables of Θ. By (7),

$$T\Theta \quad : \quad Q \subseteq P(X) \ .$$

Hence

$$T\Theta \quad : \quad \underline{q} \in Q \Rightarrow \underline{q} \subseteq X \ ,$$

so

$$T\Theta \quad : \quad \underline{r} \in \underline{q} \wedge \underline{q} \in Q \Rightarrow \underline{r} \in X$$

and (see I.3.7(5))

$$T\Theta \quad : \qquad (\exists \underline{q})(\underline{r} \in \underline{q} \wedge \underline{q} \in Q) \Rightarrow \underline{r} \in X$$

and so (see II.6.1(1))

$$T\Theta \quad : \qquad \underline{r} \in \cup Q \Rightarrow \underline{r} \in X \quad .$$

Hence (see (X) in I.3.2)

$$T\Theta \quad : \qquad (\forall \underline{r})(\underline{r} \in \cup Q \Rightarrow \underline{r} \in X) \quad .$$

Since $\underline{r}$ does not appear in $\cup Q$ or in X , this signifies (see II.2.1(1)) that

$$T\Theta \quad : \qquad \cup Q \subseteq X \quad . \qquad (8)$$

Again, by (3) and (4) ,

$$T\Theta \quad : \qquad \underline{x} \in X \Rightarrow \underline{x} \in S((\underline{x}))$$

$$T\Theta \quad : \qquad S((\underline{x})) \in Q \quad .$$

Hence

$$T\Theta \quad : \qquad \underline{x} \in X \Rightarrow \underline{x} \in S((\underline{x})) \wedge S((\underline{x})) \in Q$$

therefore

$$T\Theta \quad : \qquad \underline{x} \in X \Rightarrow (\exists \underline{q})(\underline{x} \in \underline{q} \wedge \underline{q} \in Q)$$

and so

$T\Theta$: $$\underline{x} \in X \Rightarrow \underline{x} \in \cup Q \ .$$

From this it follows (via (X) in I.3.2 and II.2.1(1)) that

$T\Theta$: $$X \subseteq \cup Q \ . \qquad (9)$$

Combining (8) and (9) and II.2.2(3) ,

$T\Theta$: $$\cup Q = X \ . \qquad (10)$$

It is a simple consequence of (3) and (7) that

$T\Theta$: $$(\forall \underline{q})(\underline{q} \in Q \Rightarrow \underline{q} \neq \emptyset \wedge \underline{q} \subseteq X) \ ; \qquad (11)$$

I leave the details of this step to the reader.

The final stage in the proof of (5) relies on the following two theorem schemas

$T\Theta$: $$S((\underline{x})) \cap S((\underline{y})) \neq \emptyset \Rightarrow (\underline{x}, \underline{y}) \in S \ , \qquad (12_1)$$

$T\Theta$: $$(\underline{x}, \underline{y}) \in S \Rightarrow S((\underline{x})) \subseteq S((\underline{y})) \ ; \qquad (12_2)$$

I leave the proof of these to the reader, with the remark that $\underline{z} \in S((\underline{x})) \Leftrightarrow (\underline{x}, \underline{z}) \in S$ is true in Θ_0 , and a reminder that S is symmetric and transitive is true in Θ . The symmetry of S allows one to proceed from the second of the above pair to

$T\Theta$: $$(\underline{x}, \underline{y}) \in S \Rightarrow S((\underline{x})) = S((\underline{y})) \ . \qquad (13)$$

By (12_1) , (13) and I.3.2(III),

$$T\Theta \quad : \qquad S((\underline{x})) \cap S((\underline{y})) \neq \emptyset \Rightarrow S((\underline{x})) = S((\underline{y})) \quad .$$

Hence

$$T\Theta \quad : \qquad \underline{p} = S((\underline{x})) \wedge \underline{q} = S((\underline{y})) \wedge \underline{p} \cap \underline{q} \neq \emptyset \Rightarrow \underline{p} = \underline{q}$$

and therefore also (see especially I.3.2(IX) or I.3.3(1))

$$T\Theta \quad : \quad (\exists\underline{x})(\exists\underline{y})(\underline{p} = S((\underline{x})) \wedge \underline{q} = S((\underline{y}))) \wedge \underline{p} \cap \underline{q} \neq \emptyset \Rightarrow \underline{p} = \underline{q} \quad .$$

By (7) , therefore,

$$T\Theta \quad : \qquad \underline{p} \in Q \wedge \underline{q} \in Q \wedge \underline{p} \cap \underline{q} \neq \emptyset \Rightarrow \underline{p} = \underline{q} \quad . \tag{14}$$

Now denote by Θ' the theory obtained by adjoining to Θ the axiom $\underline{p} \in Q \wedge \underline{q} \in Q \wedge \underline{p} \neq \underline{q}$, and by Θ'' the theory obtained by adjoining to Θ' the axiom $\underline{p} \cap \underline{q} \neq \emptyset$. Then

$$T\Theta'' : \qquad \underline{p} \in Q \wedge \underline{q} \in Q \wedge \underline{p} \cap \underline{q} \neq \emptyset$$

and so, in view of (14) ,

$$T\Theta'' : \qquad \underline{p} = \underline{q} \quad . \tag{15}$$

Since $\underline{p} \neq \underline{q}$ is true in Θ' and hence true in Θ'' , (15) makes it plain that Θ'' is contradictory. By (VII) in I.3.2, therefore,

$$T\Theta' \qquad \underline{p} \cap \underline{q} = \emptyset$$

and (VI) in I.3.2 now entails

$T\Theta$: $$(\underline{p} \in Q \wedge \underline{q} \in Q \wedge \underline{p} \neq \underline{q}) \Rightarrow \underline{p} \cap \underline{q} = \emptyset \ .$$

Since $\underline{p}$ and $\underline{q}$ are variables of Θ , a double application of (X) in I.3.2 leads to

$T\Theta$: $$(\forall \underline{p})(\forall \underline{q})((\underline{p} \in Q \wedge \underline{q} \in Q \wedge \underline{p} \neq \underline{q}) \Rightarrow \underline{p} \cap \underline{q} = \emptyset) \ . \qquad (16)$$

Combining (8) , (11) and (16) ,

$T\Theta$: $$(\cup Q = X) \wedge (\forall \underline{q})(\underline{q} \in Q \wedge \underline{q} \neq \emptyset) \wedge (\forall \underline{p})(\forall \underline{q})((\underline{p} \in Q \wedge \underline{q} \in Q \wedge \underline{p} \neq \underline{q}) \Rightarrow \underline{p} \cap \underline{q} = \emptyset) \ ,$$

and a final appeal to (VI) in I.3.2 entails (5) . □

A routine proof of (5) would be considerably shorter.

I add (without proof) that the procedure can be reversed. More specifically, given a partition Q of X , if one defines

$$S \equiv \{(\underline{x}, \underline{y}) : (\exists \underline{q})(\underline{q} \in Q \wedge \underline{x} \in \underline{q} \wedge \underline{y} \in \underline{q})\}$$

$$\equiv \{\underline{z} : (\exists \underline{x})(\exists \underline{y})(\underline{z} = (\underline{x}, \underline{y}) \wedge (\exists \underline{q})(\underline{q} \in Q \wedge \underline{x} \in \underline{q} \wedge \underline{y} \in \underline{q}))\} \ ,$$

then S can be proved to be an equivalence relation in S , and the quotient set X/S can be proved to be equal to Q . Thus equivalence relations in X and partitions of X are, in a sense, merely different aspects of the same thing.

III.2.9 Examples (i) Assuming the substance of IV.1 below, let X be a set and f a function with domain X . Then

$$S \equiv \{(\underline{x}, \underline{y}) : \underline{x} \in X \wedge \underline{y} \in X \wedge f(\underline{x}) = f(\underline{y})\}$$

$$\equiv \{\underline{z} : (\exists \underline{x})(\exists \underline{y})(\underline{x} \in X \wedge \underline{y} \in X \wedge f(\underline{x}) = f(\underline{y}) \wedge \underline{z} = (\underline{x}, \underline{y}))\}$$

may be proved to be an equivalence relation in X . The quotient set X/S is in such cases frequently abusively spoken of as the <u>quotient of</u> X <u>by</u> f . The elements of X/S are precisely those subsets of X on which f assumes a constant value, that is, the <u>sets of constancy</u> or <u>level sets</u> of f .

From what is to be said in Example IV.3.9, any quotient set of the form X/S is obtainable in the above way by a suitable choice of f : it suffices to take for f the natural mapping of X into X/S . Thus, one here has yet a third way of viewing equivalence relations and quotient sets.

(ii) A more concrete example arises if one takes X to be R and S to be the equivalence relation in R defined as

$$\{\underline{z} : (\exists \underline{x})(\exists \underline{y})(\underline{x} \in R \wedge \underline{y} \in R \wedge \underline{x} - \underline{y} \in 2\pi Z \wedge \underline{z} = (\underline{x}, \underline{y})\} ,$$

where $2\pi Z$ denotes the set of integer multiples of 2π , so that $(\underline{x}, \underline{y}) \in S$ if and only if $\underline{x}$ and $\underline{y}$ are real numbers and $\underline{x} - \underline{y} \in 2\pi Z$. The equivalence class, modulo S , of a real number $\underline{x}$ is the set

$$\{\underline{y} \in R : (\exists \underline{x})(\underline{x} \in R \wedge \underline{y} - \underline{x} \in 2\pi Z)\}$$

frequently denoted by $\underline{x} + 2\pi Z$; and the quotient set is in this case frequently denoted by $R/2\pi Z$ (another mild abus de langage).

In this particular case, the initial set R is a group under addition (see XII.2.1) and $(\underline{x}, \underline{y}) \in S$ if and only if $(\underline{x}, \underline{y}) \in R \times R$ and $\underline{x} - \underline{y}$ belongs to a certain fixed subgroup (see XII.2.2) of R . Whenever this situation obtains, the quotient set can be turned into a group in a very natural way: in the present case, the sum of two cosets $\underline{x} + 2\pi Z$ and $\underline{y} + 2\pi Z$ is defined to be the coset $(\underline{x} + \underline{y}) + 2\pi Z$. This definition (which is of a type discussed at greater length in VI.1.2) yields a unique sum of the two cosets, despite the fact that one and the same coset admits infinitely many expressions in the form $\underline{a} + 2\pi Z$ with $\underline{a} \in R$. The resulting group is naturally termed the <u>quotient</u> (or <u>factor</u>) <u>group</u> of the group

R by (or modulo) its subgroup $2\pi Z$.

This quotient group $R/2\pi Z$ arises quite naturally when one considers the so-called polar representation of complex numbers; see XII.2.4 and XII.5.4. See also VI.1 and Problems VI/2 and VI/3.

III.2.10 Concerning the definition of equivalence relations Care is needed with the terms "reflexive" , "symmetric" , "transitive" and "equivalence relation" , which are employed in slightly but significantly different senses by various authors (see, for example, Suppes (1), Sections 3.2 and 3.3; and see also Problems III/8 - III/10 below).

Our definition III.2.8(1) of " S is an equivalence relation in X " falls into four components, namely:

$$S \text{ is a relation in } X \text{ (equivalently, } S \subseteq X \times X \text{)}, \tag{1}$$

$$I_X \subseteq S , \tag{2}$$

$$\text{Inv } S = S , \tag{3}$$

$$S \circ S \subseteq S . \tag{4}$$

It is sometimes wrongly alleged that (1) , (3) and (4) together imply (2) , the suggested proof running somewhat as follows : choose any $x \in X$ and then any y such that $(x, y) \in S$; by (3) , $(y, x) \in S$; since $(x, y) \in S$ and $(y, x) \in S$, (4) implies that $(x, x) \in S$; hence $I_X \subseteq S$. This argument is fallacious because $x \in X$ does not imply that there exists y such that $(x, y) \in S$ (or, what is equivalent, that $x \in \text{Dom } S$). See (i) below and see also Suppes (1), p. 78, Exercise 1(e). The argument does, however, prove that (3) and (4) together imply that $I_{(\text{Dom } S)} \subseteq S$.

As the following simple examples show, even granted (1) , no one of (2) , (3) or (4) is implied by the remaining two. (Consistency is being assumed.)

(i) Suppose that $X \neq \emptyset$ and $S = \emptyset$. Then (1) , (3) and (4) are true and (2) is false. Likewise if $X \equiv \{0, 1\}$ and $S \equiv \{(0, 0)\}$.

(ii) Suppose that $X \equiv \{-2, -1, 0, 1, 2\}$ and

$$S \equiv \{(-2, -2), (-2, -1), (-1, -2), (-1, -1), (-1, 0), (0, -1), (0, 0), (0, 1), (1, 0), (1, 1), (1, 2), (2, 1), (2, 2)\} .$$

Then (1) , (2) and (3) are true. But (4) is false because $(0, 2) \notin S$ and yet $(0, 1) \in S$ and $(1, 2) \in S$ and therefore (see III.2.7(1')) $(0, 2) \in S \circ S$.

(iii) Suppose that $X \equiv \{0, 1\}$ and

$$S \equiv \{(0, 0), (1, 1), (1, 0)\} .$$

Then (1) , (2) and (4) are true and (3) is false.

Chapter IV. Functions

IV.1 Basic definitions and properties In informal language, a function is a relation having the special property that no two unequal ordered pairs belonging to (that is which are elements of) that relation have the same first coordinate.

To express this more formally, I write for any string f :

$$\text{Un } f \equiv_{\text{def}} (\forall \underline{x})(\forall \underline{y})(\forall \underline{z})(((\underline{x}, \underline{y}) \in f \wedge (\underline{x}, \underline{z}) \in f) \Rightarrow (\underline{y} = \underline{z})) \quad ;$$

recall the conventions of II.3.12 (according to which $\underline{x}$, $\underline{y}$, $\underline{z}$ here denote distinct letters not appearing in f). The string Un f is sometimes named " f is univocal" or " f is single-valued" . (This is similar to, but not quite the same as, Bourbaki's use of similar phrases; see Bourbaki (1), pp. 47, 81.)

The informal interpretation of Un f is usually expressed thus: if two ordered pairs belonging to f have equal first coordinates then they have equal second coordinates (and are therefore equal).

IV.1.1 Formal definition of functions

$$f \text{ is a function} \equiv_{\text{def}} (f \text{ is a relation}) \wedge \text{Un } f$$

$$\equiv (\forall \underline{z})(\underline{z} \in f \Leftrightarrow (\exists \underline{x})(\exists \underline{y})(\underline{z} = (\underline{x}, \underline{y}))) \wedge \text{Un } f \quad . \qquad (1)$$

It should be noted that $(S|\underline{u})\text{Un } f \equiv \text{Un}(S|\underline{u})f$ and that

$$(S|\underline{u})(f \text{ is a function}) \equiv (S|\underline{u})f \text{ is a function.}$$

On occasions, I will write Fn f or Fn(f) in place of " f is a function" ; see the remarks at the end of III.1.2. And I shall very frequently write " $f : X \to Y$ " or " f is a function from (or on) X into Y " in place of " f is a function and Dom f = X and Ran f $\subseteq$ Y " ; and " f is a Y-valued function" in place of " f is a function and Ran f $\subseteq$ Y " .

Some readers may wish to look at once at some of the examples listed in IV.3 below.

IV.1.2 Remarks (i) It should be clearly understood that here and everywhere in this book "function" corresponds to what is sometimes termed "single-valued function" . So-called "many-valued functions" appear frequently in conventional mathematical literature, often without any clear explanation of precisely what is meant. (A few words relating to a typical instance appear in IV.3.8.) Often they are closer to relations than to functions. In some areas, the phrase "many-valued function $X \to Y$ " refers to a function $X' \to Y$, where X' is a "covering space of X " (itself an elaborate concept); in other places, it refers to a function $F : X \to P(Y)$ and one is concerned with the existence of functions $f : X \to Y$ such that $f(x) \in F(x)$ for all $x \in X$ and possessing certain additional properties (cf. the choice functions spoken of in Problem IV/27). The matter is a complex one, but I shall say no more about it for the simple reason that the limited aims of this book make it unnecessary to do so.

(ii) As in the case of relations (see the end of III.2.2), it has also to be understood that, even at an informal level, writers differ over the details of their adopted definition of function. See the discussion in Shuard (1). Definition IV.1.1 differs in important details from that adopted by Bourbaki (for example). A careful reader's first task, after grasping the common general feature(s), is to look at the details.

To illustrate the type of difference which may be encountered, I cite the following as the informal version of a popular definition of function (cf. Bourbaki (1), p.81; Gleason (1), pp.45, 46; Hersee (1), Item 6; Stewart and Tall (1), Chapter 5; Griffiths and Hilton (1), p.42);

> A function is an ordered triplet (f, X, Y) (defined, for example, to be $((f, X), Y)$) in which S and Y are sets, f is a relation, $f \subseteq X \times Y$, $\text{Dom } f = X$, and $((x, y) \in f \wedge (x, y') \in f) \Rightarrow y = y'$.

(A function so defined is said to be <u>surjective</u> if and only if $f\langle X \rangle = Y$.)

I choose not to adopt this (perfectly workable) definition because, for most purposes I have in mind in this book, the additional complexity (compared with IV.1.1) pays no dividends (which it may well do in other contexts). The reader should ponder the differences between contending definitions, which are real and sometimes significant. For instance, according to IV.1.1,

$$\emptyset \text{ is a function}$$

is true; according to the modified definition

$$\emptyset \text{ is a function}$$

is false.

It scarcely need be said that to change from one definition to another in the course of an alleged proof immediately casts doubt (almost always justified) on the validity of the proof. Having adopted a definition, one must stick to it through thick and thin.

What has just been said illustrates a recurrent feature of informal mathematics: a term (even a technical term) is often used by different writers in essentially different (and sometimes totally inconsistent) ways. Formalisation does not, in itself, prevent this from happening, but it does make the differences more striking and thus more easily detected.

Whatever the adopted definition of "function" , the salient expected activity of a function is to "take values" . I turn to this aspect in the next subsection.

(iii) It can be proved (see Problem IV/14) that a function f is a relation with the property that, for all $\underline{x} \in \mathrm{Dom}\, f$, the image $f\langle\{\underline{x}\}\rangle$ by f of the singleton $\{\underline{x}\}$ is itself a singleton, namely $\{f(\underline{x})\}$ wherein $f(\underline{x})$ is defined as in IV.1.3 below. This helps to explain the term "single-valued" as applied to functions. See also IV.4.3(i) below.

(iv) Presumably without any intention of differing from (1) , some writers present the definition (schema)

$$f \text{ is a function} \equiv (f \text{ is a relation}) \wedge (((\underline{x}, \underline{y}) \in f \wedge (\underline{x}, \underline{z}) \in f) \Rightarrow (\underline{y} = \underline{z})) \quad , \qquad (1')$$

wherein (presumably) $\underline{x}$, $\underline{y}$, $\underline{z}$ denote distinct letters not appearing in f . (The difference between (1) and (1') lies in the omission of the universal quantifiers $(\forall \underline{x})$, $(\forall \underline{y})$, $(\forall \underline{z})$.) In spite of the fact that the right hand sides of (1) and (1') denote sentences such that (in Θ_0) the truth of one entails and is entailed by the truth of the other, they are not (in Θ_0) formally equivalent sentences. Moreover, the definition (1') is almost certainly not the intended one (which is (1)). See Problem IV/16.

(v) All of the terms "map" , "operator" , "operation" , "transformation" are at various times and places used with meanings virtually the same as "function" . Sometimes, however, they are deliberately used with other connotations; cf. (ii) above and see also Görke (1), pp.117-120. (The German word "abbildung" frequently translates mathematically into "map" or "representation" , but Görke uses it in the sense of our "relation" .) Readers have to take the trouble to search out each author's usage.

It is also customary to say that a function f maps A into (resp.

onto) B if and only if $f\langle A\rangle \subseteq B$ (resp. $f\langle A\rangle = B$). Usually (though not necessarily) this terminology is reserved for cases in which $A \subseteq \text{Dom } f$.

(vi) The basic definition IV.1.1(1) expresses the concept of function in terms of that of relation. The procedure can be reversed, provided one is granted also the concept of functional value; see Problem IV/26.

IV.1.3 <u>Formal definition of functional values</u> Informally speaking, a function f is an object which assigns to each element x of some set, called the domain of f , a certain "value" $f(x)$, called the "value of f at x " . I turn at once to the formal counterpart of this informal expectation.

For arbitrary strings f and A , I define formally and unconditionally (cf. IV.1.4(v) below)

$$f(A) \equiv_{def} \tau_{\underline{y}}((A, \underline{y}) \in f) \quad , \tag{1}$$

$\underline{y}$ denoting a letter not appearing in f or A . (The choice of $\underline{y}$ is otherwise immaterial, thanks to I.1.4(c). Here again one must bear in mind the substance of II.3.9(v); but see Problem IV/2.) $f(A)$ is termed the <u>value of</u> f <u>at</u> A or <u>the evaluation of</u> f <u>at</u> A , a concept primarily important in connection with functions.

In some contexts, one or other of the notations fA or Af is favoured in place of $f(A)$. Also, to avoid confusions, one sometimes writes $(f)(A)$ in place of $f(A)$; cf. the closing remarks in II.1.3(iii). From this point of view, a notation such as $)f, A($ might be preferable. Indeed, in certain special cases, the notation $\langle f, A\rangle$ <u>is</u> used.

Note the replacement rule

$$(S|\underline{u})f(A) \equiv ((S|\underline{u})f)((S|\underline{u})A)$$

for arbitrary strings S , f , A and an arbitrary letter $\underline{u}$. In particular,

$$(S|\underline{x})f(\underline{x}) \equiv f(S)$$

if $\underline{x}$ does not appear in f .

IV.1.4 Remarks Several remarks concerning conventional usage need to be made at once.

(i) Almost universally, f(a, b) is abusively written in place of f((a, b)) . Also, and especially in the case of sequences and families (see IV.7 and IV.11 below), f_a is often written in place of f(a) .

(ii) As has been indicated near the end of I.1.6, conventional usage of the functional value notation often pays little respect to formal niceties. In cases where one has a set f (often a function) in which a letter $\underline{x}$ appears, it may be and usually is thought to be acceptable and harmless to write f(a) to denote what is more properly denoted by $(a|\underline{x})f \equiv f[\![a]\!]$. There follows an illustration which indicates the dangers in doing this.

Suppose that B denotes a set and consider the function

$$f \equiv \{(\underline{y}, \underline{x} \cup \underline{y}) : \underline{y} \in B\}$$
$$\equiv \{\underline{z} : (\exists\underline{y})(\underline{y} \in B \wedge \underline{z} = (\underline{y}, \underline{x} \cup \underline{y}))\} \quad ;$$

here and in what follows $\underline{x}$, $\underline{y}$, $\underline{z}$, $\underline{t}$ denote distinct letters not appearing in B . It is easy to prove that f is a function with domain equal to B (cf. the general procedure discussed in IV.2.2 below). Suppose further that a denotes a set in which $\underline{x}$, $\underline{y}$, $\underline{z}$, $\underline{t}$ do not appear and such that a ∈ B is true. One may be interested in the set

$$\{(\underline{y}, a \cup \underline{y}) : \underline{y} \in B\} \qquad (1_1)$$

and it may be tempting to denote this set by f(a) . To do so, however, would be (to say the least) inviting confusion. Indeed, a moment's thought can lead one to expect a clash, inasmuch as $\underline{x}$ does not appear in $(a|\underline{x})f$ and yet presumably does appear in f(a) since it appears in f . Let us examine the matter more closely.

In the first place,

$$(a|\underline{x})f \equiv (a|\underline{x})\{\underline{z} : (\exists\underline{y})(\underline{y} \in B \wedge \underline{z} = (\underline{y}, \underline{x} \cup \underline{y}))\}$$
$$\equiv \{\underline{z} : (a|\underline{x})(\exists\underline{y})(\ldots\ldots\ldots)\} \qquad (II.1.4(r))$$

$$\equiv \{\underline{z} : (\exists \underline{y})(\underline{y} \in B \wedge \underline{z} = (\underline{y}, a \cup \underline{y}))\} \qquad \text{(I.1.8(f), II.6.2)}$$

$$\equiv \{(\underline{y}, a \cup \underline{y}) : \underline{y} \in B\} \quad . \qquad \text{(II.12.1)}$$

Thus the set (1_1) is identical with $(a|\underline{x})f$.

On the other hand, IV.1.3(1) leads to

$$f(a) \equiv \tau_{\underline{t}}((a, \underline{t}) \in f)$$

$$\equiv \tau_{\underline{t}}(((a, \underline{t})|\underline{z})(\exists \underline{y})(\underline{y} \in B \wedge \underline{z} = (\underline{y}, \underline{x} \cup \underline{y}))) \qquad \text{(S7 in II.1.2, II.3.3(3'), I.3.2(X))}$$

$$\equiv \tau_{\underline{t}}((\exists \underline{y})(\underline{y} \in B \wedge (a, \underline{t}) = (\underline{y}, \underline{x} \cup \underline{y}))) \qquad \text{(I.1.8(f))}$$

$$= \tau_{\underline{t}}((\exists \underline{y})(\underline{y} \in B \wedge a = \underline{y} \wedge \underline{t} = \underline{x} \cup \underline{y}))) \qquad \text{(S7, III.1.2(1), I.3.2(X))}$$

$$= \tau_{\underline{t}}(a \in B \wedge \underline{t} = \underline{x} \cup a) \qquad \text{(S7, Problem II/14, I.3.2(X))}$$

$$= \tau_{\underline{t}}(\underline{t} = \underline{x} \cup a) \qquad \text{(S7, I.3.7(1), I.3.2(X))}$$

$$= \underline{x} \cup a \quad . \qquad \text{(II.1.4(6))}$$

Thus

$$(a|\underline{x})f = \{(\underline{y}, a \cup \underline{y}) : \underline{y} \in B\} \qquad (1_2)$$

and

$$f(a) = \underline{x} \cup a \quad . \qquad (1_3)$$

From this it follows that sets a and B may be exhibited such that

$$f(a) \neq (a|\underline{x})f = \{(\underline{y}, a \cup \underline{y}) : \underline{y} \in B\} \qquad (1_4)$$

is true; it suffices (for example) to exhibit sets a and B such that $a \in B$, a is finite, B is finite, and $\#a > \#B$ are all true (the notation being as in V.7 below).

A similar abuse takes place in cases where U denotes a function with domain $\{0\} \cup (N \times N)$ .... say the function

$$\{(0, 0)\} \cup \{((\underline{m}, \underline{n}), 0) : \underline{m} \in N \wedge \underline{n} \in N\} \quad ,$$

where $\underline{m}$ and $\underline{n}$ denote distinct letters and 0 and N are defined as in Chapter V ...... and at the same time $U(0)$ (or U_0 ; see (i) above) is used to denote the set

$$\{(\underline{n}, U(0, \underline{n})) : \underline{n} \in N\} \quad , \qquad (1_5)$$

where $\underline{n}$ is a letter not appearing in U . The set (1_5) is (see IV.2.2 below) the function (actually a sequence) $\underline{n} \rightsquigarrow U(0, \underline{n})$ with domain N . In this case it can be shown that, if $U(0)$ is defined in accord with IV.1.3(1), then (as a consequence of the theorem $N \cap (N \times N) = \emptyset$, proved in V.3.4)

$$U(0) = \emptyset$$

is a theorem. On the other hand, the set (1_5) is nonempty.

Again, Gleason (1), p.43 writes somewhat as follows (his wording has been slightly but inessentially altered and the italics are mine) :

> For each real number x we define a function h_x from $\mathcal{R}$ to $\mathcal{R}$ by
>
> $$h_x(y) = x^2 + xy - y^3 .$$
>
> Now h itself is a function (with argument written as a subscript) from $\mathcal{R}$ to the set $\mathcal{R}^{\mathcal{R}}$ of functions from $\mathcal{R}$ to $\mathcal{R}$. Detached from any context, this construction is a pointless formality. But as one moves up the ladder of abstraction, such constructions become commonplace and *formal precision in defining them becomes the only guarantee of correctness*.

The fact is that Gleason here commits virtually the same sort of formal blunder as is mentioned above. In formal terms, his h_x is $(x|\underline{x})h$, where

$$h \equiv \{(\underline{y}, \underline{x}^2 + \underline{xy} - \underline{y}^3) : \underline{y} \in \mathcal{R}\}$$

and $\underline{x}$ and $\underline{y}$ denote distinct letters not appearing in $\mathcal{R}$ (nor in the addition, multiplication, subtraction, square and cube functions involved). The function from $\mathcal{R}$ to $\mathcal{R}^{\mathcal{R}}$ he refers to is

$$H \equiv \{(\underline{x}, h) : \underline{x} \in \mathcal{R}\} \quad ;$$

to refer to this as h is formally incorrect since the sentence

$$(\forall\underline{x})(\underline{x} \in \mathcal{R} \Rightarrow h = H) \qquad (1_6)$$

is false; see Problem IV/19.

See also the discussion in V.6.2.

Incidentally, these considerations illustrate how easily abusive notations and seemingly trivial departures from the strict formalities can produce what may appear to be (but which are not) inconsistencies in formal set theory; see the remarks in I.2.9.

(iii) One may have the intuitive feeling that, if f is a function, the domain of f is the set of $\underline{x}$ such that $f(\underline{x})$ is defined. Indeed, Spivak (1), p.38, writes:

> Note that the symbol f(x) makes sense only for x in the domain of f ; for other x the symbol f(x)

is undefined.

However, this is totally incompatible with the formal definition IV.1.3(1), according to which the metasentence " $f(\underline{x})$ is defined" , if it has any clear meaning at all, will be "true for all $\underline{x}$ " . Actually, of course, one ought to speak of "the set of $\underline{x}$ such that $\underline{\underline{A}}$ " , only in cases where $\underline{\underline{A}}$ is (or denotes) a sentence in the formal language, and " $f(\underline{x})$ is defined" is nowhere in this book attached as a name to a sentence in the formal language.

It would be more acceptable to say that useful assertions can be made about the set denoted by $f(\underline{x})$, only when the hypothesis $\underline{x} \in \mathrm{Dom}\, f$ is granted. (This corresponds to the fact that useful assertions about $\tau_{\underline{y}}(\underline{\underline{A}})$ ensue only when $(\exists \underline{y})\underline{\underline{A}}$ is known to be true.)

It may be objected that what has just been written is a feature of the particular formal scheme presented in this book. This may be true, though it is doubtful. In fact, most mathematical texts lack any precise definition of " $f(\underline{x})$ is defined" , unless it is merely a declaration that it is synonymous with " $\underline{x} \in \mathrm{Dom}\, f$ " . But in that case, it would still be better to write " $\underline{x} \in \mathrm{Dom}\, f$ " .

The phrase " $f(\underline{x})$ is defined at a " invites even more confusion. More acceptable and clearer would be " f is defined at a " ; even better would be " $a \in \mathrm{Dom}\, f$ " .

In the later and less formal parts of this book, I will sometimes follow convention and use the formally disreputable phrasing; see, for example, XI.2.2 and XI.7. This procedure may seem less offensive in informal contexts, merely because one has usually at that stage become reconciled to conditional definitions. (One formal blunder paves the way for others.)

(iv) For somewhat similar doubtful reasons, if f is a function with domain X , and if $a \notin X$, it is conventional to proceed as if one has the right to define $f(a)$ in some way which seems convenient (see XV.8.2 for an example). This, too, is forbidden by the formal definition IV.1.3(1). As soon as f and a

are specified, $f(a)$ is already determined by IV.1.3(1). To be correct, one should introduce a new function (an extension of f), say $\overline{f}$, and build the desired value at a into the definition of $\overline{f}$; this is cumbersome but necessary. Informal texts do not always follow this more lengthy procedure.

(v) In informal texts the definition of functional values is almost universally expressed in conditional form. An entirely typical example reads (cf. Gleason (1), p.41) :

> Let f be a function and let $x \in \text{Dom } f$. The unique y for which $(x, y) \in f$ is called the <u>value of</u> f <u>at</u> x and is denoted by $f(x)$.

At the corresponding point, Bourbaki (1), p.81 also adopts a conditional style; his reference to "the unique object which corresponds to x under f " presumably entails the hypothesis that x is an element of the domain of f .

The reader is referred to I.3.5(iv) for a discussion of the objections to the use of conditional definitions.

Criticism of a different sort has already been made in I.3.5(v) of the definition of $f(x)$ adopted in Suppes (1), p.87. It should now be added that although his definition (and every other known to me) agrees with ours when f is a function and x an element of $\text{Dom } f$, there may in other situations be material differences. Thus, if

$$f \equiv \{(\emptyset, \{\emptyset\}), (\emptyset, \{\{\emptyset\}\})\} ,$$

Suppes' definition ensures that $f(\emptyset) = \emptyset$ is true. According to our definition, however,

$$(f(\emptyset) = \{\emptyset\}) \vee (f(\emptyset) = \{\{\emptyset\}\})$$

is true, and it follows that our definition of $f(\emptyset)$ cannot (unless set theory is inconsistent) be equal to that prescribed by Suppes' definition.

(vi) A prevalent malpractice in informal mathematics is to confuse a function with its range. This faux pas is especially rife when the function happens to be an infinite sequence or a family (see IV.7 and IV.11), perhaps because it is from the start hallowed practice to represent the sequence (function with domain $\{1, 2, ...\}$) whose terms are $u_1, u_2, ...$ (that is, the function

$$\{(1, u_1), (2, u_2), ...\} \quad)$$

by

$$u_1, u_2, ...$$

or by

$$\{u_1, u_2, ...\} \quad .$$

This clearly invites the novice to make the objectionable confusion in question.

An infinite sequence is (as the very name suggests) always an infinite set, but its range may be either finite or infinite. In all cases one should distinguish between a sequence and its range, at least until such time as close investigation shows that the confusion is harmless in the prevailing context. It is furthermore dangerous to speak of an element of the range of a sequence u as "a member of u" ; a safer and conventional phrase is "a term of u" or "a value of u" . In this connection, see also the discussions in IV.7.1, IV.11.2 and VII.1.2(ii) below.

Strangely enough, it is much rarer for a finite sequence (that is, a sequence whose domain is a finite set) to be confused with its range - or, what is virtually the same thing - for an ordered pair to be confused with its first (or its second) coordinate, or for a complex number to be confused with its real (or its imaginary) part.

When, as is often the case, sequences (finite or infinite) are used to "represent" certain objects (as, for example, when the sequence of digits in the decimal expansion of a real number is used to represent that number), it is usually vital to avoid confusing a representative sequence with its range.

It is equally objectionable to confuse a function with its domain; for some reason, this malpractice appears to be much less common.

Further discussion appears in IV.11.2 below.

(vii) The definition schema IV.1.3(1) might be approached in a different way, along the lines discussed in II.1.3(ii). Choose and fix distinct letters $\underline{f}$ and $\underline{a}$ and make the formal definition

$$\underline{f}(\underline{a}) \equiv \tau_{\underline{y}}((\underline{a}, \underline{y}) \in \underline{f}) \quad , \tag{1_7}$$

wherein $\underline{y}$ denotes any letter different from $\underline{f}$ and $\underline{a}$; the choice of $\underline{y}$ is otherwise immaterial, thanks to I.1.4(c). Then define

$$f(A) \equiv (f|\underline{f}, A|\underline{a})\underline{f}(\underline{a}) \tag{1_8}$$

for arbitrary strings f and A . Choosing distinct letters $\underline{f}'$ and $\underline{a}'$, each different from $\underline{f}$ and $\underline{a}$ and neither appearing in f or A , and then a letter $\underline{y}$ different from $\underline{f}$, $\underline{a}$, $\underline{f}'$ and $\underline{a}'$ and not appearing in f or A , (*) in II.1.3(ii) combines with the replacement rules in I.1.4(d) and III.1.1 to lead from (1_7) and (1_8) to

$$\begin{aligned}
f(A) &\equiv (f|\underline{f}')(A|\underline{a}')(\underline{f}'|\underline{f})(\underline{a}'|\underline{a})\underline{f}(\underline{a}) \\
&\equiv (f|\underline{f}')(A|\underline{a}')(\underline{f}'|\underline{f})(\underline{a}'|\underline{a})\tau_{\underline{y}}((\underline{a}, \underline{y}) \in \underline{f}) \\
&\equiv (f|\underline{f}')(A|\underline{a}')(\underline{f}'|\underline{f})\tau_{\underline{y}}((\underline{a}', \underline{y}) \in f) \\
&\equiv (f|\underline{f}')(A|\underline{a}')\tau_{\underline{y}}((\underline{a}', \underline{y}) \in \underline{f}') \\
&\equiv (f|\underline{f}')\tau_{\underline{y}}((A, \underline{y}) \in \underline{f}') \\
&\equiv \tau_{\underline{y}}((A, \underline{y}) \in f) \quad ,
\end{aligned}$$

which agrees with IV.1.3(1).

(viii) Occasionally one finds f(g) written in place of $f \circ g$, usually when f and g are functions. A little thought at the informal level suffices to indicate that puzzlement and confusion is likely to be generated. Thus if N denotes the set of natural numbers (as defined in Chapter V) and if $g : N \to N$ and $f : N \cup \{g\} \to N$, then $f \circ g$ is a function $N \to N$ and f(g) is an element of N and informally one would not expect $f \circ g$ to be equal to f(g) , if only because $f \circ g$ is an infinite set while elements of N are expected to be finite sets. One's suspicions turn out to be formally justified. For if one supposes that $f \circ g = f(g)$, it would follow from V.3.4(35) that any element t of the nonvoid set $f \circ g$ would be an element of both N and $N \times N$, which conclusion contradicts V.3.4(37).

The use of $f(g)$ to denote $f \circ g$ may save the writer a little trouble, but it will almost certainly waste time and effort on the part of the reader.

IV.1.5 Properties of functional values Returning to formalities, I proceed to state and prove a number of theorem schemas, applicable to arbitrary sets S , T , f , g , but intuitively of greatest interest when f and g are functions.

The first is an instance of the type of theorem schema, the proof of which has hitherto been left to the reader (see the end of II.1.4), namely:

$$T! \quad : \quad (S = T \wedge f = g) \Rightarrow (f(S) = g(T)) \quad . \tag{2}$$

Proof of (2) Adjoin to Θ_0 the axiom $S = T \wedge f = g$, obtaining a theory Θ . Choose distinct letters $\underline{x}$, $\underline{y}$, $\underline{z}$ not appearing in S , T , f or g . Define $U \equiv \underline{x}(\underline{y})$, so that

$$f(S) \equiv (f|\underline{x})(S|\underline{y})U \tag{2_1}$$

and

$$g(T) \equiv (g|\underline{x})(T|\underline{y})U \quad . \tag{2_2}$$

Now

$$T\Theta \quad : \quad (S = T) \wedge (f = g) \quad . \tag{2_3}$$

By II.1.4(5), I.3.2(I) and (2_3) ,

$$T\Theta \quad : \quad (S|\underline{y})U = (T|\underline{y})U \quad ; \tag{2_4}$$

and by II.1.4(5), I.3.2(I) and (2_4) ,

$$T\Theta \quad : \quad (f|\underline{x})(S|\underline{y})U = (g|\underline{x})(T|\underline{y})U \quad . \tag{2_5}$$

By (2_1) , (2_2) , (2_5) , transitivity of equality (and the metatheorem in II.1.1 relating $\equiv$ and $=$),

$$T\Theta \quad : \quad f(S) = g(T) \quad . \tag{2_6}$$

By I.3.2(VI) and (2_6) , (2) follows. □

In the discussion of (3) - (9) below in this subsection, it is to be understood that $\underline{x}$, $\underline{y}$, $\underline{z}$ denote distinct letters not appearing in f (nor, in the case of (6) , should they appear in A ; nor, in the case of (9) , should they appear in g); cf. II.3.12.

The second theorem schema reads

$$T! : \quad \text{Fn } f \Rightarrow (((\underline{x}, \underline{y}) \in f) \Leftrightarrow (\underline{x} \in \text{Dom } f \wedge \underline{y} = f(\underline{x}))) \quad . \tag{3}$$

Informally, one may say that: if f is a function, and if $\underline{x} \in \text{Dom } f$, then $\underline{y} = f(\underline{x})$ is the unique $\underline{y}$ such that $(\underline{x}, \underline{y}) \in f$.

Proof of (3) Denote by Θ the theory obtained by adjoining to Θ_0 the axiom Fn f . By I.3.2(VI) and I.3.3(j), it suffices to prove that

$$T\Theta : \quad (\underline{x}, \underline{y}) \in f \Rightarrow \underline{x} \in \text{Dom } f \quad , \tag{3_1}$$

$$T\Theta : \quad (\underline{x}, \underline{y}) \in f \Rightarrow \underline{y} = f(\underline{x}) \quad , \tag{3_2}$$

$$T\Theta : \quad (\underline{x} \in \text{Dom } f \wedge \underline{y} = f(\underline{x})) \Rightarrow (\underline{x}, \underline{y}) \in f \quad . \tag{3_3}$$

Of these (3_1) follows at once from II.2.2(4); the details are left to the reader.

Turning to (3_2) , I.3.7(2) entails that it is enough to prove that

$$T : \quad (\text{Fn } f \wedge (\underline{x}, \underline{y}) \in f) \Rightarrow \underline{y} = f(\underline{x}) \quad . \tag{3_4}$$

To accomplish this, let Θ' denote the theory obtained by adjoining to Θ_0 the explicit axiom $\text{Fn } f \wedge (\underline{x}, \underline{y}) \in f$. Then

$T\Theta'$: $$\text{Fn } f$$

and hence (see IV.1.1)

$T\Theta'$: $$((\underline{x}, \underline{y}) \in f \wedge (\underline{x}, \underline{z}) \in f) \Rightarrow \underline{y} = \underline{z} \ . \qquad (3_5)$$

Also

$T\Theta'$: $$(\underline{x}, \underline{y}) \in f \qquad (3_6)$$

and so, by S5 in I.2.2,

$T\Theta'$: $$(\exists \underline{y})((\underline{x}, \underline{y}) \in f) \ ,$$

that is, by IV.1.3(1) and the definition of $(\exists \underline{y})$,

$T\Theta'$: $$(f(\underline{x})|\underline{y})((\underline{x}, \underline{y}) \in f) \ ,$$

that is (recall II.3.12),

$T\Theta'$: $$(\underline{x}, f(\underline{x})) \in f \ . \qquad (3_7)$$

Since $\underline{z}$ is a variable of Θ' (see II.3.12 again), (3_5) and I.3.2(II) entail

$T\Theta'$: $$((\underline{x}, \underline{y}) \in f \wedge (\underline{x}, f(\underline{x})) \in f) \Rightarrow \underline{y} = f(\underline{x}) \ . \qquad (3_8)$$

By (3_6) , (3_7) and I.3.3(j),

$T\Theta'$: $$(\underline{x}, \underline{y}) \in f \wedge (\underline{x}, f(\underline{x})) \in f \ ; \qquad (3_9)$$

hence, by (3_8) , (3_9) and I.3.2(I),

$$T\Theta' : \qquad \underline{y} = f(\underline{x}) \quad .$$

An appeal to I.3.2(VI) now entails (3_4) . (The preceding proof could be shortened by use of Problem II/25.)

Finally, consider (3_3) . By III.2.2(4),

$$T \quad : \underline{x} \in \text{Dom } f \Leftrightarrow (\exists \underline{y})((\underline{x}, \underline{y}) \in f) \equiv (f(\underline{x})|\underline{y})((\underline{x}, \underline{y}) \in f) \quad ,$$

the last step by IV.1.3(1) and the definition of $(\exists \underline{y})$. Moreover (recall II.3.12), the last-written string is identical with

$$(\underline{x}, f(\underline{x})) \in f \quad .$$

Thus

$$T \quad : \qquad \underline{x} \in \text{Dom } f \Leftrightarrow (\underline{x}, f(\underline{x})) \in f \quad . \qquad (3_{10})$$

Let Θ'' denote the theory obtained by adjoining to Θ_0 the axiom $\underline{x} \in \text{Dom } f \wedge \underline{y} = f(\underline{x})$. Then

$$T\Theta'' : \qquad \underline{y} = f(\underline{x}) \qquad (3_{11})$$

and, by use of (3_{10}) ,

$$T\Theta'' : \qquad (\underline{x}, f(\underline{x})) \in f \quad . \qquad (3_{12})$$

By (3_{11}) and S6 in II.1.2,

$$T\Theta'' : \qquad (\underline{x}, f(\underline{x})) \in f \Leftrightarrow (\underline{x}, \underline{y}) \in f \quad . \qquad (3_{13})$$

By (3_{12}) and (3_{13}) ,

$T\Theta''$: $(\underline{x}, \underline{y}) \in f$.

This, combined with I.3.2(VI), entails (3_3) . □

Remark A close look at the proof of (3_3) shows that the hypothesis Fn f has not been used, so that in fact

$$T! : (\underline{x} \in \text{Dom } f \wedge \underline{y} = f(\underline{x})) \Rightarrow (\underline{x}, \underline{y}) \in f \; . \qquad (3')$$

Other theorem schemas of importance are the following:

$$T! : \text{Fn } f \Rightarrow (\underline{x} \in \text{Dom } f \Rightarrow f(\underline{x}) \in \text{Ran } f) \; , \qquad (4)$$

$$T! : \begin{array}{l} \text{Fn } f \Rightarrow (\text{Coll}_{\underline{z}}(\exists \underline{x})(\underline{x} \in \text{Dom } f \wedge \underline{z} = (\underline{x}, f(\underline{x})) \\ \wedge\ f = \{\underline{z} : (\exists \underline{x})(\underline{x} \in \text{Dom } f \wedge \underline{z} = (\underline{x}, f(\underline{x}))\}) \; , \end{array} \qquad (5)$$

$$T! : \begin{array}{l} (\text{Fn } f \wedge A \subseteq \text{Dom } f) \Rightarrow (\text{Coll}_{\underline{y}}(\exists \underline{x})(\underline{x} \in A \wedge \underline{y} = f(\underline{x})) \\ \wedge\ f\langle A \rangle = \{\underline{y} : (\exists \underline{x})(\underline{x} \in A \wedge \underline{y} = f(\underline{x}))\}) \; ; \end{array} \qquad (6)$$

note that, if one uses the notation II.12.1(2), the last phrase of (6) is equivalent to

$$f\langle A \rangle = \{f(\underline{x}) : \underline{x} \in A\} \; .$$

I leave the proof of (4) and (6) to the reader; in the case of (6) , make use of (3) and II.2.6(1) .

Proof of (5) Let Θ denote the theory obtained by adjoining to Θ_0 the axiom Fn f . By (4) and III.2.2(6),

$$T\Theta \quad : \qquad \underline{x} \in \text{Dom } f \wedge \underline{z} = (\underline{x}, f(\underline{x})) \Rightarrow \underline{z} \in \text{Dom } f \times \text{Ran } f$$

and so, by I.3.3(l) and II.3.12,

$$T\Theta \quad : \qquad (\exists\underline{x})(\underline{x} \in \text{Dom } f \wedge \underline{z} = (\underline{x}, f(\underline{x}))) \Rightarrow \underline{z} \in \text{Dom } f \times \text{Ran } f \quad .$$

An appeal to (XII) in II.3.7 entails that

$$T\Theta \quad : \qquad \text{Coll}_{\underline{z}}(\exists\underline{x})(\underline{x} \in \text{Dom } f \wedge \underline{z} = (\underline{x}, f(\underline{x}))) \ . \qquad (5_1)$$

Now let Θ' denote the theory obtained by adjoining to Θ the axiom

$$\underline{z} \in \{\underline{z} : (\exists\underline{x})(\underline{x} \in \text{Dom } f \wedge \underline{z} = (\underline{x}, f(\underline{x})))\} \quad .$$

By (5_1) and II.3.3(3),

$$T\Theta' \ : \qquad (\exists\underline{x})(\underline{x} \in \text{Dom } f \wedge \underline{z} = (\underline{x}, f(\underline{x}))) \quad ,$$

that is,

$$T\Theta' \ : \qquad (\exists\underline{x})((\exists\underline{y})((\underline{x}, \underline{y}) \in f) \wedge \underline{z} = (\underline{x}, f(\underline{x}))) \quad .$$

Hence, by I.3.3(o),

$$T\Theta' \ : \qquad (\exists\underline{x})(\exists\underline{y})((\underline{x}, \underline{y}) \in f \wedge \underline{z} = (\underline{x}, f(\underline{x}))) \quad . \qquad (5_2)$$

By (3) and I.3.2(I),

$$T\Theta' \qquad (\underline{x}, \underline{y}) \in f \Rightarrow \underline{y} = f(\underline{x}) \quad .$$

So from (5_2) (plus various metatheorems and theorem schemas in I.3.2 and I.3.3),

$T\Theta'$: $$(\exists \underline{x})(\exists \underline{y})((\underline{x}, \underline{y}) \in f \wedge \underline{y} = f(\underline{x}) \wedge \underline{z} = (\underline{x}, f(\underline{x}))) \quad .$$

Using S6 in II.1.2 or II.1.4(5), therefore,

$T\Theta'$: $$(\exists \underline{x})(\exists \underline{y})((\underline{x}, \underline{y}) \in f \wedge \underline{z} = (\underline{x}, \underline{y}))$$

and so, using S6 again,

$T\Theta'$: $$(\exists \underline{x})(\exists \underline{y})(\underline{z} \in f)$$

that is,

$T\Theta'$: $$\underline{z} \in f \quad .$$

An appeal to I.3.2(VI) entails

$T\Theta$: $$\underline{z} \in \{\underline{z} : (\exists \underline{x})(\underline{x} \in \text{Dom } f \wedge \underline{z} = (\underline{x}, f(\underline{x}))\} \Rightarrow \underline{z} \in f \quad .$$

Since $\underline{z}$ is a variable of Θ , one concludes via II.2.1(1) that

$T\Theta$: $$\{\underline{z} : (\exists \underline{x})(\underline{x} \in \text{Dom } f \wedge \underline{z} = (\underline{x}, f(\underline{x}))\} \subseteq f \quad . \tag{5_3}$$

To complete the proof of (5) , introduce the theory Θ'' obtained by adjoining to Θ the axiom $\underline{z} \in f$. Then (since f is a relation is true in Θ , hence true in Θ'')

$T\Theta''$: $$(\exists \underline{x})(\exists \underline{y})(\underline{z} = (\underline{x}, \underline{y})) \wedge \underline{z} \in f \quad .$$

So, by I.3.3(o) again

$T\Theta''$: $$(\exists \underline{x})(\exists \underline{y})(\underline{z} = (\underline{x}, \underline{y}) \wedge \underline{z} \in f) \quad . \tag{5_4}$$

By (3) and S6 in II.1.2,

$$T\Theta \quad : \qquad \underline{z} = (\underline{x}, \underline{y}) \wedge \underline{z} \in f \Rightarrow \underline{z} = (\underline{x}, \underline{y}) \wedge \underline{x} \in \text{Dom } f \wedge \underline{y} = f(\underline{x}) \quad . \qquad (5_5)$$

By I.3.3(1), (5_4) and (5_5) ,

$$T\Theta'' \ : \qquad (\exists\underline{x})(\exists\underline{y})(\underline{x} \in \text{Dom } f \wedge \underline{y} = f(\underline{x}) \wedge \underline{z} = (\underline{x}, \underline{y})) \quad .$$

Hence, by S6 in II.1.2 and I.3.3(1) again,

$$T\Theta'' \ : \qquad (\exists\underline{x})(\exists\underline{y})(\underline{x} \in \text{Dom } f \wedge \underline{z} = (\underline{x}, f(\underline{x}))) \quad ,$$

that is,

$$T\Theta'' \ : \qquad (\exists\underline{x})(\underline{x} \in \text{Dom } f \wedge \underline{z} = (\underline{x}, f(\underline{x}))) \quad . \qquad (5_6)$$

By (5_6) , (5_1) and II.3.3(3),

$$T\Theta'' \ : \qquad \underline{z} \in \{\underline{z} : (\exists\underline{x})(\underline{x} \in \text{Dom } f \wedge \underline{z} = (\underline{x}, f(\underline{x})))\} \quad .$$

Combining this with I.3.2(VI),

$$T\Theta \quad : \qquad \underline{z} \in f \Rightarrow \underline{z} \in \{\underline{z} : (\exists\underline{x})(\underline{x} \in \text{Dom } f \wedge \underline{z} = (\underline{x}, f(\underline{x})))\} \quad .$$

Since $\underline{z}$ is a variable of Θ , one concludes via II.2.1(1) that

$$T\Theta \quad : \qquad f \subseteq \{\underline{z} : (\exists\underline{x})(\underline{x} \in \text{Dom } f \wedge \underline{z} = (\underline{x}, f(\underline{x}))\} \quad . \qquad (5_7)$$

A combination of (5_3) and (5_7) implies (via II.2.2(2))

$$T\Theta \quad : \qquad f = \{\underline{z} : (\exists\underline{x})(\underline{x} \in \text{Dom } f \wedge \underline{z} = (\underline{x}, f(\underline{x})))\} \quad . \qquad (5_8)$$

By (5_1) , (5_8) and I.3.3(j),

$$T\Theta : \quad \begin{array}{l} \mathrm{Coll}_{\underline{z}}(\exists \underline{x})(\underline{x} \in \mathrm{Dom}\ f \wedge \underline{z} = (\underline{x}, f(\underline{x}))) \\ \qquad \wedge\ f = \{\underline{z} : (\exists \underline{x})(\underline{x} \in \mathrm{Dom}\ f \wedge \underline{z} = (\underline{x}, f(\underline{x})))\} \end{array} ,$$

and an appeal to I.3.2(VI) entails (5) . □

As a corollary to (6) and III.2.6(3), one may derive

$$T! : \quad \begin{array}{l} \mathrm{Fn}\ f \Rightarrow (\mathrm{Coll}_{\underline{y}}(\exists \underline{x})(\underline{x} \in \mathrm{Dom}\ f \wedge \underline{y} = f(\underline{x})) \\ \qquad \wedge\ \mathrm{Ran}\ f = \{\underline{y} : (\exists \underline{x})(\underline{x} \in \mathrm{Dom}\ f \wedge \underline{y} = f(\underline{x}))\}) \end{array} ; \tag{8}$$

if one adopts the notation explained briefly in II.12.1, it follows from (8) that

$$T! : \quad \mathrm{Fn}\ f \Rightarrow \mathrm{Ran}\ f = \{f(\underline{x}) : \underline{x} \in \mathrm{Dom}\ f\} . \tag{8'}$$

As a corollary to (5) , one may derive

$$T! : \quad \begin{array}{l} (\mathrm{Fn}\ f \wedge \mathrm{Fn}\ g) \Rightarrow (((f = g) \Leftrightarrow ((\mathrm{Dom}\ f = \mathrm{Dom}\ g) \\ \qquad \wedge\ (\forall \underline{x})(\underline{x} \in \mathrm{Dom}\ f \Rightarrow f(\underline{x}) = g(\underline{x})))) \end{array} . \tag{9}$$

Expressed informally, (9) incorporates the vital principle that two functions are equal if and only if they have equal domains and "rules" which agree on this common domain (that is, take the same value at each point of this common domain). This stresses the fact that a function determines its domain uniquely.

Finally, the reader should prove the theorem schema

$$T! : \quad Fn\ f \Rightarrow (((a \in \mathrm{Dom}\ f) \Rightarrow (f\langle\{a\}\rangle = \{f(a)\})) \wedge ((a \notin \mathrm{Dom}\ f) \Rightarrow (f\langle\{a\}\rangle = \emptyset))) \tag{10}$$

wherein f and a denote arbitrary sets; informally expressed: if f is a function, then $f\langle\{a\}\rangle$ is equal to $\{f(a)\}$ or to $\emptyset$ according as a is or is not an element of $\mathrm{Dom}\ f$.

IV.1.6 Concerning notation On the matter of notation, it is important to stress that $f(A)$ and $f\langle\{A\}\rangle$ should not be confused; nor should $f\langle A\rangle$ and $f(A)$. For example, suppose that

$$f \equiv \{(1, 2), (2, 3), (\{1, 2\}, 4)\} ,$$

$$A \equiv \{1, 2\} ;$$

then

$$f(A) = 4 ,$$

$$f\langle\{A\}\rangle = \{4\} ,$$

$$f\langle A\rangle = \{2, 3\} ;$$

the reader should pause to prove these equations.

The risk of confusion is heightened by the sad fact that most writers use the $f(\ )$ notation to cover both what are here denoted by $f\langle\ \rangle$ and $f(\ )$. Disaster can come about as the result of this confusion, as the following amusing example (due to Bourbaki (1), p.125, Exercise *11) shows.

Fermat conjectured that (in traditional language) the "equation" (really a sentence)

$$x^n + y^n = z^n , x, y, z, n \in N , n > 2$$

is insoluble. (Here, N denotes the set of natural numbers, as in Chapter V; see also II.3.1(i) and III.2.3(viii).)

If he ever wrote out a proof, this has never been found; and all subsequent attempts at a complete proof or disproof have failed (but have been productive of an enormous amount of valuable mathematics). However, what about the following alleged disproof of Fermat's statement?

Define

$$A \equiv \{\underline{n} \in N : \underline{n} > 2 \wedge (\exists \underline{x})(\exists \underline{y})(\exists \underline{z})(\underline{x} \in N \wedge \underline{y} \in N \wedge \underline{z} \in N \wedge \underline{x}^{\underline{n}} + \underline{y}^{\underline{n}} = \underline{z}^{\underline{n}})\}$$

and consider the function $f \equiv \{(A, N)\}$. If $A = \emptyset$ (equivalently, if Fermat's statement is true), one has $N = f(\emptyset) = \emptyset$, which is a contradiction. So, on the basis of (VII) in I.3.2, Fermat's statement is false.

As the reader will no doubt perceive, the fallacy results from a confusion of $f(\emptyset)$ and $f\langle \emptyset \rangle$.

IV.1.7 Remarks concerning terminology: evaluating, locating and describing (i) Another regrettable practice hallowed by tradition, and akin to that discussed in IV.1.4(vi), is that which amounts to the use of phrases like "let $f(\underline{x})$ be a function". In almost all such instances, the function referred to is (the set denoted by) f, not (the set denoted by) $f(\underline{x})$. For another aspect of this issue, see IV.2.3(ii).

It is true that strict adherence to the legitimate terminology does seem unnatural (mainly as a result of traditional upbringing) and sometimes cumbersome, especially when speaking of specific examples ("the function $\underline{x}^2 - \underline{x} + 1$", et cetera), but this is a small price to pay at the early stages for clarity and precision. In any case, intelligent use of the $\rightsquigarrow$ notation introduced in IV.2.2 goes a long way towards answering the accusation of cumbersomeness. (Later on, when the typographical difficulties mount, infringements of the rule may be tolerated, provided that the "abus de langage" is fully understood.)

(ii) I turn next to some customary descriptive terminology which is often left totally vague - as when one writes of "evaluating" and "locating" (see, for example, X.6.7, Problem X/9, XI.1.2, XI.5.3). What is meant by these terms, is usually left unexplained or is thought to be adequately explained by reference to a few examples. (The intended meanings of new words are often left to be inferred from the context.)

The situation is akin to that in I.3.4(viii), where an attempt has been made to explain the term "exhibit" . It is difficult to be explicit and, at the same time, sufficiently inclusive.

Suppose that F is a function and a a set. It would seem that to "evaluate F(a) " is to exhibit, somewhat as described in I.3.4(viii), a set b such that $F(a) = b$. However, more than this is usually demanded (albeit not explicitly): it is intended that the evaluation shall accomplish something useful and/or significant. (For example, while 2 may be regarded as a useful evaluation of $1 + 1$, the evaluations $1 + 1 + 0$ and $3 - 1$ might be spurned because they accomplish nothing useful.) At this point, if not before, a highly subjective judgement is called for on the part of the problem-solver. It is almost impossible to describe how such judgements are to be made, or to describe what is good judgement, in terms which are both precise and of general application.

As in the case of exhibition, there is an intended ban on explicit use of the selector τ ; in particular,

$$\tau_{\underline{y}}((a, \underline{y}) \in F)$$

is not conventionally regarded as a useful evaluation of F(a) .

Again as with exhibition, an evaluation (good or bad, useful or not) involves a proof and thus has to be effected relative to, or within, a certain theory - presumably set theory, unless anything is said to the contrary. For instance

$$\text{Evaluate } F(a) \text{ for all } a \in A$$

is said to be interpreted as calling for an evaluation of $F(\underline{x})$ in the theory obtained by adjoining to set theory the axiom $\underline{x} \in A$ (where $\underline{x}$ denotes a letter not appearing in F or A), followed by the formal replacement indicated by $(a|\underline{x})$. See Problem XIV/1 for an example.

All of this is still rather unsatisfactory. The poser of a problem of the type "Evaluate ...." usually has a preconceived notion of the form the answer should take; and this form usually depends on the context. Very often the poser would be hard-pressed to explain precisely the form he expects.

Thus, after having dealt with certain aspects of integration, I might pose the problem

$$\text{Evaluate } \int_0^1 j^{\frac{1}{2}} ,$$

having in mind that which is more clearly expressed by

> Exhibit positive natural numbers a
>
> and b such that $\int_0^1 j^{\frac{1}{2}} = a^{-1}b$,

and yet being reluctant to say exactly this for fear of making the problem too easy (in the stated context). For the same reason, I might not want to express the problem as

> Exhibit a real number r
>
> such that $\int_0^1 j^{\frac{1}{2}} = r$

for fear that the problem-solver will resort to writing

$$r \equiv \lim_{n\to\infty} \sum_{k=1}^{n} (1/n)(k/n)^{\frac{1}{2}}$$

and referring to one or more general theorems appearing in Chapter XI by way of proof. I, as the problem-poser, would find it difficult to reject this solution on objective grounds, but would feel that the problem-solver has missed the point of the problem.

Again, the problem

$$\text{Evaluate } ((1 + j^2)^{-\frac{1}{2}})'$$

might cover what would be much more clearly expressed as

> Exhibit an algebraic function $f : R \to R$
>
> such that $((1 + j^2)^{-\frac{1}{2}})' = f$ (a proof being included, just to clinch matters).

The would-be solver thus often has thrown upon him the task of divining the unspoken thoughts of the poser. It is this feature which makes the wording of examination questions (on the answers to which so much may depend) such a "hot" issue.

Incidentally, one often finds reference to "the value" of a number, "the value of $f'(a)$ " , "the value of $\int_0^1 g$ " , et cetera. The phrase "the value" is, in most such cases, superfluous and confusing. Is there any distinction between a number and the value of that number? If so, what precisely is the difference? Confusion is almost unavoidable, if the formal definition IV.1.3(1) is also in mind. The only sense to be attached to "the value of $\int_0^1 g$ " comes from interpreting this as meaning "the value at g of the function $\underline{x} \rightsquigarrow \int_0^1 \underline{x}$ with domain X " , where X is some suitably chosen set of which g is an element, $\underline{x}$ being a letter not appearing in X (nor in the various sets 0, 1, R, ... involved in the definition of the integral).

It might indeed be argued that to speak of "the value of $\int_0^1 g$ " is intended to stress the distinction between the number $\int_0^1 g$ and "the process by which one arrives at that number" . However, there is no unique such process. The common features of such processes are virtually summarised in the precise concept of the function $\underline{x} \rightsquigarrow \int_0^1 \underline{x}$ with domain the set of functions $\underline{x} : [0, 1] \to R$ which are integrable over $[0, 1]$. In other words, whatever clear distinction there is to be made, is better made by distinguishing between a function and the values of that function. This brings one back to the substance of (i) above.

Finally, the terms "calculate" and "compute" are often used in the sense of "evaluate" .

(iii) Consider now the concept of "locating" elements of a set. Supposing that S denotes a nonempty set, to "locate some (maybe all) points (or elements) of S " might be taken to signify exhibiting (without explicit use of the selector τ ; see I.3.4(viii)) a positive natural number $n \in \dot{N}$ and a function $f : \{1, ..., n\} \to S$. Such a procedure "exhibits" elements $f(1), ..., f(n)$ of S . All the elements of S are thus exhibited if and only if $\operatorname{Ran} f = S$. This last can be true, only if S is finite. If S is infinite, the location of all elements of S involves replacing $\{1, ..., n\}$ by some familiar infinite set .... perhaps the set N of all natural numbers or the set R of all real numbers. (All this is closely related to the ideas dealt with in V.7 below.)

The ban on the explicit use of τ is (as in I.3.4(viii)) adopted in an attempt to rule out such trivialities as locating an element of S by merely defining

$$a \equiv \tau_{\underline{x}}(\underline{x} \in S)$$

and/or

$$f \equiv \{(1, \tau_{\underline{x}}(\underline{x} \in S))\}$$

where $\underline{x}$ is a letter not appearing in S . (These definitions do indeed lead to the theorems

$$(S \Rightarrow \emptyset) \Rightarrow a \in S \quad ,$$

$$(S \neq \emptyset) \Rightarrow (f(1) \in S) \quad ,$$

but this would rarely be regarded as significant progress.) Whether the ban succeeds in ruling out all trivialities, I do not know.

(iv) In an entirely similar vein, one sometimes speaks of "describing completely" a set S . Although this might be taken to mean defining (formally or informally) the set S , the intended meaning is usually rather that one should exhibit a set E which is in some sense known or familiar and a (preferably injective; see IV.6.1) function f with domain equal to E and range equal to S ; then one might describe S as the set of objects of the form $f(\underline{x})$ where $\underline{x} \in E$ (cf. I.3.5(vi) and II.12.1). The prescription is thus similar to and more exacting than what is required for the location of elements of S , as described in (iii) above.

Alternatively, in special cases one may be able to describe S as $\{a, b, \ldots, c\}$, where a, b, ..., c are already known or described.

See also the Remarks following VII.1.9(2).

(v) In II.3.2 we have discussed the concept of solving a condition, expressed by a sentence $\underline{\underline{A}}$, relative to a letter $\underline{x}$. Linked with this is the idea of "solving completely" relative to $\underline{x}$, or of finding the "complete solution" relative to $\underline{x}$, of $\underline{\underline{A}}$: this will be understood to signify describing completely (as in (iv)) the solution set $\{\underline{x} : \underline{\underline{A}}\}$. In particular, to find the (or a) complete solution in X relative to $\underline{x}$ of the equation $T = U$ (notation as in II.3.2), is to exhibit a known set E and a (preferably injective) function f with domain equal to E and range equal to $\{\underline{x} \in X : T = U\}$.

Both (iv) and (v) are illustrated by the following examples, which will arise naturally in due course.

(a) Suppose that $\underline{\underline{A}}$ denotes

$$(\underline{x} \text{ is a continuous function } \mathcal{R} \to \mathcal{R}) \wedge (\underline{x} \neq \underset{\sim}{0}_{\mathcal{R}})$$

$$\wedge (\forall \underline{y})(\forall \underline{z})((\underline{y} \in \mathcal{R} \wedge \underline{z} \in \mathcal{R}) \Rightarrow (\underline{x}(\underline{y} + \underline{z}) = \underline{x}(\underline{y})\cdot\underline{x}(\underline{z})) \quad ,$$

where $\underline{x}$, $\underline{y}$, $\underline{z}$ denote distinct letters not appearing in $\mathcal{R}$, 0 , $+$, $\cdot$. It will later emerge (see VIII.3.4(4)) that a complete solution relative to $\underline{x}$ of $\underline{\underline{A}}$ is obtained by taking $E \equiv]0, \to[$ (the set of all positive real numbers) and f the function

$$\underline{a} \rightsquigarrow (\text{the exponential function with base } \underline{a})$$

with domain E . (Here, and in (b) and (c) to follow, we make use of the notation $\rightsquigarrow$ explained in IV.2.2 below. Regarding the exponential function with base a , see VIII.3.2.)

(b) Suppose that $\underline{\underline{A}}$ denotes

$$(\underline{x} \text{ is a twice differentiable function } \mathcal{R} \to \mathcal{R})$$

$$\wedge (\underline{x}'' = - \underline{x}) \quad ,$$

where $\underline{x}$ denotes a letter not appearing in $\mathcal{R}$, $+$, $\cdot$. Then (see XIV.2.3) a complete solution of $\underline{\underline{A}}$ relative to $\underline{x}$ is obtained by taking $E \equiv \mathcal{R} \times \mathcal{R}$ and f the function

$$\underline{t} \rightsquigarrow (\mathrm{pr}_1\ \underline{t})\cdot\cos + (\mathrm{pr}_2\ \underline{t})\cdot\sin$$

with domain $\mathcal{R} \times \mathcal{R}$, where $\underline{t}$ is a letter not appearing in cos or sin (the trigonometric functions cos and sin are as defined in XII.5.1 below).

(c) Suppose that $\underline{\underline{A}}$ denotes the sentence

$$(\underline{x} \text{ is a differentiable function } \mathcal{R} \to \mathcal{R})$$

$$\wedge (\underline{x}' = 3\underline{x}^{2/3}) \quad ,$$

where $\underline{x}$ is as in (b) ; and that S is the corresponding solution set relative to $\underline{x}$. Some hastily written textbooks on differential equations lead their readers to think that a complete description of S , that is, a complete solution of A relative to $\underline{x}$, is obtained by taking $E \equiv \mathcal{R}$ and f the function

$$\underline{c} \rightsquigarrow (j - \underset{\sim}{\underline{c}}_{\mathcal{R}})^3$$

with domain $\mathcal{R}$, where $\underline{c}$ denotes a letter not appearing in $\mathcal{R}$, $+$, $\cdot$, $-$. However (see Problem X/11), this is false: f is indeed an injective function $E \to S$, but Ran f is a proper subset of S .

The term "complete solution" is particularly favoured in the study of differential equations, where it often suffers abuse.

(vi) Regarding the matter of "simplifying" , I can find little which may be usefully said. One reason is that the concept plainly refers to names or name symbols (rather than to mathematical objects), that there is no widely agreed list of these, and that there is no objective criterion of simplicity which is appropriate in all contexts.

While most people would agree that $\frac{1}{2}$ is a simplification of $(2\frac{1}{2} + 1\frac{1}{2})/8$, there are few who are prepared to argue with conviction as to whether $(3 + 1)/8$ and $\int_0^1 j$ are also simplifications, and if so, which is to be preferred to the other in all situations.

IV.2 Comments on the function concept

Although the current set theoretical viewpoint subordinates the concept of function to that of relation, the former has been a working tool in mathematics far longer than has the latter.

IV.2.1 The classical viewpoint It took a long while for the function concept to gain a widely accepted precise meaning. It was to begin with, closely tied to the idea of a "formula" or "rule" , but as late an epoch as the beginning of the nineteenth century witnessed acrimonious debates as to what these terms were to mean. Even today, high school and tertiary level courses can and do incorporate pronouncements typified by the following:

> A function of x is any expression that depends for its value on the value of x and is usually written $f(x)$ or $g(x)$ or $\ell(x)$.

In books treating mathematics as a service subject one may encounter somewhat different but equally vague and misleading descriptions of the function concept. The following example appears as a footnote on page 32 of Mair(1) :

> ..... to say that x is a function of y means that the relation between x and y is constant, so that when y changes, x changes.

The idea of a function as a "formula" , "rule" , or "operation" which assigns to each member of the domain X of that function a uniquely defined element, is due largely to Dirichlet (c. 1830) and remained the generally accepted notion throughout the remainder of the nineteenth century, which saw an enormous development of "function theory" (the study of special functions and important restricted classes of functions). It was not until the foundations of mathematics began to be scrutinised anew, that the current ideas began to emerge. The new ideas emerged and took the form they did, not as the result of any widespread bias in favour of logic and sets, but rather because that particular framework appeared to offer more clarity in the basic concepts.

The classical viewpoint, which tends to be suppressed in IV.1.1, is good by current standards only if one emphasises strongly that the domain to which the "rule" is to be applied must be regarded as an essential component of the function in question. One and the same rule (et cetera) can generate many different functions corresponding to different choices of domain. At the same time, apparently different rules applied to the same domain can generate the same function. For example, the two rules $\underline{y} = 0$ and $\underline{y} = \sin \pi\underline{x}$ generate the same function with domain the set of integers.

Spivak (1), p.39 writes (in connection with functions which are subsets of $R \times R$):

> unless the domain is explicitly restricted further, it is understood to consist of all numbers for which the definition (that is, the rule) makes any sense at all.

(The parenthesised insertion is mine.) The danger hidden in this is obvious: a rule may at any time and by any person be endowed with meaning or sense in cases where it hitherto had none.

The next step is to fit the "domain plus rule" concept of function into the formal system and in harmony with the definition in IV.1.3.

IV.2.2 The "domain plus rule" generation of functions and the $\rightsquigarrow$ notation In this subsection the aim is to examine from a formal point of view the "domain plus rule" generation of functions mentioned in IV.2.1.

There seems little doubt that the informal concept of "domain" is going to be subsumed under that of "set" in the formal theory. But to what is "rule" to correspond? As a plausible tentative step, one might agree that "rule" is to correspond to a choice of a set (different sets for different rules) T and a letter $\underline{x}$, which will usually appear in T, the corresponding rule being that which assigns to an arbitrary set a the "value" $(a|\underline{x})T$.

This leads one to hope that, if X and T are sets and $\underline{x}$ a letter (usually appearing in T), there will exist a function f such that Dom f = X is true and such that, for an arbitrary set a,

$$a \in X \Rightarrow f(a) = (a|\underline{x})T$$

is true (that is, that a certain sentence schema will be a theorem schema). If the letter $\underline{x}$ does not appear in X or f, this last will be secured as soon as the sentence

$$\underline{x} \in X \Rightarrow f(\underline{x}) = T$$

is true; or, what is the same thing by (X) in I.3.2, as soon as the sentence

$$(\forall\underline{x})(\underline{x} \in X \Rightarrow f(\underline{x}) = T)$$

is true. (The desired theorem schema then follows via (II) or (XI) in I.3.2, combined with the appropriate replacement rules.)

One can reduce oneself to the case in which $\underline{x}$ does not appear in X by the device of choosing a letter $\underline{x}'$ not appearing in X or T and seeking a function f, in which $\underline{x}'$ does not appear, such that Dom f = X and

$$\underline{x}' \in X \Rightarrow f(\underline{x}') = (\underline{x}'|\underline{x})T$$

is true, If this can be done, (II) in I.3.2 will yield

$$a \in X \Rightarrow f(a) = (a|\underline{x}')(\underline{x}'|\underline{x})T \equiv (a|\underline{x})T \quad ,$$

the last step by I.1.4(a).

Proceeding, then, on the assumption that $\underline{x}$ does not appear in X, it is not difficult to conjecture that if one chooses a letter $\underline{z}$, different from $\underline{x}$ and not appearing in X or T, the set

$$f \equiv \{\underline{z} : (\exists \underline{x})(\underline{x} \in X \wedge \underline{z} = (\underline{x}, T))\}$$

may well answer all requirements. (In the notation discussed in II.12.1 and §3 of the Appendix, f is the set of ordered pairs of the form $(\underline{x}, T)$ where $\underline{x} \in X$, usually denoted by $\{(\underline{x}, T) : \underline{x} \in X\}$.) Indeed, it can be shown (see equation (7) in §3 of the Appendix) to follow from the theorem schema (XIV) in II.12.1 that the sentence

$$(\exists \underline{x})(\underline{x} \in X \wedge \underline{z} = (\underline{x}, T))$$

is collectivising in $\underline{z}$. (This makes use of the assumptions that $\underline{x}$ does not appear in X and that $\underline{z}$ is different from $\underline{x}$ and does not appear in X or T.) Furthermore, there is little difficulty in proving the theorem schema

$$(f \text{ is a function}) \wedge (\text{Dom } f = X) \quad ;$$

see Problem IV/11. The remaining essential requirement is expressed by

$$T \quad : \qquad (\forall \underline{x})(\underline{x} \in X \Rightarrow f(\underline{x}) = T) \tag{1}$$

and will be proved in a moment. Meanwhile, it should be noted that the choice of $\underline{z}$, subject to the stated conditions, has no influence on f. (To verify this, make routine use of the appropriate replacement rules, especially II.3.4(q).)

All this is summarised in the

<u>Metatheorem</u> Suppose that X and T denote arbitrary sets and $\underline{x}$ a letter not appearing in X and define

$$f \equiv \{(\underline{x}, T) : \underline{x} \in X\} \equiv \{\underline{z} : (\exists \underline{x})(\underline{x} \in X \wedge \underline{z} = (\underline{x}, T))\} \quad , \qquad (1_0)$$

where $\underline{z}$ denotes any letter different from $\underline{x}$ and not appearing in X or T . Then

$$T! \quad : \qquad (\mathrm{Fn}\ f) \wedge (\mathrm{Dom}\ f = X) \wedge (\forall \underline{x})(\underline{x} \in X \Rightarrow f(\underline{x}) = T) \qquad (1')$$

and

$$T! \quad : \qquad (a \in X) \Rightarrow (f(a) = (a|\underline{x})T) \qquad (1'')$$

for an arbitrary set a .

Notice that, as a corollary of IV.1.5(9), there is (up to equality) at most one set f satisfying (1') .

Regarding notation, one may choose to write $T[\![\underline{x}]\!]$ in place of T and $T[\![a]\!]$ in place of $(a|\underline{x})T$ for arbitrary sets a ; see II.1.3(iv). Further, the function f defined by (1_0) (which is the unique f satisfying (1')) is usually described as

the function $\underline{x} \rightsquigarrow T$ (or: $\underline{x} \rightsquigarrow T[\![\underline{x}]\!]$) with domain X

or

the function with domain X and rule $\underline{x} \rightsquigarrow T[\![\underline{x}]\!]$,

and is often denoted by

$$\underline{x} \in X \rightsquigarrow T[\![\underline{x}]\!] \quad ,$$

or

$$\underline{x}(\in X) \rightsquigarrow T[\![\underline{x}]\!] \quad ,$$

or

$$\underline{x} \rightsquigarrow T[\![\underline{x}]\!] \quad \text{for (all)} \quad \underline{x} \in X \quad ;$$

some writers use an ordinary arrow $\rightarrow$ in place of $\rightsquigarrow$, but this invites confusion with other traditional usage of $\rightarrow$, as in Chapter VII for example. The replacement rules show that this function is equally well described by the phrase

the function $\underline{x}' \rightsquigarrow T[\![\underline{x}']\!]$ with domain X ,

where $\underline{x}'$ is any letter not appearing in X or T and $T[\![\underline{x}']\!] \equiv (\underline{x}'|\underline{x})T$.

Frequently

$$f : \underline{x} \rightsquigarrow T[\![\underline{x}]\!] \quad \text{with domain} \quad X$$

is written in place of

f denotes (or is defined to be) the function with rule $\underline{x} \rightsquigarrow T[\![\underline{x}]\!]$ and domain X ,

it being understood that $\underline{x}$ denotes a letter not appearing in X . The definition involved is (see (1_0) above) more explicitly expressible as

$$f \equiv \{(\underline{x}, T) : \underline{x} \in X\} \quad .$$

(A glance at texts on sex education suggests that " $\rightsquigarrow$ " might be read as "spermarrow" or "sparrow" , but this would not be conventional.)

Proof of (1) Let $\underline{y}$ denote a letter different from $\underline{x}$ and $\underline{z}$ and not appearing in X or T (hence not in f). Then, according to IV.1.3(1),

$$f(\underline{x}) \equiv \tau_{\underline{y}}((\underline{x}, \underline{y}) \in f) \quad .$$

The first step will be a proof of

$$T \quad : \qquad (\underline{x}, \underline{y}) \in f \Leftrightarrow \underline{x} \in X \wedge \underline{y} = \mathbb{T}[\![\underline{x}]\!] \quad . \tag{1_1}$$

To this end, choose a letter $\underline{x}'$ different from $\underline{x}$, $\underline{y}$ and $\underline{z}$ and not appearing in X or T (nor, therefore, in f). Since the sentence $(\exists\underline{x})(\underline{x} \in X \wedge \underline{z} = (\underline{x}, T))$ is known to be collectivising in $\underline{z}$, II.3.3(3') implies

$$T \quad : \qquad \underline{z} \in f \Leftrightarrow (\exists\underline{x})(\underline{x} \in X \wedge \underline{z} = (\underline{x}, T)) \quad .$$

On the other hand, by various replacement rules (starting with I.1.4(a)),

$$(\exists\underline{x})(\underline{x} \in X \wedge \underline{z} = (\underline{x}, T)) \equiv (\exists\underline{x}')(\underline{x}'|\underline{x})(.........)$$

$$\equiv (\exists\underline{x}')(\underline{x}' \in X \wedge \underline{z} = (\underline{x}', T')) \quad ,$$

where $T' \equiv (\underline{x}'|\underline{x})T$. Hence

$$T \quad : \qquad \underline{z} \in f \Leftrightarrow (\exists\underline{x}')(\underline{x}' \in X \wedge \underline{z} = (\underline{x}', T'))$$

and therefore, by I.3.2(II) used twice and I.3.3(j),

$$T \quad : \quad (\underline{x}, \underline{y}) \in f \Leftrightarrow ((\underline{x}, \underline{y})|\underline{z})(\exists\underline{x}')(\underline{x}' \in X \wedge \underline{z} = (\underline{x}', T'))$$

$$\equiv (\exists\underline{x}')((\underline{x}, \underline{y})|\underline{z})(\underline{x}' \in X \wedge \underline{z} = (\underline{x}', T')) \qquad \text{(I.1.8(f))}$$

$$\equiv (\exists\underline{x}')(\underline{x}' \in X \wedge (\underline{x}, \underline{y}) = (\underline{x}', T'))$$

$$\Leftrightarrow (\exists\underline{x}')(\underline{x}' \in X \wedge \underline{x}' = \underline{x} \wedge \underline{y}' = T') \qquad \text{(III.1.2(1))}$$

$$\Leftrightarrow \underline{x} \in X \wedge \underline{y} = (\underline{x}|\,\underline{x}')T' \qquad \text{(Problem II/14)}$$

$$\equiv \underline{x} \in X \wedge \underline{y} = T \quad , \qquad \text{(I.1.4(a))}$$

which proves (1_1) .

By (1_1) and Problem II/15 (applied with $\underline{y}$ in place of $\underline{x}$ and $\underline{\underline{A}} \equiv (\underline{x}, \underline{y}) \in f$),

$$T \quad : \qquad (\underline{x}, \underline{y}) \in f \Rightarrow f(\underline{x}) = \underline{y} \quad . \qquad (1_2)$$

By (1_1) and (1_2) and the transitivity of equality,

$$T \quad : \qquad (\underline{x}, \underline{y}) \in f \Rightarrow f(\underline{x}) = T[\![\underline{x}]\!] \quad .$$

Hence, by I.3.3(1) and the fact that $\underline{y}$ does not appear in the sentence $f(\underline{x}) = T[\![\underline{x}]\!]$,

$$T \quad : \qquad (\exists \underline{y})((\underline{x}, \underline{y}) \in f) \Rightarrow f(\underline{x}) = T[\![\underline{x}]\!] \quad ,$$

that is, by III.2.2(4),

$$T \quad : \qquad (\underline{x} \in \mathrm{Dom}\ f) \Rightarrow f(\underline{x}) = T[\![\underline{x}]\!] \quad ,$$

or, by the second clause of (1') (see Problem IV/11),

$$T \quad : \qquad \underline{x} \in X \Rightarrow f(\underline{x}) = T[\![\underline{x}]\!] \quad .$$

Appeal to (X) in I.3.2 now entails (1) . □

IV.2.3 Remarks (i) The procedure described in IV.2.2 is in a sense reversible. Suppose that f is a function and define $X \equiv \mathrm{Dom}\ f$ and

$$T \equiv f(\underline{x}) \equiv \tau_{\underline{y}}((\underline{x}, \underline{y}) \in f)$$

where $\underline{x}$ and $\underline{y}$ denote distinct letters not appearing in f . Then IV.1.5(5) implies

$$f = \{\underline{z} : (\exists \underline{x})(\underline{x} \in X \wedge \underline{z} = (\underline{x}, T[\![\underline{x}]\!]))\} ,$$

so that f is equal to what has been termed above the function $\underline{x} \rightsquigarrow T[\![\underline{x}]\!]$ with domain X .

It is in fact possible to prove that the definition of function given in IV.1.1 (as a special type of set of ordered pairs) is formally equivalent to that expressed by the sentence

$$f = \{\underline{z} : (\exists \underline{x})(\underline{x} \in \mathrm{Dom}\, f \wedge \underline{z} = (\underline{x}, f(\underline{x})))\} ,$$

where $\underline{x}$ and $\underline{z}$ denote distinct letters not appearing in f .

(ii) In conventional mathematics, when one has under consideration sets X , $T \equiv T[\![\underline{x}]\!]$ and a letter $\underline{x}$ not appearing in X , one sometimes introduces a function f in the following terms:

For all (or each, or every) $\underline{x} \in X$, let $f(\underline{x})$ denote (or: denote by $f(\underline{x})$) the set $T[\![\underline{x}]\!]$.

It is then taken for granted that this really does serve to bring into existence a unique function f , or describes completely an already-existent unique function f , with domain X such that $f(\underline{x}) = T[\![\underline{x}]\!]$ for all $\underline{x} \in X$. (Instances appear in the shape of what are - conventionally but rather shabbily - denoted by $u(a)$ and $E(a)$ in VII.4.2 below.) The justification for this lies in the preceding proof of the existence and uniqueness of just such a function; that is, in the metatheorem appearing in IV.2.2.

Note, however, that f and T may be (and usually are) quite different things, as also may be $f(\underline{x})$ ($= T[\![\underline{x}]\!]$) and $T(\underline{x})$. (In connection with $T[\![\underline{x}]\!]$, refer back to II.1.3(iv).) One should never confuse f with T , or $f(\underline{x})$ with $T(\underline{x})$, notwithstanding what appears in texts on informal mathematics. The situation is very similar to that discussed at some length in IV.1.4(ii).

As a simple example, define $0 \equiv \emptyset$, $1 \equiv \{\emptyset\}$ and $T \equiv \{(0, 1)\}$. Then $T[\![\underline{x}]\!] \equiv T \equiv \{(0, 1)\}$ and (making use of various theorems and metatheorems appearing in Chapters I, II, III and preceding portions of the present chapter, which the diligent reader will track down)

$$T(0) \equiv \tau_{\underline{y}}((0, \underline{y}) \in T) = \tau_{\underline{y}}(\underline{y} = 1) = 1 \quad ;$$

since $\{0\} \in 0$ is false, $T(0) = T[\![0]\!]$ is also false and, a fortiori,

$$(\forall\underline{x})(T(\underline{x}) = T[\![\underline{x}]\!])$$

is false.

(iii) One might frame the definition (schema)

$$F_{\underline{x},X,T} \equiv \{\underline{z} : (\exists\underline{x})(\underline{x} \in X \wedge \underline{z} = (\underline{x}, T))\}$$

for an arbitrary letter $\underline{x}$ and arbitrary strings X and T , $\underline{z}$ denoting any letter different from $\underline{x}$ and not appearing in X or T . (Use of I.1.8(f) and II.3.4(q) shows that the choice of $\underline{z}$, subject to these restrictions, is immaterial.) One would then have the theorem schema

$$(\mathrm{Fn}\ F_{\underline{x}, X,T}) \wedge (\mathrm{Dom}\ F_{\underline{x},X,T} = X) \wedge (\forall\underline{x})(\underline{x} \in X \Rightarrow F_{\underline{x},X,T}(\underline{x}) = T)$$

wherein X and T denote arbitrary sets and $\underline{x}$ any letter not appearing in X . When $\underline{x}$ does not appear in X , $F_{\underline{x},X,T}$ is a convenient name for the function $\underline{x} \rightsquigarrow T$ with domain X .

If $\underline{x}'$ denotes a letter not appearing in X or T , I.1.8(e) shows that

$$F_{\underline{x},X,T} \equiv F_{\underline{x}',(\underline{x}'|\underline{x})X,\ (\underline{x}'|\underline{x})T} \quad ,$$

which is identical with $F_{\underline{x}',X,(\underline{x}'|\underline{x})T}$ whenever $\underline{x}$ does not appear in X .

The notation $F_{\underline{x},X,T}$ is not conventional. If it is used, it is necessary to bear in mind that $\underline{x}$ does not appear in (the string denoted by) $F_{\underline{x},X,T}$; the situation is akin to that with $\tau_{\underline{x}}(A)$ and $(\exists\underline{x})A$, et cetera. If $\underline{x}$ and $\underline{y}$ denote distinct letters not appearing in X ,

$$(\underline{y}|\underline{x})F_{\underline{x},X,\underline{y}\backslash\underline{x}}$$

is identical with

$$F_{\underline{x},X,\underline{y}\backslash\underline{x}}$$

rather than with (the usually quite different)

$$F_{\underline{y},X,\underline{y}\backslash\underline{y}} \quad ;$$

thus, $F_{\underline{x},\{\emptyset\},\underline{y}\backslash\underline{x}} = \{(\emptyset, \underline{y})\}$ and $F_{\underline{y},\{\emptyset\},\underline{y}\backslash\underline{y}} = \{(\emptyset, \emptyset)\}$.

(Recall here the final paragraph but one in I.1.4: formal replacement for a letter in a string is not to be confused with replacement for that letter in a name for that string.)

(iv) Intuitively one tends to think of the union sign $\bigcup$ as denoting a function with rule $\underline{x} \rightsquigarrow \bigcup\underline{x}$. However, this is formally impermissible (at least in the formalism adopted in this book); cf. the discussion in III.2.3(ii) referring to the "relation of equality" .

More forcefully and precisely, if $\underline{x}$, $\underline{y}$ and f denote distinct letters,

$$T : \neg(\exists\underline{f})((\forall\underline{x})(\forall\underline{y})(\underline{y} = \bigcup\underline{x} \Rightarrow (\underline{x}, \underline{y}) \in \underline{f})) \quad . \qquad (2)$$

Here is a sketch proof of (2) . Assume the contrary (that is, adjoin the negation of (2) as an axiom) and define

$$f \equiv \tau_{\underline{f}}((\forall\underline{x})(\forall\underline{y})(.......)) \quad .$$

Then the following would be theorems (recall III.1.2(7))

$$(\forall\underline{x})((\underline{x}, \bigcup\underline{x}) \in \underline{f})$$

$$(\forall\underline{x})(\underline{x} \in \bigcup\bigcup f) \quad .$$

Thus, since $\underline{x}$ does not appear in f , $\bigcup\bigcup f$ would be the set of all sets, a contradiction (see II.3.10). Now appeal to (VII) in I.3.2.

Similar remarks apply with $\bigcap$ or the $\mathcal{P}$ in $\mathcal{P}(\ )$ replacing $\bigcup$.

IV.2.4 <u>Discussion</u> (i) The $\rightsquigarrow$ notation introduced in IV.2.2 above is a very useful addition to the semiformal vocabulary. It can be applied immediately to cases in which, for example, $T[\![\underline{x}]\!]$ is $\underline{x}$, $\underline{x} \cup a$, $\underline{x} \cap a$, $\underline{x} \times a$,

$(\underline{x}, a)$, et cetera, where a denotes a given set (in which the letter $\underline{x}$ may or may not appear). Other instances appear in IV.3.3 and IV.3.9. Many further examples appear in IV.4.4. The device is easy to use, and is widely used in informal mathematics.

(ii) In spite of (i) , if formalities are to be respected, some care is needed in the use of the $\rightsquigarrow$ notation. There is a tendency to make use of the notation in situations where closer inspection reveals unexpected complications.

A common fly in the ointment is that what one wants to use in the role of T may not, at the time one wants to use it, have been unconditionally defined for all $\underline{x}$, that is, may not have been properly presented as a set (term of the formal theory). To illustrate, consider an instance as apparently simple as that of introducing the function $\underline{x} \rightsquigarrow \underline{x}^2$ with domain N . Before this is formally justifiable one needs an unconditional definition of $\underline{x}^2$ (it is not adequate to specify the meaning of $\underline{x}^2$ subject to the condition $\underline{x} \in N$, say). If the present approach is to be justified, there seems no way out but to either rely on the substance of V.6, according to which there is a product function $\pi_N : N \times N \rightsquigarrow N$ in terms of which one may unconditionally define

$$\underline{x}^2 \equiv \pi_N((\underline{x}, \underline{x})) ,$$

or to rely on definition by recurrence; see IV.8 and V.5. In either case, more remains to be done that is at first apparent.

This example may leave the reader with the feeling that it is rare indeed to obtain something for nothing (in which he would not be wrong).

Similar remarks apply in the case of the function $\underline{x} \rightsquigarrow \underline{x}^2$ with domain R . See also IV.4.4 and VI.8.

(iii) At the risk of labouring the point, it is to be stressed again that to write

$$f(\underline{x}) = T[\![\underline{x}]\!]$$

or

$$f : \underline{x} \rightsquigarrow T[\![\underline{x}]\!]$$

does not serve to define a function f , until and unless a set X is also specified as a domain. Briefly: a function amounts to a rule ($\underline{x} \rightsquigarrow T[\![\underline{x}]\!]$), together with a set X (the domain).

In spite of this, high school teachers are sometimes subjected to official syllabus notes urging them either to adopt the "set of ordered pairs" definition of a function, or at least to stress the point that the specification of a domain is a vital part of that of a function, only to have their students later confronted with examination questions of the type (I quote):

(a) If $f(x) = \log_e [x + \sqrt{(x^2 - 1)}]$, what is the largest possible domain of the function f ?

(b) What is the natural (that is, largest possible) domain of the function $\log_e x$?

(c) Consider the function defined for all x by

$f(x) = 2 \cdot \sin(\pi x)$ ....

See also Spivak (1), p.46, Problem 3.

Regarding (a) and (b), it is not clear that any one function is in fact specified; in (b) the situation would change if $\log_e$ appeared in place of $\log_e x$. In any case, as soon as some one function f is specified, to ask

what is the largest possible (or natural) domain of f ?

is akin to enquiring after the largest possible (or natural) length of a metre rule, or the largest possible (or natural) number of sides possessed by a hexagon (previously defined as a six-sided figure).

It would be more correct and sensible to frame questions of the following type:

Say that s is a square root function if and only if s is a function, $s \subseteq R \times R$, and $s(\underline{x})^2 = \underline{x}$ for all $\underline{x} \in \text{Dom } s$. What is the largest of the domains of square root functions?

or again

Say that ℓ is a logarithmic function if and only if ℓ is a function, $\ell \subseteq R \times R$, and $\exp \ell(\underline{x}) = \underline{x}$ for all $\underline{x} \in \text{Dom } \ell$. What is the largest of the domains of logarithmic functions?

See VI.8.1 and Problems VIII/16 and XII/25.

Concerning (c) (and possibly (a) too), I can only say that I am unable to exhibit (or prove the existence of) a single function f such that $f(\underline{x}) = T[\![\underline{x}]\!]$ for all $\underline{x}$. Of course, the context often makes it virtually certain that "for all $\underline{x}$ " is intended to mean "for all $\underline{x} \in R$ " or "for all real numbers $\underline{x}$ " ; but there seems no excuse for leaving this to guesswork.

It is in any case certain (on the basis of II.3.10) that in our theory (assumed to be consistent) there is no function f such that, for all $\underline{x}$, $\underline{x} \in \text{Dom } f$ and $f(\underline{x}) = T[\![\underline{x}]\!]$. (If there were, Dom f would be the non-existent "set of all sets" . Remember yet again the substance of II.3.9(v).)

(iv) Although there is precisely one function f with domain X such that $f(\underline{x}) = T[\![\underline{x}]\!]$ for all $\underline{x} \in X$ ($\underline{x}$ denoting a letter not appearing in X), there are many functions f such that $f(\underline{x}) = T[\![\underline{x}]\!]$ for all $\underline{x} \in X$. Among such functions there are, for example, all functions

$$f \equiv \{\underline{z} : (\exists \underline{x})(\underline{x} \in X \wedge \underline{z} = (\underline{x}, T[\![\underline{x}]\!]))\} \cup g ,$$

where g denotes any function with a domain disjoint from X ; see Problem IV/12.

(v) A useful extension of the $\rightsquigarrow$ notation is as follows. Suppose that X , Y and T denote sets and that $\underline{x}$, $\underline{y}$ denote distinct letters not appearing in X or Y , though they may appear in T . One then often speaks of the function

$$(\underline{x}, \underline{y}) \rightsquigarrow T \text{ with domain } X \times Y , \tag{1}$$

or of the function

$$(\underline{x}, \underline{y})(\in X \times Y) \rightsquigarrow T ,$$

denoting thereby the function

$$\underline{z} \rightsquigarrow (\mathrm{pr}_1\, \underline{z}|\underline{x}, \mathrm{pr}_2\, \underline{z}|\underline{y})T \text{ with domain } X \times Y ,$$

where $\underline{z}$ denotes a letter different from $\underline{x}$ and $\underline{y}$ and not appearing in X , Y or T . The function thus referred to is thus identical with

$$\{\underline{t} : (\exists \underline{z})(\underline{z} \in X \times Y \wedge \underline{t} = (\underline{z}, T^*))\} ,$$

where $\underline{z}$ is as specified above,

$$T^* \equiv (\mathrm{pr}_1\, \underline{z}|\underline{x}, \mathrm{pr}_2\, \underline{z}|\underline{y})T ,$$

and $\underline{t}$ denotes a letter different from $\underline{z}$ and not appearing in X , Y or T^* . Observe that (by (*) in II.1.3(ii))

$$T^* \equiv (\mathrm{pr}_1\ \underline{z}|\underline{x}')(\mathrm{pr}_2\ \underline{z}|\underline{y}')(\underline{x}'|\underline{x})(\underline{y}'|\underline{y})T \quad ,$$

where $\underline{x}'$ and $\underline{y}'$ denote distinct letters, different from $\underline{x}$ and $\underline{y}$ and not appearing in T .

It can be proved (see Problem IV/34) that the function (1) is equal to

$$\{\underline{t} : (\exists \underline{x})(\exists \underline{y})(\underline{x} \in X \wedge \underline{y} \in Y \wedge \underline{t} = ((\underline{x}, \underline{y}), T))\} \quad , \tag{2}$$

which might (cf. II.12.1(2) and §3 of the Appendix) be otherwise denoted by

$$\{((\underline{x}, \underline{y}), T) : \underline{x} \in X \wedge \underline{y} \in Y\} \quad . \tag{3}$$

IV.3 Examples

At this point, a digression is made from the formal development in order to present some examples of functions and to illustrate the use of the domain plus rule definition of functions.

As in III.2.3, the presentation of some of the following examples involves a disregard of proper logical order; the proper place for many of them is subsequent to Chapter VI. In view of this, the reader should not expect that all the details of all the examples will be attended to here in formally satisfactory fashion. Many of the definitions connected with the examples are based upon others which are to be made in Chapters V and VI. After studying the later sections in this chapter, together with Chapters V and VI, the reader himself should be able to repair the unsatisfactory features tolerated in this section.

In the following examples, $\underline{x}$, $\underline{y}$, $\underline{z}$, $\underline{t}$ will (in the absence of other indications) denote distinct letters not appearing in R , $+$, $\cdot$, $<$; cf. VI.5.3 below.

IV.3.1 Identity functions Much of the work at high school and early tertiary levels is concerned with functions which are real-valued (that is, having ranges which are subsets of R) and whose domains are fairly simple subsets of R . Typical such subsets are

$$R_+ = \{\underline{x} \in R : \underline{x} \geq 0\} ,$$

$$\dot{R} = \{\underline{x} \in R : \underline{x} \neq 0\} ,$$

$$\text{an open interval }]a, b[= \{\underline{x} \in R : a < \underline{x} < b\} ,$$

$$\text{a closed interval } [a, b] = \{\underline{x} \in R : a \leq \underline{x} \leq b\} ,$$

the letter $\underline{x}$ being supposed not to appear in R , $<$, a or b ; see II.13 and VI.6.11. In view of this, it is natural to introduce and foster familiarity with, a special symbol for I_R (see III.2.3(ii)), which is a function with domain and range equal to R . In this book, the symbol j will play this role:

$$j \equiv_{\text{def}} I_R .$$

Note that accordingly j denotes, not a letter, but rather a set which happens to be a function $R \to R$.

More generally, if X is any set, the relation I_X is a function with domain and range each equal to X . It is, furthermore, the function $\underline{x} \rightsquigarrow \underline{x}$ with domain X , $\underline{x}$ denoting a letter not appearing in X .

IV.3.2 Constant functions and characteristic functions If X and c denote sets, the set

$$\underset{\sim}{c}_X \equiv_{\text{def}} \{\underline{z} : (\exists\underline{x})(\underline{x} \in X \wedge \underline{z} = (\underline{x}, c))\} = X \times \{c\} ,$$

where $\underline{x}$ and $\underline{z}$ denote letters not appearing in X or c , is a function. It is the function $\underline{x} \rightsquigarrow c$ with domain X and is called the constant function c

<u>with domain</u> X . (Notice that c is not necessarily a letter.) It is bad policy at this very early stage to denote this function by $\underset{\sim}{c}$, omitting explicit reference to the domain X ; and even worse to denote it by c . At this point, see Problem IV/36.

Note that this example runs counter to the definition of a function which begins "If one quantity varies with another, ..." . It also makes nonsense of a statement such as

> $y = x^2 - 5x + 1$ is a function of x , <u>because its value depends on the replacement used for</u> x

(my italics).

If X and Y are sets, the set

$$\chi_{X,Y} \equiv_{\text{def}} (X \times \{1\}) \cup ((Y \setminus X) \times \{0\}) \quad ,$$

wherein 0 and 1 denote the additive and multiplicative neutral elements of R , as named in VI.5.3, is a function called the <u>characteristic function of</u> X <u>relative to</u> Y , although usually this name is applied only when X is a subset of Y . In discussions where Y is understood, one often speaks simply of "the characteristic function of X " and uses the shortened name χ_X . (Other writers sometimes use $\phi_{X,Y}$ or $1_{X,Y}$ in place of $\chi_{X,Y}$.)

The function $\chi_{X,Y}$ has domain equal to $X \cup Y$ and range equal to a subset of $\{0, 1\}$. Moreover,

$$(\forall \underline{x})(\underline{x} \in X \Rightarrow \chi_{X,Y}(\underline{x}) = 1) \quad ,$$

$$(\forall \underline{x})(\underline{x} \in Y \setminus X \Rightarrow \chi_{X,Y}(\underline{x}) = 0)$$

and

$$\chi_{X,Y} = \underset{\sim}{1}_X \cup \underset{\sim}{0}_{Y \setminus X}$$

are theorems ($\underline{x}$ denoting a letter not appearing in X or Y); see Problem IV/10.

IV.3.3 Pointwise sums and products, et cetera The display of further examples of functions may be made to depend on general processes making use of the familiar algebraic operations on R .

In general, if c is a real number and f and g are functions: $X \to R$, functions cf or $c \cdot f$, $f + g$ and $f \cdot g$ or $fg : X \to R$ can be defined (see IV.2) as having domain X and rules $\underline{x} \rightsquigarrow c \cdot f(\underline{x})$, $\underline{x} \rightsquigarrow f(\underline{x}) + g(\underline{x})$ and $\underline{x} \rightsquigarrow f(\underline{x}) \cdot g(\underline{x})$ respectively. At greater length:

$$cf \equiv_{def} \{\underline{z} : (\exists \underline{x})(\underline{x} \in X \wedge \underline{z} = (\underline{x}, c \cdot f(\underline{x}))\} ,$$

$$f + g \equiv_{def} \underline{z} : (\exists \underline{x})(\underline{x} \in X \wedge \underline{z} = (\underline{x}, f(\underline{x}) + g(\underline{x})))\} ,$$

$$fg \equiv_{def} \{\underline{z} : (\exists \underline{x})(\underline{x} \in X \wedge \underline{z} = (\underline{x}, f(\underline{x}) \cdot g(\underline{x}))\} ,$$

where $\underline{x}$ and $\underline{z}$ denote distinct letters not appearing in c , f , g , X , R , + , $\cdot$. (All this presupposes definitions of + and $\cdot$ on the right as binary operations on R ; see the Remarks below.) The functions $f + g$ and fg thus defined are usually termed respectively the pointwise sum and pointwise product of f and g . Compare the discussion in Gleason (1), p.199, where complex-valued functions are considered. Concerning the notation for the pointwise product, see Remark (ii) at the end of III.2.7.

It is simple to prove that $cf = \underset{\sim}{c}_X \cdot f$ for all $c \in R$ and all real valued functions f with domain X .

If further $0 \notin \text{Ran } f$, it is standard procedure to define the pointwise inverse or reciprocal function $1/f$ or f^{-1} as that with domain X and rule $\underline{x} \rightsquigarrow 1/f(\underline{x})$ (the multiplicative inverse in R of $f(\underline{x})$; see VI.3.2.)

Furthermore, if $f : X \to R$ and n is a positive integer, $f^n : X \to R$ is defined recursively by writing $f^1 = f$ and $f^{n+1} = f \cdot f^n$ $(n = 1, 2, \ldots)$.

(This rests on the substance of V.5 and the subsequent identification of N with a subset of Z discussed in VI.5.3 and VI.5.4.) A similar procedure is used to define f^n in case n is a positive natural number. If also $0 \notin \text{Ran } f$ and n is a negative integer, $f^n : X \to R$ is defined to be $(1/f)^{(-n)}$. Then $f^{-1} = 1/f$ and $ff^{-1} = f^{-1}f = \underset{\sim}{1}_X$.

To complete the definition of nonnegative integer (or natural number) powers of f, if $f : X \to R$, one defines f^0 to be the constant function $\underset{\sim}{1}_X$. (In spite of the difficulty in attaching a universally convenient fixed meaning to the symbol 0^0, the preceding definition of f^0 seems on the whole to be as satisfactory as any.)

If $0 \in \text{Ran } f$ and n is a negative integer, f^n will be defined to be $(f \S (\text{Inv } f)\langle \dot{R} \rangle)^n$, in which case $ff^{-1} = f^{-1}f = \underset{\sim}{1}_S$, where $S = \text{Dom } f \cap (\text{Inv } f)\langle \dot{R} \rangle$. (If $0 \notin \text{Ran } f$, $(\text{Inv } f)\langle \dot{R} \rangle = \text{Dom } f$, and the present definition is workable and consistent with the earlier one.)

<u>The pointwise difference</u> $f - g$ may be defined to be $f + ((-1)g)$; and, if $0 \notin \text{Ran } g$, the <u>pointwise quotient</u> may be defined to be $f \cdot (1/g)$.

It is, of course quite likely that one will wish to form the pointwise sum, product, difference and quotient of two real-valued functions f and g whose domains are possibly different sets, say X and Y: making use of the notations introduced in III.2.4, the natural definitions are as above, with $f \S (X \cap Y)$ and $g \S (X \cap Y)$ in place of f and g respectively ($\S$ being as in III.2.4). (These definitions arrange that $\underset{\sim}{c}_X \cdot f = c \cdot (f \S X)$ for all $c \in R$, all real-valued functions f, and all sets X.)

<u>Remarks</u> In listing the above examples at this point, I have of course looked well ahead and assumed much that has not yet been discussed formally. The definitions appear in routine format and, from a puritanical point of view, leave much to be desired.

In the first place, they have the appearance of making definitions of pointwise sums and products and other things which are overtly conditional in nature; cf. I.3.5(iv). This can be corrected, given enough patience. <u>Some</u> improvements are indicated in IV.4.4(ii) and (iii) and VI.8, but nowhere do I discuss <u>all</u> the details (nor, I think, does anybody else!).

In the second place (and ignoring the fact that the definitions come out of logical order for the sake of illustration), the notations $f + g$ and fg involve a hideous clash with the use of $+$ and $\cdot$ to denote functions $R \times R \to R$. Thus

$f + g$ could well be (indeed: ought to be) understood to denote $\tau_{\underline{z}}(((f, g), z) \in +)$, wherein f and g are arbitrary sets in which the letter z does not appear; cf. IV.1.3. This, however, is not what is intended in the preceding definitions. In this connection, the attempted defence is usually that the intended meaning is clear from the context; but this parry is effective, only if one has already fallen from grace by adopting conditional definitions of the sort:

> If x and y are (or denote) real numbers, then $x + y = \ldots\ldots$

while

> If x and y are real-valued functions on X , then $x + y = \ldots\ldots$.

(What is illustrated here recurs frequently: one formal solecism leads to, and may even appear to demand and legitimise, still more.)

I can think of no way out of this contretemps which does not involve a good deal of labour and complexification of notations.

IV.3.4 Polynomial functions on R Starting from j (see IV.3.1) and the constant functions (see IV.3.2), repeated applications of pointwise addition and multiplication (see IV.3.3) leads to functions $R \to R$ of the form

$$p \equiv c_0\underset{\sim}{1} + c_1 j + \ldots + c_n j^n \tag{1}$$

where n is a natural number and $c_0, \ldots, c_n$ are real numbers: such functions are termed polynomial functions on R (with real coefficients).

Very frequently, the term "polynomial" is used in place of "polynomial function" : in the instance considered here (where the domain is an infinite field R), there is in practice no ultimate harm in doing this, but the situation is otherwise if, as is possible, the domain is a finite field. In the latter case it would be vital to distinguish between a polynomial (form) in the algebraic sense (for which see VI.9 below; Mulhall and Smith-White (1_3), Chapter V ; Beaumont and Pierce (1), Chapter 9; Sawyer (1), Chapter 2; Godement (1), §26) and the function it defines. This type of ambiguity would also arise if, in place of the polynomial function $p : R \to R$, one had occasion to work with the restriction of p to certain finite subsets of R .

Many specific examples of polynomials may be discussed. In doing this, the following points may be brought out. If, in (1) , $n = 0$, p is the constant function $(c_0)_{\sim R}$. If $n > 0$ and $c_n \neq 0$, two essentially different cases arise accordingly as n is odd or even.

If n is even, Ran p will be a set of the form $\{\underset{\sim}{y} \in R : \underset{\sim}{y} \geq a\} = [a, \rightarrow[$ if $c_n > 0$, or of the form $\{\underset{\sim}{y} \in R : \underset{\sim}{y} \leq a\} =]\leftarrow, a]$ if $c_n < 0$, where in each case a is a certain real number depending upon p; 0 may or may not belong to Ran p. (The letter $\underset{\sim}{y}$ is assumed not to appear in R, $\leq$, $\geq$ or a.)

If n is odd and $c_n \neq 0$, then Ran $p = R$.

These statements cannot be proved at this stage, though they are easily verified in a number of special cases, the simplest of which are the monomial functions j^n, where n is a natural number. See Problem VIII.2.6(2).

IV.3.5 Rational functions on R Pointwise sums and products of polynomial functions are again polynomial functions. However, if $q \neq \underset{\sim}{0}$ is a nonconstant polynomial function, then q^{-1} is not the restriction to (Inv q)$\langle \dot{R} \rangle$ of any polynomial function.

More generally, if p and $q \neq \underset{\sim}{0}$ are polynomial functions, one can form the pointwise quotient function $p/q = pq^{-1}$ having (Inv q)$\langle \dot{R} \rangle$ as its domain: functions obtainable in this way are termed rational functions.

One point has to be watched rather carefully: if $r \neq \underset{\sim}{0}$ is a polynomial function, the function $(pr)/(qr)$ has sometimes to be distinguished from p/q. This is because its domain, namely (Inv (qr))$\langle \dot{R} \rangle$ = (Inv q)$\langle \dot{R} \rangle \cap$ (Inv r)$\langle \dot{R} \rangle$, is sometimes a proper subset of the domain (Inv q)$\langle \dot{R} \rangle$ of p/q. In other words, $(pr)/(qr)$ will sometimes be a restriction of p/q. Whatever grounds there are for confusion of the two functions rest on the fact that $(pr)/(qr)$ can be extended by continuity so as to become equal to p/q; but this extension procedure cannot be effectively handled, and considerable caution is therefore necessary, at the present stage.

For comments on the algebraic background, see (for example) Godement (1), §29.

IV.3.6 Absolute value, positive part and integer part functions All examples appearing thus far conform tolerably well with the concept of function as it existed around the beginning of the nineteenth century, inasmuch as they are definable (at least roughly) by single algebraic-looking rules. To break the monotony, here are a few examples which are not so easily describable in this fashion.

The absolute value function with domain R, denoted by $|j|$, may be defined to be

$$\{(\underline{x}, \underline{x}) : \underline{x} \in R \wedge \underline{x} \geq 0\} \cup \{(\underline{x}, -\underline{x}) : \underline{x} \in R \wedge \underline{x} < 0\} .$$

The function $|j|$ then has range equal to R_+.

An equal (or equivalent) definition is as the function

$$\underline{x} \rightsquigarrow |\underline{x}| \text{ with domain } R ,$$

where $|\underline{x}|$ is formally defined as in VI.6.4 below.

The function $|j|$ is often informally defined by the "two part rule" (see Problem V/12) :

$$|j|(\underline{x}) = \underline{x} \quad \text{if} \quad \underline{x} \in R \quad \text{and} \quad \underline{x} \geq 0 ,$$

$$|j|(\underline{x}) = -\underline{x} \quad \text{if} \quad \underline{x} \in R \quad \text{and} \quad \underline{x} < 0 .$$

Once the square root function has been successfully defined (see IV.4.4(i) and IV.6.2), one may note that $|j|(\underline{x}) = (\underline{x}^2)^{\frac{1}{2}}$ for all $\underline{x} \in R$ and so equivalently define $|j|$ as the composite $f \circ g$, where g is the square function and f the square root function.

The positive part function j^+ may be defined to be $\frac{1}{2}(j + |j|)$; it has domain R and range R_+, and (see VI.6.8 for Max_R)

$$j^+(\underline{x}) = \mathrm{Max}_R\{\underline{x}, 0\} = \tfrac{1}{2}(\underline{x} + |\underline{x}|) \quad \text{for all} \quad \underline{x} \in R .$$

The <u>integer part function</u> $[j]$ is the function with domain R such that, for all $\underline{x} \in R$, $[j](\underline{x})$ = the integer part of $\underline{x}$, that is, the largest integer not exceeding $\underline{x}$; see VI.6.9 below. There is no very simple and natural algebraic or analytic formula which defines $[j]$, but a simple set theoretical definition is provided by

$$[j] = \{\underline{z} : (\exists \underline{n})(\exists \underline{x})(\underline{n} \in Z \wedge \underline{x} \in R \wedge 0 \leq \underline{x} < 1 \wedge \underline{z} = (\underline{n} + \underline{x}, \underline{n}))\}$$

wherein $\underline{z}$, $\underline{n}$, $\underline{x}$ denote distinct letters not appearing in (the sets denoted by) R , Z , $+$, $\leq$, $<$, 0 , 1 (to be formally defined in Chapter VI).

As a somewhat pathological monstrosity which still awaits a standard christening one can cite the function d with domain R such that $d(\underline{x}) = 1$ if $\underline{x} \in R$ is irrational and $d(\underline{x}) = 0$ if $\underline{x} \in R$ is rational, so that $\operatorname{Ran} d = \{0, 1\}$. More precisely:

$$d \equiv \{\underline{z} : (\exists \underline{x})(\underline{x} \in R \wedge \underline{x} \text{ is irrational} \wedge \underline{z} = (\underline{x}, 1))\}$$

$$\cup \{\underline{z} : (\exists \underline{x})(\underline{x} \in R \wedge \underline{x} \text{ is rational} \wedge \underline{z} = (\underline{x}, 0))\} \quad .$$

While the graphs of each of the absolute value, positive part and integer part functions are easily drawn and are useful, d is so discontinuous that drawing a useful graph of it is an impossible task.

A somewhat similar monstrosity is the function d_1 with domain R and specifying formula (or formulae) $d_1(\underline{x}) = 0$ if $\underline{x} \in R$ is irrational or zero and $d_1(\underline{x}) = 1/n$ if $\underline{x} \in \dot{R}$ is rational and has the expression m/n , where m is an integer, n a positive integer, and m and n are coprime. It can be shown that d_1 is discontinuous at each nonzero rational number and continuous at zero and at each irrational number, its graph being virtually as "wild" as that of d . (On the other hand, from some points of view ... that of integration, for example ... d_1 is decidedly less troublesome than is d ; see XI.2.4(iii).)

IV.3.7 Coordinate and polynomial functions on $R \times R$. To add a little variety of another sort, and for use in IV.3.8 to follow, two functions from $R \times R$ into R may be singled out for christening, namely:

$$j_1 \equiv_{def} \{\underline{z} : (\exists \underline{x})(\exists \underline{y})(\underline{x} \in R \wedge \underline{y} \in R \wedge \underline{z} = ((\underline{x}, \underline{y}), \underline{x}))\} ,$$

$$j_2 \equiv_{def} \{\underline{z} : (\exists \underline{x})(\exists \underline{y})(\underline{x} \in R \wedge \underline{y} \in R \wedge \underline{z} = ((\underline{x}, \underline{y}), \underline{y}))\} ;$$

these are the first and second coordinate functions (or maps, or projections) on $R \times R$.

In the notations defined in III.1.3 and IV.2, j_1 and j_2 are equal to the functions $\underline{z} \rightsquigarrow pr_1\, \underline{z}$ and $\underline{z} \rightsquigarrow pr_2\, \underline{z}$ with domain $R \times R$. Each has range equal to R.

In a similar way, one can define first and second coordinate functions $X \times Y \to X$ and $X \times Y \to Y$ for any two sets X and Y; see Problem IV/1.

The principal motive in introducing these coordinate functions on $R \times R$ is to permit the definition of a polynomial function on $R \times R$ as a function of the type (finite sums are discussed in VI.7 below)

$$P = \sum_{r=0}^{n} \sum_{s=0}^{n} c_{rs}\, j_1^r\, j_2^s \tag{2}$$

where n denotes a natural number, the c_{rs} denote real numbers, and where j_1^0 and j_2^0 are understood to denote the constant function $\underset{\sim}{1}$ with domain $R \times R$ (see IV.3.2).

(This is perhaps an especially opportune moment for the reader to glance again at the opening paragraph of III.2.3.)

IV.3.8 Algebraic functions on R. Let P be a polynomial function on $R \times R$, as defined in IV.3.7 immediately above. It is then clear that

$$F \equiv \{(\underline{x}, \underline{y}) : \underline{x} \in R \wedge \underline{y} \in R \wedge P(\underline{x}, \underline{y}) = 0\}$$

$$\equiv \{\underline{z} : (\exists\underline{x})(\exists\underline{y})(\underline{x} \in R \wedge \underline{y} \in R \wedge P(\underline{x}, \underline{y}) = 0 \wedge \underline{z} = (\underline{x}, \underline{y}))\}$$

$\underline{x}$ and $\underline{y}$ denoting distinct letters not appearing in R or P, is a relation in R. (Strictly speaking, $P(\underline{x}, \underline{y})$ ought to be written $P((\underline{x}, \underline{y}))$.) The only case of interest is that in which $P \neq 0_{\underset{\sim}{R} \times R}$, which will be assumed hereafter in this subsection. (This hypothesis is equivalent to the hypothesis that, in IV.3.7(2), at least one of the coefficients c_{rs} is nonzero.)

It was traditional to speak of F as an "algebraic function" - or, rather, of the equation $P(\underline{x}, \underline{y}) = 0$ as defining an "algebraic function" - even though it was clearly recognised that, for a given $\underline{x} \in R$, the equation $P(\underline{x}, \underline{y}) = 0$ may not uniquely determine $\underline{y}$; usually, if this equation has any sollutions $\underline{y}$, it will have several. The traditional way out of this difficulty was to speak of a "many-valued algebraic function" and to concentrate attention on certain so-called "one-valued branches" of the function.

From the point of view adopted in this book one must say simply that sometimes the relation F is not a function, but that suitable restrictions f of F are functions (the domains of which will be various subsets of R). Such a function $f : X(\subseteq R) \to R$ has the property that $P(\underline{x}, f(\underline{x})) = 0$ for all $\underline{x} \in X$. However, not all possible functions f obtained by restricting F will appear as traditional "one-valued branches".

The simplest nontrivial example will illustrate most points. For this one takes for P the polynomial function on $R \times R$ equal to $j_1 - j_2^2$. Assuming that square roots and the square root function $j^{\frac{1}{2}}$ have been added to the repertoire (see IV.4.4 and IV.6.2), one may say that

$$F = \{(\underline{x}, \underline{x}^{\frac{1}{2}}) : \underline{x} \in R_+\} \cup \{(\underline{x}, -\underline{x}^{\frac{1}{2}}) : \underline{x} \in R_+\} ,$$

the domain of F being in this case R_+. (In the above sentence, I am using the abbreviated notation mentioned in III.1.5(i).) It is thus clear that each of the functions $j^{\frac{1}{2}}$ and $-j^{\frac{1}{2}}$ are restrictions of F : these are the traditional "one-valued branches of F" having maximal domains. However, it is clear that for any function $\omega : R_+ \to \{1, -1\}$, $\omega j^{\frac{1}{2}}$ is also a restriction of F which is a function. The traditional one-valued branches are the only such restrictions which are continuous (see VIII.1.1), and it is this continuity which accounts for their privileged status.

It should perhaps be added here that the classical treatment of algebraic functions is almost always carried out in the "complex domain" , that is, one replaces R throughout by C , the set of complex numbers. In this way a much more complete and coherent picture is obtainable.

At this point the reader might also care to ponder the similarities between integers, rational numbers, and algebraic numbers (see VI.5.4 and IX.3.5) and, respectively, polynomial functions, rational functions and algebraic functions. The terminological parallel continues one step further, inasmuch as the classical functions which are not algebraic are often described as "transcendental functions" ; the logarithmic, exponential, hyperbolic, and trigonometric functions and their inverses are the most familiar examples of transcendental functions, but there are many others (the gamma and beta functions, Bessel functions, hypergeometric functions, elliptic functions, Mathieu functions, Whittaker functions, Legendre functions, spherical and ellipsoidal harmonics, sigma, theta and zeta functions, and so on). Beyond these there are unnamed categories of functions with no special names because no special role has yet been assigned to them.

IV.3.9 Projection maps By way of a more abstract example, consider the situation in which X is a set and S is an equivalence relation in X ; see III.2.8. Then the set

$$\{\underline{z} : (\exists \underline{x})(\underline{x} \in X \wedge \underline{z} = (\underline{x}, S((\underline{x}))))\} \quad ,$$

where $\underline{x}$ and $\underline{y}$ denote distinct letters not appearing in X or S , is a function with domain X and range the quotient set X/S ; it is the function $\underline{x} \rightsquigarrow S((\underline{x}))$ with domain X ; see IV.2.2. This function is usually termed the natural (or canonical) mapping or projection of X onto X/S . It would frequently be denoted by $\pi_{X,X/S}$ or some such group of symbols, or by π if X and S are understood. Note that (by Problem III/14)

$$((S \text{ is an equivalence relation in } X) \wedge (\underline{x} \in X \wedge \underline{y} \in X))$$

T! :   (1)

$$\Rightarrow ((\pi(\underline{x}) = \pi(\underline{y})) \Leftrightarrow ((\underline{x}, \underline{y}) \in S)) \quad .$$

The reader should also prove

T! : $$\text{Ran } \pi_{X,X/S} = X/S \quad .$$ (2)

Remark concerning notation In common with many other symbols, " π " is grossly overworked in informal mathematics. It is used to denote any one of many objects in the manner described in IV.3.9; it is used to denote still different objects in IV.4.4(i) and VI.3.2; and for yet another object in XII.5.2. Many other uses are made of it in other contexts. Such multiple aliases are totally at variance with the formal point of view. But it happens all the time in informal mathematics and one has to learn to accept it.

IV.3.10 Functions with domain or range a set of subsets; a theorem of Cantor

(i) Certain portions of mathematics are concerned with real (or complex) valued functions with domain a subset of $P(X)$, X denoting a set which may depend on the context.

An instance of such a function is

$\underline{x} \rightsquigarrow \#\underline{x}$ (the cardinal number of $\underline{x}$) with domain the set of all finite subsets of X ,

where $\underline{x}$ denotes a letter not appearing in X ; see V.7 below. Other instances are the functions

$\underline{x} \rightsquigarrow$ (the right endpoint of $\underline{x}$) with domain the set of all nonvoid bounded intervals in R ,

$\underline{x} \rightsquigarrow$ (the length of $\underline{x}$) with domain the set of all bounded intervals in R ,

$\underline{x} \rightsquigarrow$ (the area of $\underline{x}$) with domain the set of all bounded rectangles in $R \times R$;

see VI.6.11, VIII.2.4, XI.1.1, XI.1.3 and XI.2.2.

Extensions of the first of these functions presents a central problem of set theory; see V.7.5. (Actually, the problem here is to provide a satisfactory definition of the cardinal number of $\underline{x}$ for all sets $\underline{x}$, even though there is no function, in the strictly formal sense, which has all sets as elements of its domain.)

Extensions of the length and area functions, and of other similar functions, are of great interest and form part of what is termed "measure theory" , which is applied in constructing mathematical models of probability theory and in many other areas too.

(ii) Sometimes one is concerned with functions with range a subset of $P(X)$, where again X may vary with the context.

A trivial instance is the function

$$\underline{x} \rightsquigarrow \{\underline{x}\} \text{ with domain } X .$$

A less trivial example is the function

$$\underline{x} \rightsquigarrow S((\underline{x})) \text{ with domain } X ,$$

S denoting a given equivalence relation in X ; see III.2.8. ($\underline{x}$ denotes a letter not appearing in X or S .) Another example is (cf. the introductory remarks to II.6) the function

$\underline{x} \rightsquigarrow$ (the union of the set of all open circular

discs with centre $\underline{x}$ which are subsets
of $]0, 1[\times]0, 1[$) with domain
$R \times R$.

A conscientious reader will concern himself with the formalisation of the rules involved in the definitions expressed above. Thus the bastard phrases

the right endpoint of $\underline{x}$

the length of $\underline{x}$

the area of $\underline{x}$

each require formalisation (that is, expression as a formal and unconditional definition); cf. the general remarks in I.3.5(iv) concerning the formalisation of informal definitions and also VI.6.11 below.

(iii) A theorem of Cantor The examples in (ii) above may provoke the question : given a set X , does there exist a function f with domain X and range equal to (the whole of) $P(X)$? Cantor's theorem shows that the answer is : No, never. This theorem may be informally expressed thus:

Assume that X is a set and that f is a function $X \to P(X)$;
Define

$$S \equiv \{\underline{x} \in X : \underline{x} \notin f(\underline{x})\} .$$

Then $S \in P(X)$ (obviously !) and $S \notin \text{Ran } f$.

Herein $\underline{x}$ denotes a letter not appearing in X or f .

Proof Assume that $S \in \text{Ran } f$. By IV.1.5(8)

$$(\exists\underline{x})(\underline{x} \in X \wedge S = f(\underline{x})) .$$

Define $T \equiv \tau_{\underline{x}}(\underline{x} \in X \wedge S = f(\underline{x}))$; then

$$T \in X \wedge S = f(T) \quad . \tag{3}$$

Two cases arise:

(i) $T \in S$. Then the definition of S entails

$$T \notin f(T)$$

and therefore, by (3) , $T \notin S$, a contradiction.

(ii) $T \notin S$. Then (3) and the definition of S entail

$$T \in f(T)$$

and therefore, by (3) , $T \in S$, again a contradiction.

This completes the proof.

The above proof appears in routine form. More formally, one would make use of the metatheorem (XIII) in II.3.11, taking

$$\underline{\underline{A}} \equiv S \notin \text{Ran } f \quad ,$$

$$\underline{\underline{B}} \equiv T \in S \quad .$$

Alternatively, one might appeal to Problem I/14(iii).

The result (a theorem schema) is due to Cantor and is of great historical and theoretical interest. It implies that, whatever the set X , there is no function whose domain is a subset of X and whose range is equal to (the whole of) $P(X)$; that is, no function f such that $\text{Dom } f \subseteq X$ and $\text{Ran } f = P(X)$.

It is also the major component in the deduction that, for every set X , the cardinal number of $P(X)$ is strictly larger than that of X . (No proof of this assertion will be found in this book, partly because we nowhere discuss a sufficiently general concept of cardinal number; see V.7.5 below.)

It may also be deduced from Cantor's theorem that

$$T \quad : \qquad \neg(\exists \underline{x})(P(\underline{x}) \subseteq \underline{x}) \quad ;$$

in other words, for every set X , there exists a subset Y of X which is not an element of X . See Problem IV/32.

IV.3.11 Restrictions of functions The reader should experience no trouble in proving (see III.2.4(3) and Problem IV/17) that if f is a function and X a set, then $f \S X$ is a function with domain $X \cap \mathrm{Dom}\, f$ and that

$$(\forall \underline{x})(\underline{x} \in X \Rightarrow (f \S X)(\underline{x}) = f(\underline{x})) \quad ,$$

$\underline{x}$ denoting a letter not appearing in f or X .

More generally, every subset of a function is a function.

See also Problem IV/35.

IV.4 Implicit functions and their existence

IV.4.1 Informal discussion Situations similar to those mentioned in IV.3.8, and other more general ones, raise the following sort of question:

(1) Suppose that X and Y are given sets and that it is known that to every $x \in X$ corresponds precisely one object belonging to (element of) Y - call it provisionally α_x for want of a better name - which stands in some preassigned relation to x . Can one be sure that there is a function

$f : X \to Y$ such that $f(x) = \alpha_x$ for every $x \in X$?

The intuitively expected answer is (I suppose) "Yes" , and one anticipates by describing the function f as being that function with domain X which is defined implicitly by the said relation. (It is customary to speak of f as an "implicit function" , though it is the definition or description of f which is implicit.) Yet it is plain that the question is vaguely phrased. (What do the terms "correspond" and "relation" signify in this context?) Much may depend upon the meaning which is read into the question; that is, how it is interpreted in formal terms. I will discuss how precision can be obtained in terms of set theory, and the correspondingly precise answer.

Sometimes (though less frequently) one is concerned with cases in which the uniqueness portion of (1) is eliminated; that is, in which "at least one" is substituted for "precisely one" . Comments on this situation are delayed until IV.4.4(vi).

IV.4.2 Formal discussion; the implicit function theorem schema The meaning of the principal hypothesis of IV.4.1(1) would, upon reflection, seem to be more precisely expressed as follows: There is a sentence $\underline{\underline{A}}$ such that the following sentence is true:

$$\underline{x} \in X \Rightarrow (\text{there exists uniquely } \underline{y})(\underline{y} \in Y \wedge \underline{\underline{A}}) \quad , \tag{2}$$

where $\underline{x}$ and $\underline{y}$ denote distinct letters not appearing in X or Y , though either or both may appear in $\underline{\underline{A}}$. Accordingly (see II.1.3(iv)) I will write $\underline{\underline{A}}[\![\underline{x}, \underline{y}]\!]$ in place of $\underline{\underline{A}}$; still other letters may appear in $\underline{\underline{A}}$, of course. (The sentence $\underline{\underline{A}}$ is expected to express the preassigned relation between $\underline{x}$ and $\underline{y}$ referred to in IV.4.4(1).)

Before further progress can be made, one has to specify what is meant by the phrase "there exists uniquely $\underline{y}$ " appearing in (2) . To this end, if B denotes an arbitrary string and $\underline{y}$ an arbitrary letter, I formally define

$$(\exists_{0,1}\underline{y})B \equiv_{def} (\forall\underline{y}')(\forall\underline{y}'')(((\underline{y}'|\underline{y})B \wedge (\underline{y}''|\underline{y})B) \Rightarrow (\underline{y}' = \underline{y}'')) \quad ,$$

where $\underline{y}'$, $\underline{y}''$ denote distinct letters which are different from $\underline{y}$ and which do not appear in B . (The replacement rules show that the choice of $\underline{y}'$, $\underline{y}''$ is otherwise immaterial.)

An identical definition reads

$$(\exists_{0,1}\underline{y})B \equiv_{def} (\forall\underline{y})(\forall\underline{y}')((B \wedge (\underline{y}'|\underline{y})B) \Rightarrow (\underline{y} = \underline{y}'))$$

wherein $\underline{y}'$ denotes a letter different from $\underline{y}$ and not appearing in B .

If B denotes a sentence, the string thus defined is a sentence usually named

there exists at most one $\underline{y}$ such that B .

I also define

$$(\exists_1\underline{y})B \equiv_{def} (\exists\underline{y})B \wedge (\exists_{0,1}\underline{y})B$$

which, when B denotes a sentence, denotes a sentence usually named

there exists precisely one $\underline{y}$ such that B

or

there exists uniquely $\underline{y}$ such that B .

The conclusion of (2) is understood in this sense. (For equivalent alternative definitions of $(\exists_{0,1}\underline{y})B$ and $(\exists_1\underline{y})B$, see IV.4.3(ii) below.)

The precise meaning of IV.4.1(1) would thus presumably be expressed by

$$(\forall\underline{x})(\underline{x} \in X \Rightarrow (\exists_1\underline{y})(\underline{y} \in Y \wedge A[\![\underline{x}, \underline{y}]\!])) \quad ,$$

where $\underline{x}$ and $\underline{y}$ denote distinct letters not appearing in X or Y . At the same time, the object denoted by α_x must presumably be equal to

$$\tau_{\underline{y}}(\underline{y} \in Y \wedge \underline{\underline{A}}[\![x, \underline{y}]\!]) \quad ,$$

at least whenever $x \in X$ is true.

If there exists a function f as specified in IV.4.1(1), it is clear (from IV.1.5(9)) that the function f referred to is unique.

Thus what is needed can be expressed in formal terms as the following

Implicit function theorem schema

$$(\forall \underline{x})(\underline{x} \in X \Rightarrow (\exists_1 \underline{y})(\underline{y} \in Y \wedge \underline{\underline{A}}[\![\underline{x}, \underline{y}]\!]))$$

$$\text{T! :} \quad \Rightarrow (\exists_1 \underline{f})((\underline{f} \text{ is a function } X \to Y) \wedge (\forall \underline{x})(\forall \underline{y})((\underline{x} \in X \wedge \underline{y} \in Y)$$
$$\Rightarrow (\underline{\underline{A}}[\![\underline{x}, \underline{y}]\!] \Leftrightarrow \underline{y} = \underline{f}(\underline{x}))) \wedge (\forall \underline{x})(\underline{x} \in X$$
$$\Rightarrow (\underline{f}(\underline{x}) = \tau_{\underline{y}}(\underline{y} \in Y \wedge \underline{\underline{A}}[\![\underline{x}, \underline{y}]\!])))) \quad , \tag{3}$$

wherein X and Y denote arbitrary sets, $\underline{\underline{A}}$ an arbitrary sentence, $\underline{x}$, $\underline{y}$, $\underline{f}$ distinct letters such that $\underline{x}$ and $\underline{y}$ do not appear in X or Y (they may appear in $\underline{\underline{A}}$) and $\underline{f}$ appears in none of X , Y , $\underline{\underline{A}}$. (The reader is reminded that, in (3) and again below, $\underline{\underline{A}}[\![S, T]\!] \equiv (S|\underline{x}, T|\underline{y})\underline{\underline{A}}$ for arbitrary strings S and T ; cf. II.1.3(iv).)

The effect of this theorem schema is that, when one seeks to establish the existence of the function with domain X defined implicitly by the "relation" $\underline{y} \in Y \wedge \underline{\underline{A}}[\![\underline{x}, \underline{y}]\!]$, it suffices to prove the hypothesis

$$\underline{\underline{H}} \equiv (\forall \underline{x})(\underline{x} \in X \Rightarrow (\exists_1 \underline{y})(\underline{y} \in Y \wedge \underline{\underline{A}}[\![\underline{x}, \underline{y}]\!]))$$

in whatever theory is relevant at the time.

What is usually spoken of as the (or : an) implicit function theorem aims to announce, in special situations, sufficient conditions under which the hypothesis $\underline{\underline{H}}$ will be true. This is an elaborate matter with many ramifications. For the beginnings of the story, see Bartle and Ionescu Tulcea (1), §7.8 and also X.2.6 below, and continue with Dieudonné (1), pp.265 ff.

Two proofs of the theorem schema (3) will be given.

The first is presented in considerable detail, partly to confirm that quite elaborate theorems can be handled by systematic use of the metatheorems and theorems of I.3.2 and I.3.3, coupled with the theorems of Chapters II and III. (Not every appeal to these metatheorems and theorems is made explicit so that the following proofs are not fully semiformal; see I.3.4(ii).) This proof proceeds from "first principles" , making no explicit appeal to IV.2.2.

The second proof is shorter and less "bare-handed" than the first, since it relies on both IV.2.2 and a further lemma. A routine style proof would be shorter than either, but would not exhibit all the details of the logical structure as clearly. On the other hand, the general lines of the argument would be more easily grasped from a routine proof. The reader should compare the situation with that appearing in V.5.2.

First proof It will be convenient (in both proofs) to write (3) in the form

$$\underline{\underline{H}} \Rightarrow (\exists_1 \underline{f})\underline{\underline{Q}} \quad .$$

As has been noted, the uniqueness of $\underline{f}$ presents no problem; so I will confine myself to proving that

$$T \quad : \qquad \underline{\underline{H}} \Rightarrow (\exists \underline{f})\underline{\underline{Q}} \quad . \qquad (4)$$

To do this, denote by Θ the theory obtained by adjoining to Θ_0 the explicit axiom $\underline{\underline{H}}$, and note that $\underline{x}$, $\underline{y}$ and $\underline{f}$ are variables of Θ . By (VI) in

I.3.2, it suffices to prove that

$$T\Theta \quad : \qquad (\exists \underline{f})\underline{\underline{Q}} \ . \tag{5}$$

As to (5) , it will be deduced from

$$T\Theta \quad : \qquad (F|\underline{f})\underline{\underline{Q}} \ , \tag{6}$$

where

$$F \equiv \{\underline{z} : (\exists \underline{x})(\exists \underline{y})(\underline{x} \in X \wedge \underline{y} \in Y \wedge \underline{\underline{A}}[\![\underline{x}, \underline{y}]\!] \wedge \underline{z} = (\underline{x}, \underline{y}))\} \ , \tag{7}$$

$\underline{z}$ denoting a letter different from $\underline{x}$ and $\underline{y}$ and not appearing in X , Y or $\underline{\underline{A}}$; once (6) is proved, (5) will follow by virtue of "proof by exhibition" (see I.3.4(viii)). Notice that the replacement rules yield

$$(F|\underline{f})\underline{\underline{Q}} \equiv (F \text{ is a function } X \to Y) \wedge (\forall \underline{x})(\forall \underline{y})((\underline{x} \in X \wedge \underline{y} \in Y)$$

$$\Rightarrow (\underline{\underline{A}}[\![\underline{x}, \underline{y}]\!] \Leftrightarrow \underline{y} = F(\underline{x}))) \wedge (\forall \underline{x})(\underline{x} \in X$$

$$\Rightarrow F(\underline{x}) = \tau_{\underline{y}}(\underline{y} \in Y \wedge \underline{\underline{A}}[\![\underline{x}, \underline{y}]\!])) \ ,$$

the right hand side of which will for brevity be denoted by $\underline{\underline{L}} \wedge \underline{\underline{M}} \wedge \underline{\underline{N}}$. So (by I.3.3(j)) it suffices to prove that

$$T\Theta \quad : \qquad \underline{\underline{L}} \ , \tag{8}$$

$$T\Theta \quad : \qquad \underline{\underline{M}} \ , \tag{9}$$

and

$$T\Theta \quad : \qquad \underline{\underline{N}} \ . \tag{10}$$

To begin with, III.2.3(vii) and III.1.5(3_1) entail that F is a relation, Dom $F \subseteq X$, Ran $F \subseteq Y$ are all true in Θ_0 , and that

$$T \quad : \qquad ((\underline{x}, \underline{y}) \in F) \Leftrightarrow (\underline{x} \in X \wedge \underline{y} \in Y \wedge \underline{\underline{A}}[\![\underline{x}, \underline{y}]\!]) \ . \qquad (7_1)$$

By hypothesis $\underline{\underline{H}}$,

$$T\Theta \quad : \qquad \underline{x} \in X \Rightarrow (\exists \underline{y})(\underline{y} \in Y \wedge \underline{\underline{A}}[\![\underline{x}, \underline{y}]\!]) \ ,$$

and so

$$T\Theta \quad : \qquad \underline{x} \in X \Rightarrow (\exists \underline{y})(\underline{x} \in X \wedge \underline{y} \in Y \wedge \underline{\underline{A}}[\![\underline{x}, \underline{y}]\!]) \ .$$

Hence, by (7_1) ,

$$T\Theta \quad : \qquad \underline{x} \in X \Rightarrow (\exists \underline{y})((\underline{x}, \underline{y}) \in F)$$

and therefore

$$T\Theta \quad : \qquad \underline{x} \in X \Rightarrow \underline{x} \in \text{Dom } F \ .$$

Since $\underline{x}$ is a variable of Θ , it follows that

$$T\Theta \quad : \qquad X \subseteq \text{Dom } F \ .$$

Hence (again by the above remarks about F , now combined with II.2.2(3))

$$T\Theta \quad : \qquad X = \text{Dom } F \ .$$

Thus

$$T\Theta \quad : \qquad X = \text{Dom } F \wedge \text{Ran } F \subseteq Y \ . \qquad (11)$$

Again by hypothesis $\underline{\underline{H}}$,

$$T\Theta : \quad \underline{x} \in X \Rightarrow (\exists_{0,1}\underline{y})(\underline{y} \in Y \wedge \underline{\underline{A}}[\![\underline{x}, \underline{y}]\!]) .$$

Let Θ' denote the theory obtained by adjoining to Θ the explicit axiom $\underline{x} \in X$. Then

$$T\Theta' : \quad (\forall\underline{y}')(\forall\underline{y}'')((\underline{y}' \in Y \wedge \underline{\underline{A}}[\![\underline{x}, \underline{y}']\!] \wedge \underline{y}'' \in Y \wedge \underline{\underline{A}}[\![\underline{x}, \underline{y}'']\!]) \Rightarrow (\underline{y}' = \underline{y}'')) , \tag{12}$$

where $\underline{y}'$, $\underline{y}''$ are distinct letters different from $\underline{x}$ and $\underline{y}$ and not appearing in X , Y or $\underline{\underline{A}}$. Now, using (7_1) ,

$$T\Theta : \quad (\underline{x}, \underline{y}') \in F \Rightarrow \underline{x} \in X \wedge \underline{y}' \in Y \wedge \underline{\underline{A}}[\![\underline{x}, \underline{y}']\!] ,$$

and likewise with $\underline{y}''$ replacing $\underline{y}'$ throughout.

So

$$T\Theta' : \quad (\underline{x}, \underline{y}') \in F \wedge (\underline{x}, \underline{y}'') \in F \Rightarrow \underline{y}' \in Y \wedge \underline{\underline{A}}[\![\underline{x}, \underline{y}']\!] \wedge \underline{y}'' \in Y \wedge \underline{\underline{A}}[\![\underline{x}, \underline{y}'']\!]$$

and therefore, by (12) combined with (II) and (III) in I.3.2,

$$T\Theta' : \quad ((\underline{x}, \underline{y}') \in F \wedge (\underline{x}, \underline{y}'') \in F) \Rightarrow (\underline{y}' = \underline{y}'')$$

and so ($\underline{y}'$ and $\underline{y}''$ being variables of Θ')

$$T\Theta' : \quad (\forall\underline{y}')(\forall\underline{y}'')(((\underline{x}, \underline{y}') \in F \wedge (\underline{x}, \underline{y}'') \in F) \Rightarrow (\underline{y}' = \underline{y}'')) .$$

Accordingly, by I.3.2(VI),

$$T\Theta \quad : \quad (\underline{x} \in X) \Rightarrow ((\forall \underline{y}')(\forall \underline{y}'')(((\underline{x}, \underline{y}') \in F \wedge (\underline{x}, \underline{y}'') \in F) \Rightarrow (\underline{y}' = \underline{y}''))) \quad . \quad (13)$$

On the other hand, if Θ'' denotes the theory obtained by adjoining to Θ the explicit axiom $\underline{x} \notin X$, then (11) implies

$$T\Theta'' : \quad \neg((\underline{x}, \underline{y}') \in F \wedge (\underline{x}, \underline{y}'') \in F)$$

and so

$$T\Theta'' : \quad ((\underline{x} \in \underline{y}') \in F \wedge (\underline{x}, \underline{y}'') \in F) \Rightarrow (\underline{y}' = \underline{y}'') \quad .$$

Since $\underline{y}'$, $\underline{y}''$ are variables of Θ'' ,

$$T\Theta'' : \quad (\forall \underline{y}')(\forall \underline{y}'')(((\underline{x}, \underline{y}') \in F \wedge (\underline{x}, \underline{y}'') \in F \Rightarrow (\underline{y}' = \underline{y}'')) \quad .$$

So, by I.3.2(VI) again,

$$T\Theta \quad : \quad (\underline{x} \notin X) \Rightarrow ((\forall \underline{y}')(\forall \underline{y}'')(((\underline{x}, \underline{y}') \in F \wedge (\underline{x}, \underline{y}'') \in F \Rightarrow (\underline{y}' = \underline{y}''))) \quad . \quad (14)$$

By (13) , (14) and I.3.2(VIII),

$$T\Theta \quad : \quad (\forall \underline{y}')(\forall \underline{y}'')(((\underline{x}, \underline{y}') \in F \wedge (\underline{x}, \underline{y}'') \in F \Rightarrow (\underline{y}' = \underline{y}''))$$

and, $\underline{x}$ being a variable of Θ ,

$$T\Theta \quad : \quad (\forall \underline{x})(\forall \underline{y}')(\forall \underline{y}'')(((\underline{x}, \underline{y}') \in F \wedge (\underline{x}, \underline{y}'') \in F) \Rightarrow (\underline{y}' = \underline{y}'')) \quad .$$

Thus (see IV.1.1 and (11))

$$T\Theta \ : \quad F \text{ is a function } X \to Y \ ,$$

which is (8) .

I now pass to the proof of (9) . Let Θ_3 denote the theory obtained by adjoining to Θ the explicit axiom $\underline{x} \in X \wedge \underline{y} \in Y$, and Θ_4 that obtained by adjoining to Θ_3 the explicit axiom $\underline{\underline{A}}[\![\underline{x}, \underline{y}]\!]$. Then

$$T\Theta_4 \ : \quad \underline{x} \in X \wedge \underline{y} \in Y \wedge \underline{\underline{A}}[\![\underline{x}, \underline{y}]\!]$$

and so, by (7_1) ,

$$T\Theta_4 \ : \quad (\underline{x}, \underline{y}) \in F \ .$$

Hence, by (8) and IV.1.5(3)

$$T\Theta_4 \ : \quad \underline{y} = F(\underline{x}) \ .$$

Thus, by I.3.2(VI),

$$T\Theta_3 \ : \quad \underline{\underline{A}}[\![\underline{x}, \underline{y}]\!] \Rightarrow \underline{y} = F(\underline{x}) \ . \tag{15}$$

Further, if Θ_5 denotes the theory obtained by adjoining to Θ_3 the explicit axiom $\underline{y} = F(\underline{x})$, then (8) , (11) and IV.1.5(3) imply

$$T\Theta_5 \ : \quad (\underline{x}, \underline{y}) \in F$$

and so (by (7_1)) also

$$T\Theta_5 \ : \quad \underline{\underline{A}}[\![\underline{x}, \underline{y}]\!] \ .$$

Hence

$$T\Theta_3 : \quad \underline{y} = F(\underline{x}) \Rightarrow \underline{\underline{A}}[\![\underline{x}, \underline{y}]\!] \quad . \tag{16}$$

By (15) and (16) ,

$$T\Theta_3 : \quad \underline{\underline{A}}[\![\underline{x}, \underline{y}]\!] \Leftrightarrow \underline{y} = F(\underline{x}) \quad .$$

Hence, by I.3.2(VI),

$$T\Theta \ : \quad (\underline{x} \in X \wedge \underline{y} \in Y) \Rightarrow (\underline{\underline{A}}[\![\underline{x}, \underline{y}]\!] \Leftrightarrow \underline{y} = F(\underline{x})) \quad . \tag{17}$$

Since $\underline{x}$ and $\underline{y}$ are variables of Θ , (17) and I.3.2(X) entail (9) .

Considering (10) , let Θ' be as above (Θ with $\underline{x} \in X$ adjoined), and let Θ_6 be the theory obtained by adjoining to Θ' the explicit axiom $\underline{y} \in Y \wedge \underline{\underline{A}}[\![\underline{x}, \underline{y}]\!]$. Then

$$T\Theta_6 : \quad \underline{x} \in X \wedge \underline{y} \in Y \wedge \underline{\underline{A}}[\![\underline{x}, \underline{y}]\!] \quad .$$

Also, by (17) and I.3.7(2),

$$T\Theta_6 : \quad (\underline{x} \in X \wedge \underline{y} \in Y \wedge \underline{\underline{A}}[\![\underline{x}, \underline{y}]\!]) \Rightarrow \underline{y} = F(\underline{x}) \quad .$$

Hence

$$T\Theta_6 : \quad \underline{y} = F(\underline{x})$$

and so, by I.3.2(VI),

$$T\Theta' : \quad (\underline{y} \in Y \wedge \underline{\underline{A}}[\![\underline{x}, \underline{y}]\!]) \Rightarrow \underline{y} = F(\underline{x}) \quad . \tag{18}$$

Now let Θ_7 denote the theory obtained by adjoining to Θ' the explicit axiom $\underline{y} = F(\underline{x})$. Then

$$T\Theta_7 : \qquad \underline{x} \in X \wedge \underline{y} = F(\underline{x}) \quad .$$

So, by (8) , (11) and IV.1.5(3) ,

$$T\Theta_7 : \qquad \underline{y} \in Y \wedge (\underline{x}, \underline{y}) \in F$$

and therefore, by (7_1) ,

$$T\Theta_7 : \qquad \underline{y} \in Y \wedge \underline{A}[\![\underline{x}, \underline{y}]\!] \quad .$$

Thus, by I.3.2(VI),

$$T\Theta' : \qquad \underline{y} = F(\underline{x}) \Rightarrow (\underline{y} \in Y \wedge \underline{A}[\![\underline{x}, \underline{y}]\!]) \quad . \tag{19}$$

Combining (18) and (19) ,

$$T\Theta' : \qquad (\underline{y} \in Y \wedge \underline{A}[\![\underline{x}, \underline{y}]\!] \Leftrightarrow \underline{y} = F(\underline{x})$$

and so, $\underline{y}$ being a variable of Θ' ,

$$T\Theta' \qquad (\forall \underline{y})((\underline{y} \in Y \wedge \underline{A}[\![\underline{x}, \underline{y}]\!]) \Leftrightarrow \underline{y} = F(\underline{x})) \quad . \tag{20}$$

Using S7 in II.1.2, (20) implies that

$$T\Theta' : \qquad \tau_{\underline{y}}(\underline{y} \in Y \wedge \underline{A}[\![\underline{x}, \underline{y}]\!]) = \tau_{\underline{y}}(\underline{y} = F(\underline{x})) \quad ;$$

hence, by Problem II/13 (note that $\underline{y}$ does not appear in $F(\underline{x})$)

$$T\Theta' : \qquad F(\underline{x}) = \tau_{\underline{y}}(\underline{y} \in Y \wedge \underline{\underline{A}}[\![\underline{x}, \underline{y}]\!]) \quad .$$

By I.3.2(VI),

$$T\Theta \quad : \qquad \underline{x} \in X \Rightarrow (F(\underline{x}) = \tau_{\underline{y}}(\underline{y} \in Y \wedge \underline{\underline{A}}[\![\underline{x}, \underline{y}]\!]))$$

and ($\underline{x}$ being a variable of Θ) (10) follows. □

The second proof of (3) is based upon a lemma (schema), itself of some general interest; see also Problem IV/13.

Lemma Suppose that $\underline{\underline{B}}$ is a sentence and $\underline{y}$ a letter. Then

$$T! \quad : \qquad (\exists_{0,1}\underline{y})\underline{\underline{B}} \Rightarrow (\forall\underline{y})(\underline{\underline{B}} \Rightarrow (\underline{y} = \tau_{\underline{y}}(\underline{\underline{B}}))) \quad , \qquad \text{(i)}$$

and

$$T! \quad : \qquad (\exists_{1}\underline{y})\underline{\underline{B}} \Rightarrow (\forall\underline{y})(\underline{\underline{B}} \Leftrightarrow (\underline{y} = \tau_{\underline{y}}(\underline{\underline{B}}))) \quad . \qquad \text{(ii)}$$

Proof of lemma (i) Let Θ denote the theory obtained by adjoining to Θ_0 the explicit axiom $\underline{\underline{B}} \wedge (\exists_{0,1}\underline{y})\underline{\underline{B}}$; and define

$$t \equiv \tau_{\underline{y}}(\underline{\underline{B}}) \quad .$$

Then

$$T\Theta \quad : \qquad \underline{\underline{B}} \quad , \qquad (20)$$

$$T\Theta \quad : \qquad (\exists_{0,1}\underline{y})\underline{\underline{B}} \quad . \qquad (21)$$

By (20) , S5 in I.2.2 and I.3.2(I),

$T\Theta \quad : \qquad (\exists \underline{y})\underline{\underline{B}} \quad ,$

that is,

$$T\Theta \quad : \qquad (t|\underline{y})\underline{\underline{B}} \quad . \tag{22}$$

By (20) , (22) and I.3.3(j),

$$T\Theta \quad : \qquad (\underline{y}|\underline{y})\underline{\underline{B}} \wedge (t|\underline{y})\underline{\underline{B}} \quad . \tag{23}$$

By (21) and (23) and I.3.2(I),

$$T\Theta \quad : \qquad \underline{y} = t \quad . \tag{24}$$

By (24) and I.3.2(VI),

$$T \quad : \qquad (\underline{\underline{B}} \wedge (\exists_{0,1}\underline{y})\underline{\underline{B}}) \Rightarrow (\underline{y} = t) \quad . \tag{25}$$

By (25) , I.3.7(2) and I.3.2(I),

$$T \quad : \qquad (\exists_{0,1}\underline{y})\underline{\underline{B}} \Rightarrow (\underline{\underline{B}} \Rightarrow (\underline{y} = t)) \quad .$$

Hence, by I.3.3(l),

$$T \quad : \qquad (\exists_{0,1}\underline{y})\underline{\underline{B}} \Rightarrow (\forall \underline{y})(\underline{\underline{B}} \Rightarrow (\underline{y} = t)) \quad ,$$

which is identical with (i) .

(ii) Let Θ' denote the theory obtained by adjoining to Θ_0 the explicit axiom $(\exists_1\underline{y})\underline{\underline{B}}$. Then, by I.3.3(j) and I.3.2(I),

$T\Theta'$: $$(\exists_{0,1}\underline{y})\underline{\underline{B}} \qquad (26)$$

$T\Theta'$: $$(\exists\underline{y})\underline{\underline{B}} \quad . \qquad (27)$$

By (i) , (26) , I.3.2(I) and the fact that Θ' is stronger than Θ_0 ,

$T\Theta'$: $$\underline{\underline{B}} \Rightarrow (\underline{y} = t) \quad . \qquad (28)$$

On the other hand, by (27) ,

$T\Theta'$: $$(t|\underline{y})\underline{\underline{B}} \quad . \qquad (29)$$

By S6 in II.1.2,

T : $$(\underline{y} = t) \Rightarrow (\underline{\underline{B}} \Leftrightarrow (t|\underline{y})\underline{\underline{B}}) \quad ,$$

and so (again since Θ' is stronger than Θ_0)

$T\Theta'$: $$(\underline{y} = t) \Rightarrow (\underline{\underline{B}} \Leftrightarrow (t|\underline{y})\underline{\underline{B}}) \quad . \qquad (30)$$

If Θ'' denotes the theory obtained by adjoining to Θ' the explicit axiom $\underline{y} = t$, then (since Θ'' is stronger than Θ') (29) implies

$T\Theta''$: $$(t|\underline{y})\underline{\underline{B}} \quad ; \qquad (31)$$

and (by the same token) I.3.2(I) and (30) entail

$T\Theta''$: $$(t|\underline{y})\underline{\underline{B}} \Leftrightarrow \underline{\underline{B}} \quad . \qquad (32)$$

By (31) , (32) , I.3.3(j) and I.3.2(I),

$T\Theta''$: $\underline{\underline{B}}$.

Hence, by I.3.2(VI),

$T\Theta'$: $(\underline{y} = t) \Rightarrow \underline{\underline{B}}$. (33)

By (28) , (33) and I.3.3(j),

$T\Theta'$: $\underline{\underline{B}} \Leftrightarrow (\underline{y} = t)$

and, by I.3.2(VI) again,

T : $(\exists_1 \underline{y})\underline{\underline{B}} \Rightarrow (\underline{\underline{B}} \Leftrightarrow (\underline{y} = t))$.

Finally, by I.3.3(l) again,

T : $(\exists_1 \underline{y})\underline{\underline{B}} \Rightarrow (\forall \underline{y})(\underline{\underline{B}} \Leftrightarrow (\underline{y} = t))$,

which is identical with (ii) . □

Second proof of (3) One proceeds as before down to (5) , but this time (6) will be proved with

$$F \equiv \text{the function } \underline{x} \rightsquigarrow \tau_{\underline{y}}(\underline{\underline{B}}) \text{ with domain } X ,$$

wherein

$$\underline{\underline{B}} \equiv \underline{y} \in Y \wedge \underline{\underline{A}}[\underline{x}, \underline{y}] .$$

By IV.2.2(1),

T : $$\underline{x} \in X \Rightarrow F(\underline{x}) = \tau_{\underline{y}}(\underline{\underline{B}})$$

and so, by I.3.2(X),

T : $$(\forall \underline{x})(\underline{x} \in X \Rightarrow F(\underline{x}) = \tau_{\underline{y}}(\underline{\underline{B}})) \equiv \underline{\underline{N}} \quad . \tag{34}$$

As before, let Θ denote the theory obtained by adjoining to Θ_0 the explicit axiom

$$\underline{\underline{H}} \equiv (\forall \underline{x})(\underline{x} \in X \Rightarrow (\exists_1 \underline{y})\underline{\underline{B}}) \quad ;$$

note that $\underline{x}$ and $\underline{y}$ are variables of Θ . Then

TΘ : $$\underline{x} \in X \Rightarrow (\exists_1 \underline{y})\underline{\underline{B}} \quad . \tag{35}$$

Again as before, let Θ' denote the theory Θ with the explicit axiom $\underline{x} \in X$ adjoined; then, by (35) and I.3.2(I),

TΘ' : $$(\exists_1 \underline{y})\underline{\underline{B}} \quad .$$

This, together with part (ii) of the above lemma and I.3.2(I), implies

TΘ' : $$\underline{\underline{B}} \Leftrightarrow (\underline{y} = \tau_{\underline{y}}(\underline{\underline{B}})) \quad ; \tag{36}$$

and then, by (34) , the fact that $\underline{x} \in X$ is true in Θ' , S6 in II.1.2 and I.3.2(I), it follows that

TΘ' : $$\underline{\underline{B}} \Leftrightarrow (\underline{y} = F(\underline{x})) \quad . \tag{37}$$

Further, by (34) , (35) , S6 in II.1.2 and I.3.2(I),

TΘ' : $$(F(\underline{x})|\underline{y})\underline{\underline{B}} \quad . \tag{38}$$

By (38), the definition of $\underline{\underline{B}}$, I.3.3(j) and I.3.2(I),

$$T\Theta' : \quad F(\underline{x}) \in Y \quad ,$$

which, in view of I.3.2(VI), entails

$$T\Theta : \quad \underline{x} \in X \Rightarrow F(\underline{x}) \in Y \quad . \tag{39}$$

By (39) and the definition of F (plus the theorem schemas in IV.1.5),

$$T\Theta : \quad (F \text{ is a function } X \to Y) \equiv \underline{\underline{L}} \quad . \tag{40}$$

Next, by (37) and I.3.2(VI),

$$T\Theta : \quad (\underline{x} \in X) \Rightarrow (\underline{\underline{B}} \Leftrightarrow \underline{y} = F(\underline{x})) \quad ,$$

that is,

$$T\Theta : \quad (\underline{x} \in X) \Rightarrow ((\underline{y} \in Y \wedge \underline{\underline{A}}[\![\underline{x}, \underline{y}]\!]) \Leftrightarrow \underline{y} = F(\underline{x})) \quad . \tag{41}$$

Let Θ_3 be as before (namely, Θ with $\underline{x} \in X \wedge \underline{y} \in Y$ adjoined as an explicit axiom). By (41) and I.3.2(I),

$$T\Theta_3 : \quad (\underline{y} \in Y \wedge \underline{\underline{A}}[\![\underline{x}, \underline{y}]\!]) \Leftrightarrow \underline{y} = F(\underline{x}) \quad . \tag{42}$$

But, since $\underline{y} \in Y$ is true in Θ_3, I.3.7(1) implies that

$$T\Theta_3 : \quad (\underline{y} \in Y \wedge \underline{\underline{A}}[\![\underline{x}, \underline{y}]\!]) \Leftrightarrow \underline{\underline{A}}[\![\underline{x}, \underline{y}]\!] \quad . \tag{43}$$

By (42), (43) and I.3.2(I) (or I.3.7(1) again)

$$T\Theta_3 : \qquad \underline{\underline{A}}[\![\underline{x}, \underline{y}]\!] \Leftrightarrow \underline{y} = F(\underline{x}) \quad ,$$

which, by I.3.2(VI), entails

$$T\Theta \quad : \qquad (\underline{x} \in X \wedge \underline{y} \in Y) \Rightarrow (\underline{\underline{A}}[\![\underline{x}, \underline{y}]\!] \Leftrightarrow \underline{y} = F(\underline{x})) \quad . \tag{44}$$

So, by I.3.2(X) used twice (recall that $\underline{x}$ and $\underline{y}$ are variables of Θ),

$$T\Theta \quad : \qquad (\forall \underline{x})(\forall \underline{y})(44) \equiv \underline{\underline{M}} \quad . \tag{45}$$

By (34) , (40) , (45) and I.3.3(j) (recall also I.3.7(6)),

$$T\Theta \quad : \qquad \underline{\underline{L}} \wedge \underline{\underline{M}} \wedge \underline{\underline{N}} \equiv (F|f)\underline{\underline{Q}} \quad ,$$

which is (6) . The rest is as before. □

It is worth noting as a corollary of (3) that, if one defines f to be the function

$$\underline{x} \rightsquigarrow \tau_{\underline{y}}(\underline{y} \in Y \wedge \underline{\underline{A}}[\![\underline{x}, \underline{y}]\!])$$

with domain X , then (see Problem IV/8)

$$T! \quad : \qquad \begin{aligned} &\underline{\underline{H}} \Rightarrow (\forall \underline{x})(\forall \underline{y})((\underline{x} \in X \wedge \underline{y} \in Y) \\ &\Rightarrow (\underline{\underline{A}}[\![\underline{x}, \underline{y}]\!] \Leftrightarrow \underline{y} = f(\underline{x}))) \quad . \end{aligned} \tag{46}$$

IV.4.3 Remarks concerning notation

(i) Notice that $\mathrm{Un}\ f$ and $\mathrm{Fn}\ f$, as defined in IV.1, are identical with

$$(\forall \underline{x})(\exists_{0,1}\underline{y})((\underline{x}, \underline{y}) \in f)$$

and

$$(f \text{ is a relation}) \wedge (\forall \underline{x})(\exists_{0,1}\underline{y})((\underline{x}, \underline{y}) \in f)$$

respectively, where $\underline{x}$ and $\underline{y}$ denote distinct letters not appearing in f .

(ii) Some writers use $\exists!$ or $E!$ to denote equivalents of our $\exists_1$. Thus, Kleene (2), p.154, uses $\exists! yB(y)$ as an abbreviation for what in our formalism is indicated by

$$(\exists!\underline{y})\underline{\underline{B}} \equiv (\exists\underline{y})(\underline{\underline{B}} \wedge (\forall\underline{z})((\underline{z}|\underline{y})\underline{\underline{B}} \Rightarrow (\underline{z} = \underline{y}))) \quad ,$$

where $\underline{z}$ denotes a letter different from $\underline{y}$ and not appearing in B . It is left to the reader to prove the theorem schemas

$$(\exists!\underline{y})\underline{\underline{B}} \Leftrightarrow (\exists_1\underline{y})\underline{\underline{B}} \quad ,$$

$$(\exists_1\underline{y})\underline{\underline{B}} \Leftrightarrow (\exists\underline{z})(\forall\underline{y})(\underline{\underline{B}} \Leftrightarrow (\underline{y} = \underline{z})) \quad ,$$

$$(\exists_{0,1}\underline{y})\underline{\underline{B}} \Leftrightarrow (\exists\underline{z})(\forall\underline{y})(\underline{\underline{B}} \Rightarrow (\underline{y} = \underline{z})) \quad ,$$

where $\underline{z}$ is as above; see Problems IV/3, IV/13 and IV/18.

IV.4.4 <u>The notation $\rightsquigarrow$ again</u> Reverting to the situation described at the outset of IV.4.1, mathematicians frequently would refer to the function F defined in IV.4.2(7) as that defined by the rule $\underline{x} \rightsquigarrow \alpha_{\underline{x}}$ and having domain equal to X : in symbols,

$$F : \underline{x}(\in X) \rightsquigarrow \alpha_{\underline{x}} \quad ;$$

or, more informally, as the function

$$\underline{x}(\in X) \rightsquigarrow (\text{the unique } \underline{y} \in Y \text{ such that } A[\![\underline{x}, \underline{y}]\!]) \quad .$$

In this they may be a little less than (formally) correct, because they would usually operate in a way which leaves the definition of $\alpha_{\underline{x}}$ conditional (conditional upon $\underline{x} \in X$, for one thing); cf, the remarks on this issue in IV.3.10. However, this may be rectified by making use of IV.4.2(3) to replace $\alpha_{\underline{x}}$ by (the unconditionally defined) term

$$\tau_{\underline{y}}(\underline{y} \in Y \wedge \underline{A}[\![\underline{x}, \underline{y}]\!]) \quad .$$

Of course, this would not be the end of the matter: it would still require satisfactory proof that this definition of F secures (granted the hypothesis of (3) in IV.4.2) the truth of

$$(\forall\underline{x})(\forall\underline{y})((\underline{x} \in X \wedge \underline{y} \in Y) \Rightarrow (\underline{A}[\![\underline{x}, \underline{y}]\!] \Leftrightarrow \underline{y} = F(\underline{x}))) \quad .$$

As the following three examples show, the procedure in IV.4.2 contributes towards formalising the definitions of the square root function (sketched in IV.3.8) and dispelling some of the formally unsatisfactory features mentioned in the Remarks in IV.3.3. A little more will be said on both these issues in VI.8.

(i) Suppose that $X \equiv Y \equiv R_+$ (the set of nonnegative real numbers; see VI.6.11) and

$$A[\![\underline{x}, \underline{y}]\!] \equiv \underline{x} = \underline{y}^2 \quad .$$

This presupposes the developments in Chapter VI below (cf. the remarks in IV.2.4(ii)); in particular, $\underline{y}^2 \equiv \pi((\underline{y}, \underline{y}))$, where π is as in VI.3.2. Assuming that $\underline{x}$ and $\underline{y}$ denote distinct letters not appearing in R_+ or π , the hypothesis of IV.4.2(3) is then a theorem: this may be inferred by using VIII.1.2, VIII.2.1 and the fact that the function $\underline{y} \rightsquigarrow \underline{y}^2$ with domain R_+ is strictly increasing (see IV.6.5). The function F whose existence and uniqueness is guaranteed by IV.4.2(3) is the square root function $\underline{x} \rightsquigarrow \underline{x}^{\frac{1}{2}}$ with domain R_+ , usually denoted by $j^{\frac{1}{2}}$; cf. IV.3.6, IV.3.8 and IV.6.2. Here $\underline{x}^{\frac{1}{2}}$ is understood to denote

$$\tau_{\underline{y}}(\underline{y} \in R_+ \wedge \underline{x} = \underline{y}^2) \quad ;$$

beware the possible confusion when the notation $z^{\frac{1}{2}}$ is used in connection with complex numbers z (see XII.5.4).

(ii) Again, if $Y \equiv R$ and f and g are as in IV.3.3 and

$$\underline{\underline{A}}[\![\underline{x}, \underline{y}]\!] \equiv \underline{y} = \sigma(f(\underline{x}), g(\underline{x})) \quad ,$$

where σ is as in VI.3.2 and $\underline{x}$, $\underline{y}$ are distinct letters not appearing in σ , f or g , one would obtain the pointwise sum function $f + g$ referred to in IV.3.3.

(iii) If f is as in IV.3.3, X is taken to be $(\text{Inv } f)\langle \dot{R} \rangle$, Y to be R (or $\dot{R}$) and

$$\underline{\underline{A}}[\![\underline{x}, \underline{y}]\!] \equiv \pi((f(\underline{x}), \underline{y})) = 1 \quad ,$$

where $\underline{x}$ and $\underline{y}$ denote distinct letters not appearing in π or f , one obtains the pointwise inverse function f^{-1} referred to in IV.3.3. (That the hypothesis of IV.4.2(3) is true in this case, follows from the field properties of R discussed in VI.3.2.)

(iv) A less basic and perhaps more typical use of IV.4.2 appears in VII.6.1 and VII.6.8 in connection with the exponential function.

(v) It is also conventional, when one has established a theorem of the type

$$\underline{\underline{H}} \equiv (\forall \underline{x})(\underline{x} \in X \Rightarrow (\exists_1 \underline{y})(\underline{y} \in Y \wedge \underline{\underline{A}}[\![\underline{x}, \underline{y}]\!])) \quad ,$$

to write something like:

> For all (or each, or every) $\underline{x} \in X$, let $f(\underline{x})$ denote the unique element $\underline{y}$ of Y such that $A[\![\underline{x}, \underline{y}]\!]$,

and this is taken as adequate definition of the appropriate function f (which will have domain X and range a subset of Y). The justification for this lies, of course, in IV.4.2(3). As has been indicated, a typical instance is provided by VII.6.1 and VII.6.8.

(vi) It is worth noting that there is a companion to IV.4.2(3) in which both hypothesis and conclusion are weakened. Reference to Y is eliminated and, in place of $\underline{\underline{H}}$, one assumes that, for each $\underline{x} \in X$, there exists (at least one) $\underline{y}$ such that $\underline{\underline{A}}[\![\underline{x}, \underline{y}]\!]$; uniqueness of such a $\underline{y}$ is not assumed. The appropriate theorem schema reads (see Problems II/50 and IV/33)

$$\mathrm{T!} : \quad \begin{aligned} &(\forall \underline{x})(\underline{x} \in X \Rightarrow (\exists \underline{y})\underline{\underline{A}}[\![\underline{x}, \underline{y}]\!]) \\ &\Rightarrow (\exists \underline{f})(\mathrm{Fn}\ \underline{f} \wedge \mathrm{Dom}\ \underline{f} = X \wedge (\forall \underline{x})(\underline{x} \in X \Rightarrow \underline{\underline{A}}[\![\underline{x}, \underline{f}(\underline{x})]\!])) \end{aligned} \tag{1}$$

wherein X denotes an arbitrary set, $\underline{\underline{A}}$ an arbitrary sentence, $\underline{x}$, $\underline{y}$ and $\underline{f}$ distinct letters, $\underline{x}$ and $\underline{y}$ not appearing in X and $\underline{f}$ not appearing in X or $\underline{\underline{A}}$.

For an f whose existence is asserted one may take the function

$$\underline{x} \rightsquigarrow \tau_{\underline{y}}(\underline{\underline{A}}[\![\underline{x}, \underline{y}]\!]) \text{ with domain } X .$$

Now, however, one can say nothing a priori about $\mathrm{Ran}\ f$; and one will not know that $\underline{f}$ is unique.

In practice, however, one usually has in mind a set Y for which the hypothesis of IV.4.2(3) is true: this is why that particular variant has been examined first and in detail.

(vii) Observe also that, if f is a function and

$$X \equiv \operatorname{Dom} f \quad , \quad Y \equiv \operatorname{Ran} f \quad , \quad \underline{\underline{A}}[\![\underline{x}, \underline{y}]\!] \equiv (\underline{x}, \underline{y}) \in f \quad ,$$

$\underline{x}$ and $\underline{y}$ denoting distinct letters not appearing in f, then f is equal to the function

$$\underline{x} \rightsquigarrow \tau_{\underline{y}}(\underline{y} \in Y \wedge \underline{\underline{A}}[\![\underline{x}, \underline{y}]\!]) \text{ with domain } X$$

and to the function

$$\underline{x} \rightsquigarrow \tau_{\underline{y}}(\underline{\underline{A}}[\![\underline{x}, \underline{y}]\!]) \text{ with domain } X \quad .$$

(To confirm this, refer back to IV.2.3(i).) In other words, every function is obtainable by appeal to a suitable instance of the implicit function theorem schema (though this procedure may be unnecessarily circuitous).

(viii) There is another, rather similar theorem schema which is frequently used, often without explicit mention. Informally expressed, the situation in question is that in which X and T are given sets and $\underline{\underline{A}}$ a given sentence ($\underline{\underline{A}}$ is regarded as expressing a relation between $\underline{x}$ and $\underline{y}$ and $\underline{y}$ may appear in T); it is supposed that, for all $\underline{x} \in X$, there exists $\underline{y}$ in the relation $\underline{\underline{A}}$ to $\underline{x}$; and that, for all $\underline{x} \in X$, if $\underline{y}$ and $\underline{y}'$ are both in the relation $\underline{\underline{A}}$ to $\underline{x}$, then $T = (\underline{y}'|\underline{y})T$. The conclusion is that there is a unique function f with domain X such that, for all $\underline{x} \in X$, $f(\underline{x}) = T$ for all $\underline{y}$ in the relation $\underline{\underline{A}}$ to $\underline{x}$; that is, the value $f(\underline{x})$ of f at $\underline{x} \in X$ is found by choosing any $\underline{y}$ in the relation $\underline{\underline{A}}$ to $\underline{x}$ and taking the corresponding T to be the said value.

One of the simplest illustrations arises when one has a function

$g : R \to R$ which has period 2π, that is,

$$g(\underline{y} + 2\pi) = g(\underline{y}) \text{ for all } \underline{y} \in R \ ;$$

there is then a unique function f with domain the unit circle Γ in the complex plane such that

$$f(\underline{x}) = g(\underline{y})$$

for all $\underline{x} \in \Gamma$ and all $\underline{y} \in R$ such that $\underline{x} = e^{i\underline{y}}$; see Chapter XII. In this instance, $X \equiv R$, $T \equiv g(\underline{y})$ and $\underline{\underline{A}} \equiv (\underline{y} \in R \wedge e^{i\underline{y}} = \underline{x})$.

Another illustration appears in Problem XI/10.

The appropriate theorem schema is as follows (see Problem IV/23 below) :

$$((\forall\underline{x})(\underline{x} \in X \Rightarrow (\exists\underline{y})\underline{\underline{A}}) \wedge (\forall\underline{x})(\forall\underline{y})(\forall\underline{y}')((\underline{x} \in X \wedge \underline{\underline{A}} \wedge (\underline{y}'|\underline{y})\underline{\underline{A}}) \Rightarrow (T = (\underline{y}'|\underline{y})T)))$$

$$\Rightarrow (\exists_1\underline{f})(\text{Fn } \underline{f} \wedge \text{Dom } \underline{f} = X \wedge (\forall\underline{x})(\forall\underline{y})((\underline{x} \in X \wedge \underline{\underline{A}}) \Rightarrow (\underline{f}(\underline{x}) = T)))$$

wherein X and T denote arbitrary sets, $\underline{\underline{A}}$ an arbitrary sentence and $\underline{x}$, $\underline{y}$, $\underline{y}'$, $\underline{f}$ denote distinct letters, $\underline{x}$ not appearing in X or T (it may appear in $\underline{\underline{A}}$), $\underline{y}$ not appearing in X (it may appear in $\underline{\underline{A}}$ and/or T), and $\underline{y}'$ and $\underline{f}$ not appearing in X , T or $\underline{\underline{A}}$.

IV.5 The Axiom of Choice

IV.5.1 Informal discussion A theorem schema akin to those stated in IV.4.2 and IV.4.4(vi) might be expressed informally as follows : Suppose that X and Y denote sets and $\underline{\underline{A}}[\![\underline{x}, \underline{y}]\!]$ a sentence (regarded as expressing that $\underline{y}$ bears a certain relation to $\underline{x}$); suppose also that, for every $\underline{x} \in X$, there exists at least one $\underline{y} \in Y$ satisfying $\underline{\underline{A}}[\![\underline{x}, \underline{y}]\!]$; then there exists at least one function f with domain X such that, for every $\underline{x} \in X$, $f(\underline{x})$ bears the said relation to $\underline{x}$.

One might visualise the production of such a function by imagining the elements $\underline{x}$ of X laid out on display; for each such element $\underline{x}$ one may (by assumption) "choose" an object $\underline{y} \in Y$ such that $\underline{A}[\![\underline{x}, \underline{y}]\!]$; suppose such a choice made for every $\underline{x} \in X$. Surely (one is inclined to say), there will be a function f with domain X such that, for every $\underline{x} \in X$, $f(\underline{x})$ is equal to the corresponding chosen object? Cannot one simply define f by these choices?

Well, this visualised procedure is (in outcome, if not in every pictorial detail) formalised by resort to the selector τ in the manner described in IV.4 : for each $\underline{x} \in X$ the chosen object may as well be taken to be $\tau_{\underline{y}}(\underline{y} \in Y \wedge \underline{A}[\![\underline{x}, \underline{y}]\!])$ and a function with the desired properties is the function f with domain X and rule

$$\underline{x} \rightsquigarrow \tau_{\underline{y}}(\underline{y} \in Y \wedge \underline{A}[\![\underline{x}, \underline{y}]\!]) \quad .$$

Thus, all is well in our formalism.

However, as has been indicated in II.12.3, several versions of formal axiomatic set theory lack a selector. In such a version of set theory, one is left facing the problem of formalising the intuitive multi-choice procedure described above. The apparent need for making possibly infinitely many "choices" , some of which may be dependent upon preceding ones, was seen by many mathematicians as a stumbling block. Arguments of this sort intruded quite frequently, even into relatively elementary mathematics, where they were often glossed over; examples appear in V.8.3(ii), Problem V/23, VII.3.4, VII.3.5 and VIII.2.2.

In certain versions of set theory lacking a selector, the most strenuous efforts finally showed that there is no way of sanctioning all such "infinitely many choice" arguments without the introduction of an axiom for that special purpose, the so called Axiom of Choice; see II.12.3, IV.5.3 and §4 of the Appendix. The version of set theory briefly described in this book is different inasmuch as it has this axiom built-in in the guise of the signs τ and $\square$ and the rules for their use. In spite of this, and in deference to tradition, I will continue to describe as an axiom what is in our theory a theorem (schema); and I will occasionally indicate "local" appeals to the "axiom". (Such indications are most appropriate later on, when the presentation has become more informal and one tends to lose sight of the fundamental role the "axiom" plays again and again.)

One of many brief informal discussions of the axiom appears in Section 11-5 of Gleason (1). Gleason's discussion covers numerous other points of interest bearing on the contact between intuitive and formal approaches. See also Stoll (2), pp. 112, 302 and Problems IV/27 - IV/29. Concerning the early history (or pre-history) of the axiom, See Fraenkel et al. (1), pp. 56-58, 80-86.

IV.5.2 <u>The axiom</u> There are many equivalent versions of the axiom; see, for example, Suppes (1), Chapter 8; Fraenkel et al. (1), pp. 53-56; Rubin and Rubin (1); Jech (1). For our purposes perhaps the most convenient version is that which expresses that the following is a theorem schema (proved in §4 of the Appendix):

> If R is a relation, there is a function f which is contained in R (as a subset) and which has domain equal to that of R .

This is, of course, an informal statement of the axiom.

This version of the axiom makes no overt reference to the difficulties attendant upon infinitely many choices illustrated by example in IV.5.1 (and used there to assemble the ingredients of a function with certain desired properties). Nevertheless, this version resolves the matter satisfactorily. One has in fact merely to take for R the relation defined to be the set of all ordered pairs $(\underline{x}, \underline{y}) \in X \times Y$ such that $\underline{A}[\![\underline{x}, \underline{y}]\!]$. The principal hypothesis in the situation discussed in IV.5.1, namely

$$(\forall \underline{x})(\underline{x} \in X \Rightarrow (\exists \underline{y})(\underline{y} \in Y \wedge \underline{A}[\![\underline{x}, \underline{y}]\!])) ,$$

implies that $\mathrm{Dom}\, R = X$; and the axiom implies the existence of a function f with the desired properties.

IV.5.3 Remarks As a matter of passing interest, I remark that the precise status of the Axiom of Choice in relation to the remaining axioms of other versions of set theory was for decades an outstanding problem. Part of the solution came in 1938 as the result of work by Gödel, but the complete solution came nearly 30 years later (in 1963) and was the work of Paul Cohen. Meantime, the majority of mathematicians adopted the axiom in every day work (though not always without unease) since it was not known to lead to any contradictions. (Others voiced their opposition, some even denying that the axiom had any clear meaning.) For a few more details concerning the work of Gödel, Cohen and Solovay, see §4 of the Appendix.

Although mathematicians make frequent (and often unremarked) appeals to the axiom in situations (like that described above) where what is desired seems intuitively obvious, the Axiom of Choice is known to lead to some truly staggering theorems which are unprovable without it. For instance, Banach and Tarski used it in 1924 to show that, if S and T denote any two solid spheres in $\mathbb{R}^3$, there is a natural number n and partitions

$$S = S_1 \cup S_2 \cup \ldots \cup S_n \quad ,$$

$$T = T_1 \cup T_2 \cup \ldots \cup T_n \quad ,$$

where the S_i are pairwise disjoint, the T_i are pairwise disjoint, and where S_i is congruent to T_i for all $i \in \{1, 2, \ldots, n\}$. When one considers the volumes of the S_i and T_i, this conclusion seems quite incredible, since no restriction is placed on the radii of S and T : S may be as small as one pleases and T as large as one pleases. It has been suggested that the Banach-Tarski theorem caps anything to be found in the tales of the Arabian nights!

The following example is much more routine in nature.

IV.5.4 Example: Right inverses If f is a function, there exists a function g such that $\mathrm{Dom}\, g = \mathrm{Ran}\, f$, $\mathrm{Ran}\, g \subseteq \mathrm{Dom}\, f$ and

$$\underline{y} \in \mathrm{Ran}\, f \Rightarrow f(g(\underline{y})) = \underline{y} \quad ,$$

where $\underline{y}$ denotes a letter not appearing in f or g. Any such function g is what is usually termed a right inverse of f; cf. IV.6 below.

I will describe a routine proof of this; the reader should consider how to convert it into a semiformal proof. Apply the Axiom of Choice (as stated in IV.5.2) to the relation $R \equiv \mathrm{Inv}\, f$ to deduce that there is a function g such that

$$(\text{Dom } g = \text{Dom }(\text{Inv } f)) \wedge (g \subseteq \text{Inv } f) \quad .$$

Hence, by III.2.5(3), Dom g = Ran f . Assume $\underline{y} \in$ Ran f . Then $\underline{y} \in$ Dom g and so $(\underline{y}, g(\underline{y})) \in g$ as a corollary of IV.1.5(3'). Hence, by III.2.5(5),

$$(g(\underline{y}), \underline{y}) \in f \quad . \tag{1}$$

Furthermore, $g(\underline{y}) \in$ Ran g by IV.1.5(4), and hence $g(\underline{y}) \in$ Dom f by III.2.2(12) and III.2.5(4). Therefore, by IV.1.5(3),

$$(g(\underline{y}), f(g(\underline{y}))) \in f \quad . \tag{2}$$

Since f is a function, (1) and (2) imply $\underline{y} = f(g(\underline{y}))$, and the proof is complete.

Remarks (i) There is no hope of proving that there is necessarily a left inverse of f , that is, a function h such that Dom h = Ran f and

$$\underline{x} \in \text{Dom } f \Rightarrow h(f(\underline{x})) = \underline{x} \tag{3}$$

where $\underline{x}$ denotes a letter not appearing in f or h . (These conditions imply that Ran h $\subseteq$ Dom f .) Indeed, were such a function h to exist, it would follow that

$$((\underline{x}, \underline{z}) \in f \wedge (\underline{y}, \underline{z}) \in f) \Rightarrow (\underline{x} = \underline{y}) \quad ,$$

where $\underline{y}$ and $\underline{z}$ denote distinct letters different from $\underline{x}$ and not appearing in f , that is, that f is injective in the sense described in IV.6. But this is not necessarily the case. Cf. the remark following IV.6.1(6).

(ii) The reader should pause to reflect on the routine proof sketched in

IV.5.4. For example, what is the status of (1) ? Plainly, the sentence labelled (1) is not a theorem of Θ_0 , but rather a theorem of a certain theory (not described explicitly in the routine proof) obtained from Θ_0 by the adjunction of certain axioms (of which $\underline{y} \in \text{Ran } f$ is but one). Close scrutiny will indicate how much there is in a routine proof which is left non-explicit and which has to be carried in the reader's mind. With practice, of course, it becomes easier to do this than to write out the much longer semiformal proof. Cf. the substance of IV.6.1.

(iii) The theorem schema in IV.5.4 is a pure "existence theorem" ; there is no hint of how g is to be obtained in practice. Sometimes (see IV.6.1(3)) one can without further effort be more specific and avoid any appeal to the Axiom of Choice at this particular point. This is not always possible, however.

(iv) The precise meanings assigned to the terms "right inverse" and "left inverse" are not fully standardised, and the reader must be wary. For example, " g is a right inverse of f " might be taken to signify that $f \circ g = I_{(\text{Ran } f)}$. If one makes use of IV.6.6(3) and IV.1.5(9), one can see that this last is implied by the definition given at the outset of this subsection. However, the converse is not true: it can happen that $f \circ g = I_{(\text{Ran } f)}$ and yet Dom g be a proper superset of Ran f . Indeed, suppose that g is a right inverse of f according to the definition at the outset of this subsection; choose $b \notin \text{Ran } f = \text{Dom } g$ and $a \notin \text{Dom } f$; define $g_1 \equiv g \cup \{(b, a)\}$; then $f \circ g_1 = f \circ g = I_{(\text{Ran } f)}$ and $\text{Dom } g_1 = \text{Dom } g \cup \{b\} = \text{Ran } f \cup \{b\}$ is a proper subset of Ran f .

IV.5.5 <u>Lifting</u> Let X be a set and S an equivalence relation in X (see III.2.8). It follows from IV.5.4 and IV.3.9 that the function

$$\pi \equiv \pi_{X,X/S} : X \to X/S$$

has at least one right inverse ψ , that is, a function $X/S \to X$ such that

$$\pi(\psi(\underline{q})) = \underline{q} \text{ for all } \underline{q} \in X/S \quad . \tag{1}$$

Any such function ψ is called a <u>lifting of</u> X/S <u>into</u> (or <u>up to</u>) X . In effect, it chooses, for each coset $\underline{q}$ of X modulo S , an element $\underline{x} \equiv \psi(\underline{q})$ of X whose coset modulo S is equal to $\underline{q}$. Usually, of course, there will be more than one such element $\underline{x}$ of X (that is, π is usually not an injective function in the sense of IV.6.1 below); and then there will be more than one right inverse of π .

It is important to record the theorem schema

$$\begin{aligned} T! \quad : \qquad & ((S \text{ is an equivalence relation in } X) \\ & \wedge (\psi \text{ is a lifting of } X/S \text{ into } X) \\ & \wedge (\underline{x} \in X)) \Rightarrow (\underline{x}, \psi(\pi(\underline{x}))) \in S \quad , \end{aligned} \tag{2}$$

where $\underline{x}$ denotes a letter not appearing in X , S or ψ (hence not in $\pi \equiv \pi_{X,X/S}$ either).

<u>Proof</u> Let Θ denote the theory obtained by adjoining to Θ_0 the axiom

$$\begin{aligned} \underline{A} \equiv {} & (S \text{ is an equivalence relation in } X) \\ & \wedge (\psi \text{ is a lifting of } X/S \text{ into } X) \quad . \end{aligned}$$

By IV.6.6(2) (no circularity is involved) and the replacement rule in IV.1.3,

$$T \quad : \qquad \underline{x} \in X \Rightarrow ((\pi \circ \psi)(\pi(\underline{x})) = \pi(\psi(\pi(\underline{x}))) \quad . \tag{3}$$

By (3) , (1) , IV.1.5(9) and IV.3.9(2),

$$T\Theta \quad : \qquad \pi \circ \psi = I_{X/S} \quad . \tag{4}$$

Also, by IV.3.9, IV.3.1 and IV.2.2(1),

$$T \quad : \qquad \underline{x} \in X \Rightarrow \pi(\underline{x}) \in X/S \Rightarrow I_{X/S}(\pi(\underline{x})) = \pi(\underline{x}) \quad . \tag{5}$$

By (3) , (4) and (5) ,

$$T\Theta \quad : \qquad \underline{x} \in X \Rightarrow (\pi(\underline{x}) = \pi(\psi(\pi(\underline{x}))) \quad . \tag{6}$$

On the other hand, by IV.3.9(1),

$$T\Theta \quad : \qquad (\underline{x} \in X \wedge \underline{y} \in X) \Rightarrow (\pi(\underline{x}) = \pi(\underline{y}) \Leftrightarrow (\underline{x}, \underline{y}) \in S) \quad , \tag{7}$$

where $\underline{y}$ denotes a letter different from $\underline{x}$ and not appearing in X or S . By I.3.7(2) and (7) ,

$$T\Theta \quad : \qquad (\underline{x} \in X \wedge \underline{y} \in X \wedge \pi(\underline{x}) = \pi(\underline{y})) \Rightarrow (\underline{x}, \underline{y}) \in S \quad . \tag{8}$$

But $\underline{x} \in X \Rightarrow \psi(\pi(\underline{x})) \in X$ is true in Θ (prove this) and so, by (8) ,

$$T\Theta \quad : \qquad (\underline{x} \in X \wedge (\pi(\underline{x}) = \pi(\psi(\pi(\underline{x}))))) \Rightarrow (\underline{x}, \psi(\pi(\underline{x})))) \in S \quad . \tag{9}$$

By (9) , (6) and I.3.7(1),

$$T\Theta \quad : \qquad \underline{x} \in X \Rightarrow (\underline{x}, \psi(\pi(\underline{x}))) \in S \quad . \tag{10}$$

By (10) and I.3.2(VI),

$$T \quad : \qquad \underline{\underline{A}} \Rightarrow ((\underline{x} \in X) \Rightarrow (\underline{x}, \psi(\pi(\underline{x}))) \in S) \quad . \tag{11}$$

Finally, (2) follows from (11) and I.3.7(2). $\square$

The existence of a right inverse ψ of π is useful when one seeks to derive from a binary operation in X a corresponding binary operation in X/S, a procedure instanced in III.2.9(ii) and again in VI.1.2. In the former situation, the sum of elements q and q' of $R/2\pi Z$ may be defined to be $\pi(\psi(q) + \psi(q'))$, wherein $+$ denotes addition in R ; and $+_{R/2\pi Z}$ (addition in $R/2\pi Z$) is the function

$$(\underline{q}, \underline{q}') \rightsquigarrow \pi(\psi(\underline{q}) + \psi(\underline{q}') \quad \text{with domain} \quad (R/2\pi Z) \times (R/2\pi Z)$$

or, more properly expressed, the function

$$\underline{z} \rightsquigarrow \pi(\psi(pr_1\ \underline{z}) + \psi(pr_2\ \underline{z})) \quad \text{with domain} \quad (R/2\pi Z) \times (R/2\pi Z) \quad ,$$

where $\underline{z}$ denotes a letter not appearing in R , Z or ψ . (One has still to consider how the above operation of addition in $R/2\pi Z$ is influenced by the choice of ψ : it turns out to have no influence; cf. Problem VI/2.)

Sometimes one is concerned with the existence of a lifting ψ with additional properties (continuity, for example). This can be a difficult problem to settle, but no such problems will arise in this book.

IV.6 Inverse and composite functions

In this subsection the principal concern rests with functions which are "invertible" . In its simplest sense, to invert a function f is to carry back each element y of the range of f to the element x of the domain of f to which it corresponds under f (that is, for which $f(x) = y$). This specification may be ambiguous, because there may be at least one element of the range which corresponds under f to more than one element of the domain. Nevertheless, this is the key idea.

IV.6.1 Injective functions and their inverses Although the primary interest rests with functions rather than relations in general, it is worth taking a first step which preserves maximal symmetry in comparison with IV.1.1. This is exhibited in the following simple theorem schema:

$$T \ : \quad (\text{Inv } S \text{ is a function}) \Leftrightarrow (\forall \underline{x})(\forall \underline{x}')(\forall \underline{y})(((\underline{x}, \underline{y}) \in S \wedge (\underline{x}', \underline{y}) \in S) \Rightarrow (\underline{x} = \underline{x}')) \tag{1}$$

where $\underline{x}$, $\underline{x}'$, $\underline{y}$ denote distinct letters not appearing in S . The proof of (1) may be effected by use of the theorems in III.2.5; see Problem IV/4.

Recall also from III.2.5 tnat the domain and range of Inv S are (whether or not S is a function) respectively equal to Ran S and Dom S .

From this point on to the end of this subsection, I shall be dealing with the case in which S is assumed to be a function f . In the arguments below, $\underline{x}$, $\underline{x}'$, $\underline{y}$ denote distinct letters not appearing in f . One task is to correspondingly reformulate (1) in the more convenient form it takes in the new situation. This reformulation and the sentence schemas (2) - (5) displayed in this subsection will be theorem schemas of the theory Θ obtained by adjoining to Θ_0 the explicit axiom Fn f ($\equiv f$ is a function).

These major theorem schemas are as follows:

$$T\Theta \ : \quad (\text{Inv } f \text{ is a function}) \Leftrightarrow (\forall \underline{x})(\forall \underline{x}')((\underline{x} \in \text{Dom } f \wedge \underline{x}' \in \text{Dom } f \wedge f(\underline{x}) = f(\underline{x}')) \Rightarrow (\underline{x} = \underline{x}')) \ . \tag{2}$$

The conjunction of " f is a function" with the sentence appearing on the right of (2) is usually expressed in English as " f <u>is an injective</u> (or <u>one to one</u>) <u>function</u>" or " f <u>is an injection</u>" ; thus

$$f \text{ is an injection } \equiv_{def} \text{Fn } f \wedge (\forall \underline{x})(\forall \underline{x}')((\underline{x} \in \text{Dom } f \wedge \underline{x}' \in \text{Dom } f \wedge (f(\underline{x}) = f(\underline{x}') \Rightarrow (\underline{x} = \underline{x}')) ,$$

$\underline{x}$ and $\underline{x}'$ denoting distinct letters not appearing in f .

(The phrase "one to one" is intended to contrast with "many to one" , which is descriptive of what happens with general functions, or indeed with general relations: many elements of the domain may correspond to one element of the range.)

$$T\Theta! \quad : \qquad f \circ (\text{Inv } f) = I_{(\text{Ran } f)} . \tag{3}$$

$$T\Theta! \quad : \qquad \begin{aligned} (\text{Inv } f) \circ f &= \{(\underline{x}, \underline{x}') \in \text{Dom } f \times \text{Dom } f : f(\underline{x}) = f(\underline{x}')\} \\ &\supseteq I_{(\text{Dom } f)} . \end{aligned} \tag{4}$$

As a corollary to (2) and (4) :

$$T\Theta! \quad : \qquad \begin{aligned} (\text{Inv } f \text{ is a function}) &\Leftrightarrow (\text{Inv } f) \circ f \subseteq I_{(\text{Dom } f)} \\ &\Leftrightarrow (\text{Inv } f) \circ f = I_{(\text{Dom } f)} . \end{aligned} \tag{5}$$

The customary presentation of these results would, of course, be somewhat as follows:

<u>Theorem</u> If f is a function, then:

..................................; (2')

..................................; (3')

..................................; (4')

<u>Corollary</u> If f is a function, then

.... (5')

In the line numbered (2') there would conventionally appear a more informal stand-in for (2) , possibly something like

> Inv f is a function, if and only if $x = x'$ whenever x and x' are elements of the domain of f such that $f(x) = f(x')$.

Similar remarks apply to the lines numbered (3') , (4') and (5') .

Turning to proofs, I will sketch those of (2) and (4) , leaving the remainder to the reader; see Problem IV/5.

Proof of (2) In view of (1) , it is enough to prove that

$$\mathrm{T}\Theta : \quad (\forall \underline{x})(\forall \underline{x}')(\forall \underline{y})(((\underline{x}, \underline{y}) \in f \wedge (\underline{x}', \underline{y}) \in f) \Rightarrow (\underline{x} = \underline{x}'))$$
$$\Leftrightarrow (\forall \underline{x})(\forall \underline{x}')((\underline{x} \in \mathrm{Dom}\ f \wedge \underline{x}' \in \mathrm{Dom}\ f \wedge f(\underline{x}) = f(\underline{x}')) \Rightarrow (\underline{x} = \underline{x}')) \ .$$

This may be achieved by making use of IV.1.5(3), as a corollary of which

$$\mathrm{T}\Theta : \qquad (\underline{x}, \underline{y}) \in f \Leftrightarrow \underline{x} \in \mathrm{Dom}\ f \wedge \underline{y} = f(\underline{x}) \ .$$

The reader should elucidate the details. □

Proof of (4) By III.2.7(1'), the associated collectivising theorems, III.2.5(5), and various theorem schemas and metatheorems in Chapter I and II.3,

$$\mathrm{Inv}\ f \circ f = \{(\underline{x}, \underline{x}') : (\exists \underline{y})((\underline{x}, \underline{y}) \in f \wedge (\underline{y}, \underline{x}') \in \mathrm{Inv}\ f)\}$$

$$= \{(\underline{x}, \underline{x}') : (\exists \underline{y})((\underline{x}, \underline{y}) \in f \wedge (\underline{x}', \underline{y}) \in f)\}$$

$$= \{(\underline{x}, \underline{x}') : (\exists \underline{y})(\underline{x} \in \mathrm{Dom}\ f \wedge \underline{y} = f(\underline{x}) \wedge \underline{x}' \in \mathrm{Dom}\ f \wedge \underline{y} = f(\underline{x}'))\} \quad (\mathrm{IV.1.5(3)})$$

$$= \{(\underline{x}, \underline{x}') : \underline{x} \in \text{Dom } f \wedge \underline{x}' \in \text{Dom } f \wedge f(\underline{x}) = f(\underline{x}')\}$$

(fill in the gap in this step!)

$$= \{(\underline{x}, \underline{x}') \in \text{Dom } f \times \text{Dom } f : f(\underline{x}) = f(\underline{x}')\} \supseteq I_{(\text{Dom } f)} \quad . \qquad \square$$

It is also worth noting the following (informally stated) theorem (schema)

<u>Theorem</u> If f and g are functions, and if

$$g \circ f = I_{(\text{Dom } f)} \quad ,$$

then Inv f is a function and

$$\text{Inv } f = g \mathbin{\S} \text{Ran } f \quad . \tag{6}$$

<u>Proof</u> Proceeding as in the proof of (4) , one first proves that

$$g \circ f = \{(\underline{x}, \underline{x}') : \underline{x} \in \text{Dom } f \wedge f(\underline{x}) \in \text{Dom } g \wedge \underline{x}' = g(f(\underline{x}))\}$$

which by hypothesis is equal to $I_{(\text{Dom } f)}$. This implies

$$\underline{x} \in \text{Dom } f \wedge f(\underline{x}) \in \text{Dom } g \wedge \underline{x}' = g(f(\underline{x})) \Leftrightarrow \underline{x} = \underline{x}' \in \text{Dom } f \quad .$$

In particular,

$$\underline{x} \in \text{Dom } f \Rightarrow g(f(\underline{x})) = \underline{x} \quad . \tag{6_1}$$

By (2) and (6_1) , it follows that Inv f is a function. Thus, to prove (6) , it suffices (see IV.1.5(9)) to prove that $g(\underline{y}) = (\text{Inv } f)(\underline{y})$ for all $\underline{y} \in \text{Ran } f$.

Assume $\underline{y} \in \text{Ran } f$, then $\underline{y} = f(\underline{x})$ for some $\underline{x} \in \text{Dom } f$ (see IV.1.5(8)). Then, by (6_1) (and S6 in II.1.2),

$$g(\underline{y}) = g(f(\underline{x})) = \underline{x} \quad .$$

So, since f is a function, IV.1.5(3) and III.2.5(5) imply

$$(\underline{y}, g(\underline{y})) = (f(\underline{x}), \underline{x}) \in \text{Inv } f \quad . \tag{6_2}$$

Also, III.2.5(3) and IV.1.5(3') imply (since $\underline{y} \in \text{Ran } f$ by hypothesis)

$$(\underline{y}, (\text{Inv } f)(\underline{y})) \in \text{Inv } f \quad . \tag{6_3}$$

Hence, since Inv f is a function, (6_2) and (6_3) combine with IV.1.1 to imply $g(\underline{y}) = (\text{Inv } f)(\underline{y})$. □

Remarks (i) The reader should note carefully that the analogous sentence (schema)

$$(f \text{ is a function} \wedge g \text{ is a function} \wedge f \circ g = I_{(\text{Ran } f)})$$
$$\Rightarrow (\text{Inv } f \text{ is a function})$$

is not true (cf. IV.5.4). The reader should consider the example in which f is the function $\underline{x} \rightsquigarrow \underline{x}^2$ with domain R and range R_+ and g is the function $\underline{y} \rightsquigarrow \underline{y}^{\frac{1}{2}}$ with domain R_+ ; then $f \circ g = I_{R_+}$ (this may be proved directly or deduced from the result of substitution in the theorem schemas IV.6.6(1) - (3)); but f is not injective and so Inv f is not a function.

(ii) The preceding proof of (6) omits many references to theorem schemas and metatheorems in Chapters I and II. In this respect it falls far short of being semiformal according to the initial standards described in I.3.4(ii).

However, a recovery of those standards presents no difficulty in principle.

The following theorem schemas are also useful (compare (7) and (8) with (9) and (11) in III.2.6):

$$\text{T!} : \quad f \text{ is an injective function} \Rightarrow f\langle A \cap B \rangle = f\langle A \rangle \cap f\langle B \rangle \quad ; \tag{7}$$

$$\text{T!} : \quad f \text{ is an injective function} \Rightarrow f\langle A \setminus B \rangle = f\langle A \rangle \setminus f\langle B \rangle \quad ; \tag{8}$$

$$\text{T!} : \quad (f \text{ is an injective function} \wedge g \text{ is an injective function}) \Rightarrow f \circ g \text{ is an injective function}; \tag{9}$$

see Problems IV/6 and IV/7.

IV.6.2 Inverse of j^n when $n \in \dot{N}$; partial inverses The substance of this and the next two subsections properly belongs after Chapter VI, inasmuch as certain definitions related to and properties of $\dot{N}$ and R are taken for granted. $\dot{N}$ denotes the set of positive (that is, nonzero) natural numbers; see V.2.1.

Suppose first that $n \in \dot{N}$ is odd. It may then be proved that $a^n < b^n$ for all real numbers a and b such that $a < b$. Thus the function $j^n : R \to R$ is an injection. Moreover (see VIII.2.6(2)), $j^n\langle R \rangle = R$. Consequently, the inverse of j^n is a function with domain R and range R ; this function is denoted by $j^{1/n}$.

Suppose next that $n \in \dot{N}$ is even, Then j^n is not an injection since in fact $(-a)^n = a^n$ for every $a \in R$. However, if R_+ denotes the set of nonnegative real numbers, it may be proved that $j^n \S R_+$ is an injection and that the range of $j^n \S R_+$ is equal to R_+ . In this case, $j^{1/n}$ is defined to be $\mathrm{Inv}(j^n \S R_+)$ and as such has domain and range both equal to R_+ . For the case $n = 2$, see also IV.4.4(i).

Remark More generally, if f is a function and X a subset of Dom f such that $f \S X$ is injective, $\mathrm{Inv}(f \S X)$ is often termed a partial inverse of

f . An example is $j^{1/2}$ as defined immediately above; other examples appear in XII.5.6 and XII.6.1.

IV.6.3 Definition of j^r when $r \in Q$ Here Q denotes the set of rational numbers; see VI.5.4.

If r is a positive rational number, it is uniquely expressible in the form m/n , where m is a positive integer, n is a positive integer, and m and n are coprime (see Problem VI/28). Accordingly, j^r is defined to be $(j^{1/n})^m$; it has domain R if n is odd, R_+ if n is even.

If r is a negative rational number, j^r is defined to be $(j^{(-r)} \S X)^{-1}$, where $X \equiv (\text{Dom } j^{(-r)}) \cap \dot{R}$, and $\dot{R}$ denotes the set of nonzero real numbers.

In order to preserve as much as possible of the index laws, j^0 has to be defined to be $\underset{\sim}{1}$, in agreement with IV.3.3.

IV.6.4 The functions f^r Having thus defined j^r for rational exponents r , f^r can be defined (for any $f : X \to R$) as $(j^r) \circ f$. The domain of f^r will depend quite crucially on r : it may be any one of

$$(\text{Inv } f)\langle R \rangle \quad , \quad (\text{Inv } f)\langle R_+ \rangle \quad , \quad (\text{Inv } f)\langle \dot{R} \rangle \quad , \quad (\text{Inv } f)\langle P \rangle$$

A satisfactory discussion of irrational exponents is not possible at this stage: it is best left until after the exponential and logarithmic functions have been dealt with; see VIII.3 below.

Note that the foregoing definition of f^r is overtly conditional; cf. the Remarks in IV.3.3.

IV.6.5 Monotone functions After certain properties of continuous real-valued functions have been examined, it will be proved (see VIII.2.7) that a continuous real-valued function f with domain an interval in R is injective if and only if either

$$(\forall \underline{x})(\forall \underline{y})((\underline{x} \in \text{Dom } f \wedge \underline{y} \in \text{Dom } f \wedge \underline{x} < \underline{y}) \Rightarrow (f(\underline{x}) < f(\underline{y}))) \quad (1)$$

or

$$(\forall \underline{x})(\forall \underline{y})((\underline{x} \in \text{Dom } f \wedge \underline{y} \in \text{Dom } f \wedge \underline{x} < \underline{y}) \Rightarrow (f(\underline{x}) > f(\underline{y}))) \quad (2)$$

(The "only if" part of this assertion is that which is relatively difficult to prove, the "if" part being almost evident and true without the assumption of continuity.)

A function $f \subseteq R \times R$ which satisfies (1) (resp. (2)) is said to be <u>strictly increasing</u> (resp. <u>strictly decreasing</u>); and a function $f \subseteq R \times R$ is said to be <u>strictly monotone</u> if and only if it is either strictly increasing or strictly decreasing. The word "monotonic" is often used in place of "monotone".

If the strict inequalities between $f(\underline{x})$ and $f(\underline{y})$ featuring in (1) and (2) are replaced by the associated weak (or wide) inequalities ($<$ by $\leq$ and $>$ by $\geq$), one obtains the definitions of <u>increasing</u>, <u>decreasing</u> and <u>monotone</u> functions.

Expressed another way, a function $f \subseteq R \times R$ is increasing (resp. strictly increasing) if and only if it "preserves" the weak (resp. strong) order relation $\leq$ (resp. $<$) in R (see Problem VI/11). Decreasing (resp. strictly decreasing) functions, on the other hand, "reverse" the weak (resp. strong) order relation.

Most functions $f \subseteq R \times R$ are not monotone, nor even expressible as the pointwise difference of two monotone functions; this remains true, even if the functions f are assumed to be continuous.

If f is a function which is a subset of $R \times R$, and if $E \subseteq R$, f is said to be <u>monotone</u> (or <u>monotonic</u>) <u>on</u> E if and only if the restriction $f \S E$ is monotone. Analogously for <u>strictly monotone on</u> E, <u>increasing on</u> E, <u>strictly increasing on</u> E, et cetera.

The above definitions appear in conventional conditional form. Formal and unconditional versions might read f is (a) strictly increasing (function)

$$\equiv_{def} \text{Fn } f \wedge f \subseteq R \times R$$

$$\wedge\ (\forall \underline{x})(\forall \underline{y})((\underline{x} \in \text{Dom } f \wedge \underline{y} \in \text{Dom } f \wedge \underline{x} < \underline{y}) \Rightarrow (f(\underline{x}) < f(\underline{y})))$$

et cetera, $\underline{x}$ and $\underline{y}$ denoting distinct letters not appearing in f or P (see VI.3 below).

IV.6.6 Properties of composite functions In practice, the handling of composite functions is based upon the following theorem (schema), presented as it might conventionally appear; cf. IV.6.1. (On the matter of notation, see Remark (ii) at the end of III.2.7.)

Theorem If f and g are functions, then:

$$f \circ g \text{ is a function } ; \tag{1}$$

$$\underline{x} \in \text{Dom } (f \circ g) \Rightarrow f \circ g(\underline{x}) = f(g(\underline{x})) \quad ; \tag{2}$$

$$\text{Dom } (f \circ g) = \{\underline{x} \in \text{Dom } g : g(\underline{x}) \in \text{Dom } f\} \quad ; \tag{3}$$

$$\text{Ran } (f \circ g) = f\langle \text{Ran } g \rangle \quad . \tag{4}$$

(In (2) and (3) , $\underline{x}$ denotes a letter not appearing in f or g .)

Proof Assume throughout that f and g are functions. Let $\underline{x}$, $\underline{y}$, $\underline{y}'$, $\underline{z}$ denote distinct letters not appearing in f or g .

Dealing with (1) , assume also that

$$(\underline{x}, \underline{y}) \in f \circ g \wedge (\underline{x}, \underline{z}) \in f \circ g \quad . \tag{5}$$

By III.2.7(2'), the first clause in (5) implies that

$$(\exists \underline{y}')((\underline{x}, \underline{y}') \in g \wedge (\underline{y}', \underline{y}) \in f) \quad .$$

Hence, using IV.1.5(3) (and I.3.3(1)),

$$(\exists \underline{y}')(\underline{y}' = g(\underline{x}) \wedge \underline{y} = f(\underline{y}')) \quad .$$

Using Problem II/14, one deduces that

$$\underline{y} = f(g(\underline{x})) \quad . \tag{5'}$$

Similarly, the second clause of (5) implies

$$\underline{z} = f(g(\underline{x})) \quad . \tag{5''}$$

By (5') and (5") (combined with I.3.3(j), II.1.4(4) and I.3.2(I)), $\underline{y} = \underline{z}$. Thus (by hidden appeal to I.3.2(VI)) $f \circ g$ is a function and (1) is proved.

Turning to (2) , assume that $\underline{x} \in \text{Dom}\,(f \circ g)$. Then, by IV.1.5(3'),

$$(\underline{x}, f \circ g(\underline{x})) \in f \circ g \quad .$$

By the proof of (1) above,

$$(\underline{x}, \underline{y}) \in f \circ g \Rightarrow \underline{y} = f(g(\underline{x})) \quad .$$

These last two theorems together imply that $f \circ g(\underline{x}) = f(g(\underline{x}))$. This entails (via I.3.2(VI)) (2) .

Turning to (3) , by III.2.7(3),

$$\text{Dom}\,(f \circ g) = (\text{Inv}\ g)\langle \text{Dom}\ f \rangle$$

$$= \{\underline{y}' : (\exists \underline{x}')(\underline{x}' \in \text{Dom}\ f \wedge (\underline{x}', \underline{y}') \in \text{Inv}\ g)\} \quad .$$

So, making repeated implicit appeals to II.3.6(18),

$$\text{Dom}\,(f \circ g) = \{\underline{y}' : (\exists \underline{x}')(\underline{x}' \in \text{Dom}\ f \wedge (\underline{y}', \underline{x}') \in g)\} \qquad \text{(III.2.5(5))}$$

$$= \{\underline{y}' : (\exists \underline{x}')(\underline{x}' \in \text{Dom}\ f \wedge \underline{y}' \in \text{Dom}\ g \wedge \underline{x}' = g(\underline{y}'))\} \quad \text{(IV.1.5(3))}$$

$$= \{\underline{y}' : g(\underline{y}') \in \text{Dom}\ f \wedge \underline{y}' \in \text{Dom}\ g\} \qquad \text{(Problem II/14)}$$

$$= \{\underline{y}' \in \text{Dom}\ g : g(\underline{y}') \in \text{Dom}\ f\} \quad ,$$

which is identical with (3) .

Finally, (4) is obtained by substitution in the theorem schema III.2.7(4) . □

<u>Corollary</u> If f and g are functions, then ($\underline{x}$ and $\underline{z}$ denoting distinct letters not appearing in f or g):

$$f \circ g = \{\underline{z} : (\exists \underline{x})(\underline{x} \in \text{Dom}\ g \wedge g(\underline{x}) \in \text{Dom}\ f \wedge \underline{z} = (\underline{x}, f(g(\underline{x}))))\} \quad . \qquad (6)$$

<u>Proof</u> Combine the above theorem with IV.1.5(5). □

This corollary incorporates the practical way of computing the composite of any two given functions, so much so that it may well assume the role of a definition. It might otherwise be expressed by saying that $f \circ g$ is the function defined by the rule $\underline{x} \rightsquigarrow f(g(\underline{x}))$ with domain $\text{Dom}\ g \cap (\text{Inv}\ g)\langle \text{Dom}\ f \rangle$ (the latter being equal to the set of all $\underline{x}$ belonging to Dom g such that

$g(\underline{x}) \in \text{Dom } f$).

IV.7 Sequences

The material of this section is for the most part presented in informal style and is again out of proper logical order, which would place it after Chapters V and VI. At that stage formalisation would present no difficulties of principle. As always on such occasions, the derangement is as explained in the Foreword, namely, to make contact with topics which are supposedly familiar at an informal level to the reader and thereby to encourage and stimulate his continued interest in the underlying formalities.

O'Brien (1) provides a brief introductory informal discussion of sequences, with an emphasis on interesting special cases.

IV.7.1 Definition of sequences Granted the general definition of function, that of sequence is the result of specifying the domain.

If X is a set, an X-valued sequence (sequence in X, sequence of elements of X) is simply a function $u : S \to X$, where S denotes either N or $\dot{N}$. The differences between the cases $S \equiv N$ and $S \equiv \dot{N}$ are almost always trivial; for definiteness the case $S \equiv \dot{N}$ will be considered in this section.

Granted the definition of N and $\dot{N} \equiv N \setminus \{\emptyset\}$ to appear in V.2.1 below, the preceding informal definitions may be formalised thus:

$$u \text{ is a sequence with domain } N \equiv_{\text{def}} (\text{Fn } u) \wedge (\text{Dom } u = N) \quad ,$$

$$u \text{ is a sequence with domain } \dot{N} \equiv_{\text{def}} (\text{Fn } u) \wedge (\text{Dom } u = \dot{N}) \quad ,$$

$$u \text{ is an X-valued sequence with domain } N \text{ (resp. } \dot{N} \text{)}$$

$$\equiv_{\text{def}} (u \text{ is a sequence with domain } N \text{ (resp. } \dot{N} \text{)}) \wedge (\text{Ran } u \subseteq X) \quad ,$$

wherein u and X denote arbitrary sets.

If u is a sequence, it is traditional to write u_n in place of $u(n)$ for the value of u at $n \in \mathring{N}$; u_n or $u(n)$ is termed the <u>n-th term</u> of the sequence u . Sometimes the sequence u is denoted by some such notation as (u_n) , $\{u_n\}$, $(u_n)_{n=1}^{\infty}$, $\langle u_n \rangle$; but extreme care should be taken not to confuse the sequence u with its range, which is equal to what would often be denoted by $\{u_n : n \in \mathring{N}\}$. (The reader is reminded that the use of this notation is still somewhat premature; it is an instance of the notation introduced in II.12.1(2) and still to be discussed in §3 of the Appendix; see also IV.1.5(8) and IV.1.5(8').) In general work, it would seem best to use a notation consistent with that in general use for functions. From the point of view adopted in this book, the parentheses notation ((u_n) , et cetera) is preferable to the braces notation ($\{u_n\}$, et cetera) for denoting sequences.

Particular sequences are frequently otherwise denoted in the form $(u_1, u_2, u_3, \ldots)$: $(1, 1/2, 1/3, \ldots)$, $(0, 1, 0, -1, \ldots)$ are examples. The ellipsis is taken to indicate an injunction "continued ad infinitum according to the pattern (or rule) displayed (or specified) by what precedes" . The objections to this are evident. It is an entirely subjective matter to decide upon what pattern is exhibited by a finite number of explicitly written sets (or names of sets), enclosed between parentheses or braces and separated by commas. It is thus asking for trouble to pose the question:

> What are the next three numbers in the following sequences
>
> $(3, 12, 14, 192, \ldots)$; $(3, 1, 5, 2, 7, 3, \ldots)$?

In any case, there are infinitely many possible answers which, if any is deemed correct, are all equally correct. If u and v are any sequences and $m \in \mathring{N}$, the sequence w defined by the rule

$$\underline{n} \rightsquigarrow u(n) + (\underline{n} - 1)(\underline{n} - 2) \ldots (\underline{n} - m)v(m)$$

has its first m terms agreeing with the first m terms of u. Rarely are there adequate objective grounds for picking and choosing among the various sequences w in terms of simplicity, a defensive stance often adopted by the originator of such a question when he is challenged. (Of course, if the desired extra terms are explicitly stated to be such that one is to obtain a segment of an arithmetic progression or a geometric progression, the problem becomes either insoluble or uniquely soluble as soon as the first two terms are displayed. However, one cannot expect the student to divine the presence of unstated assumptions. He should, on the contrary, be awarded full marks for writing:

This problem is not uniquely soluble ,

when such is the case.)

If one wishes to refer to the sequence whose $\underline{n}$-th term is (say) $1/\underline{n}^2$, the notation

$$u : \underline{n} \rightsquigarrow 1/\underline{n}^2$$

is more precise, and certainly no more trouble to write, than

$$(1, 1/2^2, 1/3^2, \ldots) \quad .$$

A proposed definition such as

> A Sequence is a set of numbers obeying a rule such that the next element of the set can be predicted

is so misleading that it is difficult to know where to begin putting it right. This alleged definition, which does indeed appear in at least one popular high school text book, refers to real-valued sequences only (a restriction which is fair enough in the context in which it appears). But, while a sequence (being a function) is indeed a set - in this case a set of ordered pairs $(\underline{n}, \underline{x})$ in which

$n \in \mathring{N}$ and $\underline{x} \in R$ - the alleged definition clearly indicates the confusion of a sequence with its range, to do which is a cardinal error. Again, the phrase "obeying a rule ...." (does this refer to the aforesaid set, or to its members?) presumably refers to an unspecified one of several stock methods (for example, the domain plus rule procedure, or definition by recurrence; see IV.7.2) of defining individual sequences; it forms no part of the definition of the general concept. Finally, the phrase "next element of the set" (next to what, one might ask) might, in the case of real-valued sequences, be interpreted as that member of Ran u (if there is one) which lies nearest to the right (or the left?) of the chosen member of Ran u : this is, presumably, not intended. The confusion here is between the essential order of $\mathring{N}$ (the domain) and a (here irrelevant) order on R (a convenient superset of the range). The proposed definition might also be taken to suggest (wrongly) that every sequence is definable by a suitable one-term recurrence formula; see IV.8.3.

Care is necessary not to give the impression that (for example) a sequence of real numbers is a subset of R . It is in fact a subset of $N \times R$. Yet, if one uses any of the standard realisations of R known to me, R and $N \times R$ are disjoint sets. (This is because every element of $N \times R$ is an ordered pair and therefore a finite set, whereas every standard realisation of R known to me arranges that every real number is an infinite set; see VI.1.3.)

More generally, it is often vital that a sequence be not confused with its range, a point which recurs in VII.1.2(ii); see also IV.1.4(vi). In an informal introductory account of the concept of sequence, one may well (cf. O'Brien (1), p.1) explain that a sequence is (or is the result of) associating with (or assigning to) the natural number 1 an object a_1 , with the natural number 2 an object a_2 , and so on indefinitely. However, there are potential dangers in attempting to further clarify the concept by adding (cf. O'Brien, loc. cit.) something of the sort

> the infinite ordered set $a_1, a_2, a_3, \dots$
> is called a <u>sequence</u>.

For one thing, this addition may be seen as an invitation to confuse the sequence in question with the set $\{a_1, a_2, a_3, \ldots\}$, which is the range of the said sequence (and which, incidentally, may not be infinite). For another, the term "ordered set" is attached to a relatively advanced and fairly complicated general concept (see Problem VI/11), and its use in this context is a doubtful advantage, especially in connection with sequences which are not injective (that is, in which $a_m = a_n$ for at least one pair (m, n) of natural numbers satisfying $m \neq n$); cf. the informal discussion at the outset of III.1. In brief, the description in terms of association or assignment is a relatively undistorted description of the essence of the concept of sequence, and might be better left without elaboration of the type described.

All operations applying to real-valued functions (see IV.3.2 and IV.3.3) apply ipso facto to real-valued sequences; one need not start again ab initio.

The question of the convergence of sequences will be discussed in Chapter VII.

IV.7.2 <u>The role of sequences</u> The significance of sequences in mathematics is, naturally, something which comes to be appreciated only gradually; it can scarcely be conveyed in a few lines immediately after the definition has been posed. (For one thing, it is linked with the substance of V.13.) However, some indication of their usefulness might well be given at this stage by choosing an example or examples which tie up with problems of which the student is (or should be) already aware. One example of this nature will be indicated now.

Although earlier work may have involved no precise construction of any but a few particular irrational numbers, the student doubtless will be aware that every irrational number is approximable by rational numbers, but he may not have seen discussed in any detail any algorithm for the construction of approximants. For the case of those irrational numbers which are square roots of positive rational numbers, such an algorithm can be formulated in terms of sequences defined by suitable so-called "recurrence formulas" (see IV.8).

The following description is entirely conventional in format, adopted to make the main ideas clear, while leaving the formal details aside. The substratum formal mathematics would involve defining a sequence u in which a letter $\underline{x}$ appears ($\underline{x}$ being subject to no a priori mathematical restriction such as " $\underline{x}$ is a positive rational number" , though it may be subject to metamathematical restrictions), and making use of the general technique discussed in IV.8. To do this would be extremely lengthy and tedious, and the resulting account would be inappropriate to the immediate aim. One would resort to some such lengths, only in a more appropriate context (perhaps in expanded and more formal versions of Chapters VI and VII below).

Let x denote any positive rational number; the only case of interest is that in which x is not the square of a rational number, but the following process works perfectly well without this assumption. Define a real-valued sequence u in the following way: u_1 is a freely chosen positive rational number; then $u_2, u_3, \ldots$ are defined by recurrence via the recurrence formula

$$u_{n+1} = \tfrac{1}{2}(u_n + x/u_n) \quad \text{for all} \quad n \in \dot{N} \; ; \tag{1}$$

u denotes the sequence $n \leadsto u_n$. (It is for the moment assumed that a unique sentence is defined in this way; see the substance of IV.8. In keeping with informality, n appears in place of $\underline{n}$.) The aim is to prove that the sequence u is convergent to a limit which is equal to $x^{\frac{1}{2}}$, that is, equal to the unique positive real number y satisfying $y^2 = x$. This means (as will be explained again later in Chapter VII) that the absolute value $|y - u_n|$ of the difference $y - u_n$ is less than any preassigned positive number ε for all $n \in \dot{N}$ bigger than some suitably chosen $n_0(\varepsilon) \in \dot{N}$. The inequalities to be established below make this assertion plainer.

First, however, some calculations, throughout which n denotes a positive natural number. It is easily proved by induction that $u_n > 0$ for all $n \in \dot{N}$. Next, (1) implies

$$u_{n+1} - u_n = \tfrac{1}{2}(x/u_n - u_n) \tag{2}$$

and

$$u_{n+1}^2 - x = (u_n - x/u_n)^2/4 \quad . \tag{3}$$

From (3) it is deducible that $u_n \geq x^{\frac{1}{2}}$ at least for $n \geq 2$; and then from (2) that $u_{n+1} \leq u_n$ for $n \geq 2$. To avoid having to continue to write the condition $n \geq 2$, it may as well and will be assumed that u_1 has been chosen at least as large as $x^{\frac{1}{2}}$, and then it is true that $u_{n+1} \leq u_n$ and $x^{\frac{1}{2}} \leq u_n$ for all $n \in \dot{N}$.

A general theorem in analysis (see VI.3.4 and VII.4.1 below) implies that the decreasing sequence u of positive real numbers is convergent to a limit y ; and in the present case this limit satisfies $y \geq x^{\frac{1}{2}}$.

Other general theorems to be discussed in Chapter VII show that, granted the convergence of the sequence u to the limit y , the recurrence relation (1) implies the equation

$$y = \tfrac{1}{2}(y + x/y) \quad .$$

Since y is positive, it follows at once that $y = x^{\frac{1}{2}}$.

What can be said about how close u_n comes to y for large values of n ? Define $e_n \equiv u_n - y$, so that $e_n \geq 0$ for all $n \in \dot{N}$. From (3)

$$u_{n+1}^2 - x = (u_n^2 - x)^2/4u_n^2 \quad ,$$

which implies

$$e_{n+1} = e_n^2(\tfrac{1}{2} + y/2u_n)^2(u_{n+1} + y)^{-1}$$

and therefore (since $u_n \geq y$)

$$e_{n+1} \leq re_n^2 \quad , \text{where} \quad r = 1/2y \quad . \tag{4}$$

If one is interested in the foregoing as a means of calculating approximants to $x^{\frac{1}{2}}$, one may as well assume that $x > 1$ (since $(x^{-1})^{\frac{1}{2}} = (x^{\frac{1}{2}})^{-1}$ and the computation of reciprocals presumably involves a little trouble) and then

r is small. This having been done, (4) implies that the error e_n ultimately tends to zero with extreme rapidity. Thus,

$$e_{n+p} \leq (r^{\frac{1}{2}}e_n)^{2^p} \quad \text{for all} \quad p \in \mathring{N} \quad . \tag{5}$$

There are many other similar "iterative" processes of approximation which depend on the construction of sequences (whose terms may be numbers or functions). Another fairly simple example, which might be made a source of illustrations, is the following one designed to approximate roots (that is, solutions) of an equation

$$f(x) = x \quad . \tag{6}$$

Suppose that f is a given real-valued function whose domain is an interval $X = [x_0 - h, x_0 + h]$, where $x_0 \in R$, $h \in R$, $h > 0$; suppose too that there is a number $r < 1$ such that

$$|f(x) - f(x')| \leq r|x - x'| \quad \text{for all} \quad x, \quad x' \in X \quad ;$$

suppose finally that

$$|f(x_0) - x_0| < (1 - r)h \quad .$$

Choose any $x_1 \in X$ and define a sequence by the recurrence formula

$$x_{n+1} = f(x_n) \quad \text{for all} \quad n \in \mathring{N} \quad . \tag{7}$$

It can be proved that the equation (6) has a solution $t \in X$ and that the sequence whose n-th term is x_n converges quite rapidly to the limit t:

$$|x_n - t| \leq 2hr^{n-1} \quad \text{for all} \quad n \in \mathring{N} \quad . \tag{8}$$

(This iterative process may not work unless some a priori restrictions like those listed above are imposed; as a counterexample consider the case in which $f = j^2$ and $x_1 > 1$. For another approximation method, see Problem X/46.)

The use of sequences is not confined to any particular branch of mathematics. Although the foregoing instances belong to analysis, the purely algebraic construction of polynomial (forms) may very well be done in terms of sequences; see for example VI.9 below and Beaumont and Pierce (1), pp.312 ff.

IV.8 Sequences defined by recurrence formulas

The opening remarks of the preceding section apply equally well here: the material of this section is presented in relatively informal style and formalisation must wait until after V.5 below, when it could then proceed without any difficulty in principle.

IV.8.1 Simple recurrence formulas The situation, illustrated by the examples in IV.7.2 may be described as follows: given a set X, a function $f : X \to X$, and $a \in X$, one is concerned with the existence and uniqueness of an X-valued sequence u such that

$$u(1) = a \quad , \tag{1}$$

$$u(\underline{n} + 1) = f(u(\underline{n})) \quad \text{for all} \quad \underline{n} \in \mathring{N} \quad . \tag{2}$$

Assuming the existence and uniqueness of u, this sequence is traditionally spoken of as that defined by the initial value a and (reverting to the more traditional notation $u_{\underline{n}}$ in place of $u(\underline{n})$) the recurrence (or recursion) formula $u_{\underline{n}+1} = f(u_{\underline{n}})$.

Intuitively, the existence and uniqueness of u may appear to the reader to be quite clear; but what is, or is not, intuitively clear is a subjective matter, and it is desirable to minimise the subjective element. (One has at the very least to prove the existence of a set u such that $(1, a) \in u$, $(2, f(a)) \in u$, $(3, f(f(a))) \in u$, et cetera; cf. the discussion in II.5.7 and II.6.4.) In fact, the question of definition by recurrence cannot be satisfactorily treated until the set N of natural numbers has been embraced more or less formally into mathematics. Further and more formal discussion is therefore deferred until V.5 below.

IV.8.2 Other recurrence formulas In IV.8.1 I have mentioned "one-term" recurrence formulas. It is sometimes necessary to consider " k-term" recurrence formulas, where $k \in \dot{N}$ is given, and sequences defined by them. In such a case, one is given a set X, an element a of X^k (see IV.9.3 below), and a function $f : X^k \to X$, and one is concerned with the existence and uniqueness of an X-valued sequence u such that

$$u(\underline{n}) = a(\underline{n}) \quad \text{for all} \quad \underline{n} \in \{1, 2, \ldots, k\} \tag{3}$$

and

$$u(\underline{n} + k) = f(u(\underline{n}), u(\underline{n} + 1), \ldots, u(\underline{n} + k - 1)) \quad \text{for all} \quad \underline{n} \in \dot{N} . \tag{4}$$

Here again one may prove the existence and uniqueness of such a sequence u.

IV.8.3 Remarks Care should be taken to guard against the impression that every X-valued sequence u is definable by (1) and (2) for a suitable choice of $a \in X$ and $f : X \to X$; or by (3) and (4) for a suitable choice of the natural number $k \geq 1$, $a \in X^k$ and $f : X^k \to X$. A counterexample is provided by taking $\dot{N}$ for X and u to be the sequence whose successive terms are (suggested by the scheme)

$$1,1,2,2,2,3,3,3,3,4,4,4,4,4,5,5,5,5,5,5,\ldots\ldots\ldots;$$

that is, the sequence

$$u \equiv \{(\underline{n}, \underline{m}) \in \dot{N} \times \dot{N} : \underline{m} \cdot (\underline{m} + 1)$$

$$\leq 2 \cdot \underline{n} < (\underline{m} + 1) \cdot (\underline{m} + 2)\} ,$$

where $\underline{n}$ and $\underline{m}$ denote distinct letters; or (what is equal) the sequence

$$\underline{n} \rightsquigarrow \mathrm{Max}_N \{\underline{m} \in \dot{N} : \underline{m} \cdot (\underline{m} + 1) \leq 2 \cdot \underline{n}\} \text{ with domain } \dot{N} ;$$

see Chapter V, especially V.6 and V.8.1.

Again, it can be proved (see Problem V/27) that the sequence u with domain N and rule

$$\underline{n} \rightsquigarrow 2^{2^{\underline{n}^2}}$$

satisfies no recurrence relation

$$u_{\underline{n}+1} = P(u_n) ,$$

where P is a polynomial function. (If it did, there would exist natural numbers A and a (both independent of $\underline{n}$; cf. V.11.5 and VI.10.1) such that

$$u_{\underline{n}} \leq A \cdot 2^{a^{\underline{n}}} \text{ for all } \underline{n} \in \dot{N} ,$$

which is false.)

On the other hand, every sequence $u : \dot{N} \to X$ satisfies a recursion formula

$$u_{\underline{n}+1} = F(\underline{n}, u_{\underline{n}}) \text{ for all } \underline{n} \in \dot{N}$$

for a suitably chosen function $F : \dot{N} \times X \to X$, where F may depend upon u . To prove this (by exhibition), it suffices to define

$$F \equiv \{\underline{z} : (\exists \underline{n})(\exists \underline{x})(\underline{n} \in \dot{N} \wedge \underline{x} \in X \wedge \underline{z} = ((\underline{n}, \underline{x}), u(\underline{n} + 1)))\} ,$$

where $\underline{n}$, $\underline{x}$, $\underline{z}$ are distinct letters not appearing in X or u . (This function F is equal to the function

$$(\underline{n}, \underline{x}) \rightsquigarrow u(\underline{n} + 1) \text{ with domain } \dot{N} \times X ,$$

$\underline{n}$ and $\underline{x}$ denoting distinct letters not appearing in u or X ; cf. IV.2.4(v).)

IV.9 Ordered triplets, et cetera

It is not long before high school mathematics makes use of ordered triplets (for example, in three dimensional coordinate geometry), and ordered quadruplets, ordered quintuplets, et cetera, are not slow in following. This circumstance presents an opportunity to present other, somewhat less obvious, uses of the function concept.

Taking ordered triplets first, their characteristic property is that two ordered triplets $(\underline{x}, \underline{y}, \underline{z})$ and $(\underline{x}', \underline{y}', \underline{z}')$ are equal if and only if $\underline{x} = \underline{x}'$, $\underline{y} = \underline{y}'$ and $\underline{z} = \underline{z}'$ (cf.III.1.2). Now it is easy to see that this property would be secured by defining $(\underline{x}, \underline{y}, \underline{z})$ to be $((\underline{x}, \underline{y}), \underline{z})$. But the same end would be achieved by defining $(\underline{x}, \underline{y}, \underline{z})$ to be $(\underline{x}, (\underline{y}, \underline{z}))$, or $((\underline{x}, \underline{z}), \underline{y})$, and so forth. This superabundance of more or less arbitrary choices, which would only multiply when one came to speak of ordered quadruplets and quintuplets, suggests that one seek a definition which is a little less arbitrary in appearance.

As in sections IV.7 and IV.8, some of the material in Chapter V is being presupposed.

IV.9.1 <u>A new look at ordered pairs</u> A pointer to a more satisfactory definition will result from adopting a slightly different outlook on ordered pairs. For this, one will adopt a standard two-element set: instead of taking $2 \equiv 0^{++} \equiv \{\emptyset, \{\emptyset\}\}$ (cf. V.2), it is traditional in this connection to take the set $\{1, 2\}$ (cf. V.7.2), which will here be denoted by S. This being done, let X and Y be sets, and let P be the set of all functions $\underline{f}$ with domain S, range a subset of $X \cup Y$, and such that $\underline{f}(1) \in X$ and $\underline{f}(2) \in Y$; that is,

$$P \equiv \{\underline{f} \in \mathcal{P}(S \times (X \cup Y)) : \underline{f} \text{ is a function}$$

$$\wedge \text{ Dom } \underline{f} = S \wedge \underline{f}(1) \in X \wedge \underline{f}(2) \in Y\} ,$$

where $\underline{f}$ denotes a letter not appearing in X or Y. Denote by F the function

$$\underline{z} \rightsquigarrow \tau_{\underline{f}}(\underline{f} \in P \wedge \underline{f}(1) = \text{pr}_1\ \underline{z} \wedge \underline{f}(2) = \text{pr}_2\ \underline{z})$$

$$\text{with domain } X \times Y ,$$

where $\underline{z}$ and $\underline{f}$ denote distinct letters not appearing in X or Y (and hence not in P either). It can be proved (see Problem IV/9) that F is an injective function with domain $X \times Y$ and range P. In other words $X \times Y$ is in a one-to-one correspondence with P. This is usually taken to be a good enough reason for developing this theme into a definition of ordered triplets.

IV.9.2 <u>Ordered triplets</u> Thus, choose a standard three-element set, namely $T = \{1, 2, 3\}$. Given sets X, Y, and Z, the cartesian product $X \times Y \times Z$ is defined ab initio as the set of functions g with domain T, range a subset of $X \cup Y \cup Z$, and such that $g(1) \in X$, $g(2) \in Y$, $g(3) \in Z$. The ordered triplet (x, y, z) could then be defined to be the unique t such that $\{x\} \times \{y\} \times \{z\} = \{t\}$, namely, $\{(1, x), (2, y), (3, z)\}$.

The definition of ordered quadruplets (and of ordered k-uplets for any $k \in \dot{\mathcal{N}}$) is made in an exactly analogous fashion.

IV.9.3 Definition of Y^X The definitions in IV.9.2 suggest defining Y^X to be the set of functions with domain X and ranges which are subsets of Y : this is what is usually done and what we shall do, the possibility of "identification" spoken of in IV.9.4 being borne in mind.

In particular, if k is a positive natural number and

$$\{1, \ldots, k\} \equiv \{\underline{n} \in N : 1 \leq \underline{n} \leq k\} \quad ,$$

where $\underline{n}$ is a letter not appearing in k , $X^{\{1,\ldots,k\}}$ is what is more informally denoted by $X \times X \times \ldots \times X$ (k "factors"); it is also sometimes denoted by X^k (though consistency is then obtained only at the expense of identifying k with $\{1, \ldots, k\}$, which, in accord with the conventions explained in IV.9.4, is sanctioned by V.7.2(8) below). $X^{\{1,\ldots,k\}}$ is the set of all ordered k-uples of elements of X . See also Problem V/60.

Furthermore,

$$(f \text{ is a function } X \to Y) \Leftrightarrow f \in Y^X$$

and

$$(u \text{ is an } X\text{-valued sequence}) \Leftrightarrow u \in X^N \quad (\text{or } u \in X^{\mathring{N}}) \quad .$$

The notation Y^X is itself not altogether happy, since it involves a clash with that used for exponentiation of natural numbers as defined in V.6.4. It might be better to use, say $\mathrm{Fns}[\![X, Y]\!]$, but this would be unconventional. Compare with the notation $F(X, Y)$ in Bourbaki (1), p. 102.

Remarks The definitions in IV.9.2 and IV.9.3 are informal in style. At the appropriate stage of the formal development (subsequent to Chapter V), formalisation would present no problem. For example:

$$X \times Y \times Z \equiv_{\text{def}} \{\underline{g} : \text{Fn } \underline{g} \wedge \text{Dom } \underline{g} = \{1, 2, 3\} \wedge \underline{g}(1) \in X$$

$$\wedge\ \underline{g}(2) \in Y \wedge \underline{g}(3) \in Z\} \quad ,$$

$\underline{g}$ denoting a letter not appearing in X , Y or Z ;

$$(x, y, z) \equiv_{\text{def}} \tau_{\underline{t}}(\{\underline{t}\} = \{x\} \times \{y\} \times \{z\}) \quad ,$$

$\underline{t}$ denoting a letter not appearing in x , y or z ;

$$X^k \equiv_{\text{def}} \{\underline{h} : \text{Fn } \underline{h} \wedge \text{Dom } \underline{h} = \{1, \ldots, k\}$$

$$\wedge\ (\forall \underline{n})(\underline{n} \in \{1, \ldots, k\} \Rightarrow \underline{h}(\underline{n}) \in X\} \quad ,$$

$\underline{h}$ and $\underline{n}$ denoting distinct letters not appearing in X or k .

IV.9.4 Identifications Having proved (see IV.9.1) that there is an injective function F with domain $X \times Y$ and range P , it would be quite customary in routine mathematics to "identify" $X \times Y$ and P and proceed as if they were equal sets (which usually they are not). Behind this particular identification is, of course, a definite function F realising an injective mapping of $X \times Y$ into P .

Such identifications are made very frequently. A very familiar instance is that of identifying R with $R \times \{0\}$ via the "natural" injection $\underline{x} \rightsquigarrow (\underline{x}, 0)$: the result of this particular identification is that one thereby causes R to appear as a subset of $R \times R$. I return to this example in IV.9.6. Another familiar instance is the identification of polynomial forms with polynomial functions; see IV.3.4 and VI.9.5.

Again, whenever there is a "natural" injection of Y_1 onto Y_2 , it is usual to make an identification of X^{Y_1} and X^{Y_2} . For instance, there is an obvious "natural" injection of $2 \equiv \{\emptyset, \{\emptyset\}\}$ onto $\{1, 2\}$ and so X^2 is usually identified with $X^{\{1,2\}}$. (Note however, that there is no justification for identifying X^2 with $X^{\{2\}}$, since there is no injective function with domain 2 and range $\{2\}$.) Other examples appear in Chapter VI. (Regarding the term "natural" , see the Remark below.)

Identifications are logically indefensible and have no place in the formal theory (sets A and B are often identified in cases where $A \neq B$ is a theorem, and one must not proceed as if the identification amounts to adjoining $A = B$ as a new axiom). They amount to the deliberate use of multiple aliases of name symbols (cf. I.3.5). It might appear at first sight that they cannot fail to generate intolerable chaos. However, the procedure seems to work tolerably well in informal mathematics - indeed, to work better than one has any right to expect (provided one ignores a few "awkward" questions; see IV.9.6 below). What is more, the practice of identifying certain sets is often judged to be almost a practical necessity as a device for keeping notational complexities within bounds. It may even on occasions be positively fruitful for new developments, perhaps because it forces one to look at things in new ways.

Remark The term "natural" , as employed above, is admittedly vague (though experience can clothe it with some significance). A precise meaning is based on the concepts explained in the Appendix to the original French version of Bourbaki (1), Chapter IV, where the term "canonique" (usually Anglicised as "canonical") is employed, and where identifications are also discussed.

IV.9.5 When does one identify? Although there are no agreed rules, it would seem that it is regarded as legitimate (though not always desirable or profitable) to identify two sets X and Y whenever there is an injective function with domain equal to X and range equal to Y (that is, whenever X and Y are equipotent in the sense of V.7.1). In such a situation, whether or not the identification is made, depends largely on a subjective judgement, which may be made in a limited and temporary context. If such an identification is legitimately adopted and experience shows it to be helpful, it is likely to receive common assent; if not, it is likely to be abandoned.

For example, it is legitimate to identify N and $N \times N$; in some contexts the identification *is* made (usually as a local and temporary expedient), but in most contexts it is not. The outcome is that this identification has not the almost universal assent granted to the identification of R with $R \times \{0\}$ in the context of complex numbers (see IV.9.6 below).

Again, if f is a given function, the function

$$f^* : \underline{x} \rightsquigarrow (\underline{x}, f(\underline{x})) \quad \text{with domain} \quad \text{Dom } f$$

is an injective function with domain equal to Dom f and range equal to f . It is thus legitimate to identify f and Dom f . But few mathematicians would dream of doing so; and some might be hard-pressed to give a totally convincing reason why they refrain in this particular instance.

IV.9.6 Awkward questions To illustrate what is meant by "awkward" questions, consider the sets of real and complex numbers (see Chapters VI and XII). In Chapter VI it is explained how at one stage agreement is reached on a definition of the set R of real numbers. In XII.1.1 one constructs from R the set $C \equiv R \times R$ of complex numbers. There is an injective "natural (or canonical) mapping" $\beta : \underline{x} \rightsquigarrow (\underline{x}, 0)$ with domain R and for various reasons one identifies R with its image $R \times \{0\}$ under this mapping. The intended outcome of this is that one "replaces" or "identifies" every real number x by or with the complex number (x, 0) , and so comes to view R as a subset of C : from an informal point of view, this picture is almost universal. However, if (as is often the case) the above procedure is carried out rather hastily, the student can often be led to ask querulously: "I am being encouraged to write - and often see written - $R \subset C$, but is a real number ever really a complex number?" If one abides by the original definitions, the answer must be an emphatic "No" : with the original definitions, in fact, $R \cap C = \emptyset$. If one makes the said identification (that is, if one uses R as a name for a different set), the answer is "Yes" . The question may be described as "awkward" (though "fruitless" or "pointless" might be thought to be more apt adjectives), but it is quite reasonable.

As has been indicated in IV.9.5, to say that sets X and Y may be identified often means no more and no less than that they are equipotent (have the same cardinal number); see V.7 below. This may happen, even though the elements of X are of quite a different "nature" from those of Y . For instance, although R and $R \times \{0\}$ are often identified, most definitions of R ensure that every real number is an infinite set (but see VI.4.1), whereas every element of $R \times \{0\}$ is a set with at most two elements. Again, if X is any set, it is identifiable with a suitably chosen set Y , all of whose elements are singletons (sets with exactly one element). If one attends to the "nature" of the elements of sets which have been legitimately identified, awkward questions are very likely to arise.

IV.10 The use of the term "variable"

In I.2.6 there is introduced a concept of "variable" (or "free letter"), quite precisely defined in relation to a given theory Θ (or formal system). In particular, there is a concept of "variable" appropriate to set theory.

However, the term "variable" is traditional in mathematics, predating any precise concept of formal theory. In this traditional guise it is ill-defined when taken out of such contexts as "function of a real variable" , "function of two variables" , "partial differentiation (or integration) with respect to the first variable" . Naturally, text books follow this usage.

When it is used in such contexts as those just mentioned, it is a convenient way of expressing what can usually be expressed (probably at greater length) without use of the word "variable" at all. For instance,

"function of a real variable" means "function with domain R (or perhaps with domain an interval in R)" ;

"function of two variables" means "function with domain $X \times Y$ (X , Y being usually two specified sets understood in the context)" ;

"partial derivation of f with respect to the first variable" means "derivative of the partial function $f_{*\underline{y}} : \underline{x} \rightsquigarrow f((\underline{x}, \underline{y}))$ " ;

and so on. There is little one may object to in this.

However, <u>some</u> uses of the term "variable" and some alleged definitions thereof, are less happy. For instance, to say (I quote)

> x is a variable over a set X if it denotes an element from the set X

is plainly confusing, unless "element" is understood as "variable element" , in which case the circularity becomes plain. Again (and likewise a quote)

> For a given set (x, y) , we frequently call x and y variables .....

is confusing. See also Görke (1), p.126.

Especially reprehensible is the common practice of using the term in such a way as to suggest that to every function f is linked one (or more) variables, of any one of which f is said to be a function. This is the picture conveyed when one adopts the inexcusable habit of speaking about "the function $f(x)$ " when reference is really being made to "the function f " . One must, for the sake of clarity, insist on preserving a distinction between f and $f(x)$.

In the above, I have been talking about pure mathematics. In applied mathematics (theoretical physics, et cetera) there may be practical need for a concept of variable x which is somewhat akin to that of an equivalence class of functions (two functions going into the same class if and only if they have the same range); when this is done $f(x)$ becomes closely akin to the composite function $f \circ x$. See the comments in XIV.2.1. However, I do not intend to pursue this matter in detail here.

See also the Foreword to Volume 2.

IV.11 Families

IV.11.1 Families and index sets Sometimes the term "family" is used as a synonym for "set" , the sole reason for the replacement being to assist the reader in forming an informal picture and to avoid frequent and tiresome repetition of the word "set" . (The term "collection" is also used in like fashion; see, for example, the opening paragraph in II.6.) This usage is not adopted in this book, and I am here concerned with a different use of the term "family" .

This second use is that explained in Halmos (1), Section 9 and is more technical. In this sense, and formally speaking, "family" is a synonym for "function" . What differences there are, are informal; these affect the conventional terminology. Halmos explains the difference by writing:

> There are occasions when the range of the

> function is deemed to be more important than the function itself.

This embodies some truth, but it should not be interpreted as an invitation to confuse the function with its range; see IV.11.2 below. It is perhaps nearer the truth to say that there are occasions when it is convenient, in formulating a definition or in proving a statement about a set S, to introduce a function f such that $\text{Ran } f = S$ (that is, to represent S as an indexed set); and in such a situation one's ultimate concern is with the range of f (namely, the set S) rather than with f itself. (Instances arise in V.7 below.)

At all events, a <u>family</u> is simply a function. The domain of a family f is usually spoken of as the <u>index set</u> of the family, and an element of the domain as an <u>index</u>; if i is an index, the value $f(i)$ is usually written f_i ("index notation") and is spoken of as the <u>i-th element</u> or <u>i-th term</u> of the family f. (The former name is bad insofar as sometimes f_i is not an element of the set f; see IV.11.2.) If $\text{Ran } f \subseteq X$, the family is often spoken of as a <u>family of elements of</u> X. If I is the index set of f, $\text{Ran } f$ is termed an <u>indexed set</u> with index set I. (A given set S may be indexed in many ways and with various index sets.)

For example, an X-valued sequence, as defined in IV.7.1, would be spoken of as a family of elements of X with index set N (or $\dot{N}$). In view of this particular case, a family f with index set I is often denoted by (f_i) or $(f_i)_{i \in I}$. The notation $\{f_i\}$ or $\{f_i\}_{i \in I}$ is also sometimes used, but this invites a sometimes dangerous confusion between f and $\text{Ran } f$.

IV.11.2 <u>Some prevalent confusions</u> Attention has been drawn in IV.1.4(vi) to the common malpractice of confusing a function with its range. This is especially tempting in the case of sequences and families, perhaps because of the feeling expressed in the passage from Halmos (1) quoted in IV.11.1 immediately above.

As has already been stressed, one should not confuse a family with its range. For example, since the usual definitions of real numbers (see Chapter VI) agree in making every real number an infinite set, $f \cap \text{Ran } f = \emptyset$ for every family f of real numbers.

For the same reason, if f is a family with index set I and $i \in I$, one should not (as is often done) refer to $f(i)$ as an element of f. (The elements of f are exactly the ordered pairs of the form $(i, f(i))$ where $i \in I$.) Either of the phrases

i-th term of f

or

i-th value of f

are much to be preferred.

In what follows, other prevalent confusions are discussed briefly.

(i) If f is a family with index set I, it is customary to say that f is a (pairwise) disjoint family if and only if $f_i \cap f_j = \emptyset$ for any two unequal indices i and j belonging to I.

This concept must be distinguished from " f is disjoint" as defined in II.9. Consider, for example, the set

$$f \equiv \{(\emptyset, \{\emptyset\}), (\{\emptyset\}, \{\emptyset\})\} ,$$

which (see Chapter V) is equal to $\{(0, 1), (1, 1)\}$; then

f is a family (with index set $\{\emptyset, \{\emptyset\}\}$),
f is disjoint,
f is not a disjoint family,
$\text{Ran } f$ is disjoint

are all true sentences.

(ii) In VI.7 we shall define and study sums of finite families of real numbers. We do <u>not</u> initially attempt to define sums of finite sets of real numbers. If one subsequently feels the need of such a definition, one would define the sum of a finite set S of real numbers as the common sum of every injective family with range equal to S .

The circuitous approach is necessary. Consider the two families

$$f \equiv \{(1, 0), (2, 1), (3, 1)\} \quad , \qquad g \equiv \{(1, 0), (2, 1)\} \quad .$$

These are both finite families of real numbers, and each has a range equal to $S \equiv \{0, 1\}$. But the sum of the family f is $0 + 1 + 1 = 2$ and that of the family g is $0 + 1 = 1$. (Note that g is injective and f is not.)

(iii) Define $C \equiv R \times R$; elements of C may here be thought of as vectors in a plane. Define the inner (scalar) product of vectors x and y by

$$\langle x|y \rangle \equiv \sum_{k=1}^{2} x(k)y(k) \quad ,$$

and the length $\|x\|$ of a vector x as the nonnegative square root of $\langle x|x \rangle$. Vectors x and y are said to be orthogonal if and only if $\langle x|y \rangle = 0$ and a set $S \subseteq C$ is said to be orthogonal if and only if $\langle x|y \rangle = 0$ for all $x \in S$ and all $y \in S$ such that $x \neq y$.

Sometimes (see, for example, Hewitt and Stromberg (1), (16.10), p.236) one encounters the following alleged extension of Pythagoras' theorem:

If $\{z_1, z_2, \ldots, z_n\}$ is an orthogonal subset of C , then

$$\|z_1 + \ldots + z_n\|^2 = \|z_1\|^2 + \ldots + \|z_n\|^2 \quad .$$

This is simply not true, as is seen by taking $n = 2$ and $z_1 = z_2 = z$, where $z \in C$ and $z \neq (0, 0)$. The alleged extension becomes a theorem, if "orthogonal subset" is replaced by "orthogonal family" , a family $(z_i)_{i \in I}$ of elements of C being defined to be orthogonal if and only if $\langle z_i | z_j \rangle = 0$ for every two unequal indices $i, j \in I$. See also Halmos (3), p. 124, Theorem 1.

(iv) Similar confusions often arise in connection with linearly dependent sets versus linearly dependent families; see Problem IV/37.

IV.12 Review of the semiformal language

IV.12.1 Provisional survey I have now covered all that is needed in this book of what might be called basic general logic and set theory. Everything else to be discussed in this book can be expressed, in not too indirect and cumbersome fashion, in terms of the concepts and notations discussed up to this point. (The last sentence should perhaps be qualified by the phrase "in theory" : in practice, the expression is sometimes very cumbersome and is made tolerable only by the introduction of abbreviating names.) For this reason it may be helpful to summarise the corresponding basic semiformal vocabulary, which is vastly less cumbersome to use than would be the formal language. Even so, by most standards, the semiformal language is still very formal; and for most purposes it would be regarded as itself an instrument of adequate precision. Almost always, one would be content to clarify a bastard phrase or sentence by translating it into this semiformal language and stop there.

Unlike the formal language (which is immutable once the formal system or theory has been decided upon), the semiformal language is allowed to grow in parallel with formal developments, though the growth is (or should be) carefully controlled.

For the reader's convenience, the next subsection lists the semiformal language as it stands right now.

IV.12.2 Semiformal vocabulary This vocabulary comprises the following:-

(i) Logical (formal) letters: though these will henceforth often appear without underlines and be handled like conventional "variables" .

(ii) The simple logical connectives and constants: $\vee$, $\neg$, $\wedge$, $\Rightarrow$, $\Leftrightarrow$; explicit appearances of τ are relatively rare (mainly in formal definitions), its use being for most of the time hidden by use of the quantifiers $\exists$ and $\forall$.

(iii) The simpler set theoretical connectives and constants: $\in$, $=$, $\subseteq$, $\subset$, $\supseteq$, $\supset$, $\emptyset$.

(iv) More elaborate groups of symbols in which there appear one or more set symbols and sentence symbols:

$$\{\underline{x} : \underline{\underline{A}}\} \ , \ \{T : \underline{x} \in X\} \ ,$$

$$\{X\} \ , \ \{X, Y\} \ ,$$

$$\bigcup X \ , \ \bigcap X \ , \ X \cup Y \ , \ X \cap Y \ , \ \mathcal{P}(X) \ , \ X \setminus Y \ ,$$

$$(X, Y) \ , \ X \times Y \ , \ \mathrm{Dom}\ X \ , \ \mathrm{Ran}\ X \ ,$$

$$X \S Y \ , \ \mathrm{Inv}\ X \ , \ X \circ Y \ , \ X\langle Y \rangle \ , \ X/Y \ ,$$

$$f(X) \ , \ f : X \to Y \ , \ \underline{x}(\in X) \rightsquigarrow T[\![\underline{x}]\!] \ , \ X^Y \ .$$

Whether or not bastard phrases such as " f is a function" or " f is an injection" are included in the list, is largely a matter of taste.

Additions to the semiformal vocabulary to be made later include the following:

$$0 \ , \ \mathcal{N} \ , \ + \ , \ \cdot \ , \ < \ , \ > \ , \ \#X \ , \ \mathcal{R} \ , \ \mathcal{P} \ ,$$
$$\mathcal{Q} \ , \ Z \ , \ \lim u \ , \ e \ , \ \exp \ , \ \ln \ , \ \mathcal{C} \ , \ i \ ,$$
$$\pi \ , \ \sin \ , \ \cos \ , \ \tan \ , \ \ldots\ldots\ldots\ldots \ .$$

Many of these are no more than generally accepted names for (more or less specific) sets. Most of them denote strings of astronomical length.

IV.12.3 General remarks The reader should occasionally practice his "semiformal language" : there are occasional problems which encourage him to do this. In these, he may be asked to render a bastard sentence or phrase into the semiformal language. He may (with justification) object that there is very often room for debate over such translations. The existence of such ambiguities is a fact of informal mathematical life, and also one of the arguments in favour of an adequate degree of formalisation. All who deal with informal mathematics have to come to terms with the situation.

In this connection, refer back to I.2.9(ix). There is no guarantee that an author who writes the bastard sentence

> there exists a natural number m such that, if n is a natural number and $n > m$, then $u(n) = 0$

intends this to be a translation of what is expressed semiformally as

$$(\exists \underline{m})(\underline{m} \in \mathcal{N} \wedge ((\underline{n} \in \mathcal{N} \wedge \underline{n} > \underline{m}) \Rightarrow (u(\underline{n}) = 0))) \quad .$$

Indeed , although this almost certainly the "nearest" semiformal expression, it is also almost certain that (as was discussed in I.2.9(ix)) the said author intends to refer to what is expressed semiformally as

$$(\exists \underline{m})(\underline{m} \in \mathcal{N} \wedge (\forall \underline{n})((\underline{n} \in \mathcal{N} \wedge \underline{n} > \underline{m}) \Rightarrow (u(\underline{n}) = 0))) \quad ,$$

which is materially different. A genuine novice could be caused much (unnecessary) grief by the author's laxity. A more experienced mathematician (that is, in this context, one who is in some measure already acquainted with this particular situation or with analogous ones) would "see through" the laxity and (at least mentally) make repairs.